Advances in

Photochemistry

Volume 8

Advances in Photochemistry

Volume 8

Editors

J. N. PITTS, JR., *Department of Chemistry,*
University of California, Riverside, California, and
Director Statewide Air Pollution Research Center,
University of California

GEORGE S. HAMMOND, *Department of Chemistry,*
California Institute of Technology, Pasadena, California

W. ALBERT NOYES, JR., *Department of Chemistry,*
University of Texas, Austin, Texas

1971

Wiley–Interscience

a division of John Wiley & Sons, Inc., New York · London · Sydney · Toronto

Copyright © 1971 by John Wiley & Sons, Inc.

All rights reserved. No part of this book may be reproduced by any means nor transmitted, nor translated into a machine language without the written permission of the publisher.

Library of Congress Catalog Card Number 63-13592
ISBN 0-471-69091-0

1 2 3 4 5 6 7 8 9 10

PRINTED IN THE UNITED STATES OF AMERICA

Foreword

Volume 1 of this series, *Advances in Photochemistry*, appeared in 1963. The stated purpose of this series was to explore the frontiers of photochemistry through the medium of chapters written by pioneers who are experts. Frontiers in photochemistry are at times full of confusion and for this reason they are fascinating!

In the first six volumes the editors have not acted as censors. We have followed the principle set forth in the original introduction to this series, that is, "the authors are free to make any statement they wish which cannot be proven wrong." We have solicited articles from experts who have strong personal points of view concerning fundamental aspects of photochemistry and spectroscopy. We have asked these authors to present their points of view through critical discussion and evaluation of existing data. In no sense have the articles been simply literature surveys although in some cases they have fulfilled both purposes.

We trust that during the past few years, when photochemistry has truly come of age, that this series has proven to be stimulating and has served our original purpose of bringing together points of views of photochemists and spectroscopists who only a short five to ten years ago did not even use the same vocabulary.

We intend to continue the policies as established in the previous volumes; the change in order of the editors simply reflects a redistribution of responsibilities among them.

James N. Pitts, Jr.
George S. Hammond
W. Albert Noyes, Jr.

Contents

ELECTRONICALLY EXCITED HALOGEN ATOMS
By D. Husain and R. J. Donovan, *Department of Physical Chemistry, Cambridge University, and Gonville and Caius College, Cambridge* . 1

THE PHOTOCHEMISTRY OF α-DICARBONYL COMPOUNDS
By Bruce M. Monroe, *Explosives Department, E. I. duPont de Nemours, Wilmington, Delaware* 77

PHOTO-FRIES REARRANGEMENT AND RELATED PHOTOCHEMICAL [1,j]-SHIFTS (j=3,5,7) OF CARBONYL AND SULFONYL GROUPS
By Daniel Bellus, *J. R. Geigy AG, Basel, Switzerland* . . . 109

PHOTOASSOCIATION IN AROMATIC SYSTEMS
By Brian Stevens, *Department of Chemistry, University of South Florida, Tampa* 161

PHOTOCHEMISTRY IN THE METALLOCENES
By R. E. Bozak, *Department of Chemistry, California State College, Hayward* 227

COMPLICATIONS IN PHOTOSENSITIZED REACTIONS
By Paul S. Engel and Bruce M. Monroe, *National Institutes of Health, Bethesda, Maryland, and Explosives Department, E. I. duPont de Nemours, Wilmington, Delaware.* 245

PHOTOCHEMICAL AND SPECTROSCOPIC PROPERTIES OF ORGANIC MOLECULES IN ADSORBED OR OTHER PERTURBING POLAR ENVIRONMENTS
By Colin H. Nicholls and Peter A. Leermakers, *Hall-Atwater Laboratories, Wesleyan University, Middletown, Connecticut.* . 315

AUTHOR INDEX 337

SUBJECT INDEX 357

CUMULATIVE INDEX (VOLUMES 1–8) 364

Advances in

Photochemistry

Volume 8

Electronically Excited Halogen Atoms

D. HUSAIN, *The Department of Physical Chemistry, The University of Cambridge, Cambridge, England*

and

R. J. DONOVAN, *Department of Chemistry, University of Edinburgh, Edinburgh, Scotland.*

CONTENTS

I. Introduction. 2
II. Consideration of Atomic States 4
   A. Spin Orbit Coupling 4
   B. Spontaneous Emission and the Mean Radiative Lifetimes of the Halogen Atoms in the np^5 $^2P_{1/2}$ States 5
   C. Atomic Transitions 6
III. Experimental Methods for Studying Electronically Excited Halogen Atoms 9
   A. Time-Resolved Absorption Spectroscopy in the Ultraviolet and Vacuum Ultraviolet Following Flash Photolytic Initiation 10
   B. Time-Resolved Atomic Emission Following Flash Photolytic Initiation . 12
   C. Stimulated Emission from Excited Halogen Atoms Following Flash Photolysis 14
   D. Atomic Emission in Flow Systems 15
   E. Electron Spin Resonance Studies of Electronically Excited Halogen Atoms 17
   F. Molecular Halogen Emission in Flow Discharge Systems. 17
   G. Molecular Halogen Emission in Shock-Heated Gases 19
   H. Time-Resolved Molecular Halogen Emission Following Flash Photolysis 20
   I. Time-Resolved Mass Spectrometric Studies. 22
   J. Classical Photochemical Studies 23
IV. Primary Photochemical Processes Resulting in the Production of Electronically Excited Halogen Atoms. 24
   A. General Considerations 24
   B. Primary Processes, the Halogens Cl_2, Br_2, and I_2 25
   C. Primary Processes, the Interhalogens 26
   D. Primary Processes, the Hydrogen Halides HCl, HBr, and HI . . . 29
   E. Primary Processes, the Polyatomic Halides 32
V. Secondary Processes Giving Rise to the Production of $^2P_{1/2}$ Halogen Atoms via Metathetical Reactions 34
VI. Stimulated Emission 36
   A. Population Inversion and the Atomic Photodissociation Laser . . . 36
   B. Quenching of Stimulated Emission 37
   C. Future Possibilities, the Bromine and Chlorine Atom Lasers. . . . 38

VII.	Spin Orbit Relaxation	39
	A. General Considerations	39
	B. General Equation for the Removal of Electronically Excited Halogen Atoms	41
	C. Diffusion of $I(5^2P_{1/2})$ in Inert Gases	42
	D. Experimental Determination of the Mean Radiative Lifetime of $I(5^2P_{1/2})$	44
	E. Collisional Quenching of $X(^2P_{1/2})$ by X_2 where X = Halogen	45
	F. Collisional Quenching by Atoms	47
	G. Collision with Diatomic Molecules	48
	H. Quenching by Polyatomic Molecules	49
VIII.	$X(^2P_{1/2})$ in Atomic Recombination	51
	A. Molecular Emission Accompanying Atomic Recombination	51
	B. Halogen Atom Recombination in Flow Discharges	51
	C. Halogen Atom Recombination in the Shock Tube	52
	D. Halogen Atom Recombination in Flash Photolysis Experiments	53
IX.	Chemical Reactions of Electronically Excited Halogen Atoms	55
	A. The Thermochemistry of Some Reactions of $X(^2P_{1/2})$	55
	B. Spin Orbit Relaxation	56
	C. H Atom Abstraction from Paraffins by $I(5^2P_{1/2})$ by Classical Photochemical Investigation	57
	D. Reactions of $I(5^2P_{1/2})$ with Alkyl Iodides	58
	E. Reactions of $I(5^2P_{1/2})$ and $I(5^2P_{3/2})$ with Halogens and Interhalogens	61
	F. Some Reactions of $Br(4^2P_{1/2})$ and $Cl(3^2P_{1/2})$	64
	G. $I(5^2P_{1/2})$ with Polyatomic Molecules	65
	1. Nitrosyl Halides	65
	2. Nitrous Oxide, Ozone, and Nitrogen Dioxide	66
	3. Olefins	67
	4. Allyl Halides	68
X.	General conclusions	68
References		70

I. INTRODUCTION

This review on "electronically excited" halogen atoms is confined to the higher of the two spin orbit states associated with the np^5 ground state electronic configuration. The lower state is designated $^2P_{3/2}$ and the higher $^2P_{1/2}$, the difference in energy arising only from the coupling of the spin and orbital motions of the electrons. The energies of the $^2P_{1/2}$ atoms are shown in Table I and vary from 1.15 kcal for fluorine to 21.7 kcal for iodine. There now exists a large body of data on the chemistry of the ground state halogen atoms, as indicated in the review of Fettis and Knox[1] on this subject in 1964. While it has often been appreciated in studies involving halogen atoms that the $^2P_{1/2}$ state may play a significant role in a given system, it is only since 1964 that direct studies of $^2P_{1/2}$ halogen atoms have been carried out and this now provides a basis for more detailed discussions.

TABLE I
Spin Orbit Energies of Electronically Excited Halogen Atoms and Boltzmann Fraction at 300°K

$$\frac{[X^2P_{1/2}]}{[X^2P_{1/2}] + [X^2P_{3/2}]}$$

	F	Cl	Br	I
$E(^2P_{1/2} - {}^2P_{3/2})$, kcal	1.155	2.518	10.533	21.733
Boltzmann fraction	6.7×10^{-2}	7.3×10^{-3}	1.1×10^{-8}	7.4×10^{-17}

The large range of experimental conditions that have been employed in studies involving halogen atoms requires that the population of the $^2P_{1/2}$ state under any given condition be considered in detail. Table I shows the Boltzmann fractions of halogen atoms in the $^2P_{1/2}$ state at 300°K which for fluorine and chlorine atoms could make a significant kinetic contribution at this temperature. The thermal fraction for Br($4^2P_{1/2}$) at 2500°K, the highest temperature to which molecular bromine was shock heated by Palmer[2] in his investigation of the accompanying molecular emission, is 5.7% and controls the emission from the state correlating with these atoms. However, it is clearly in photochemical systems, where the primary species may be produced in markedly nonequilibrium distributions, that the reactions of electronically excited halogen atoms are most important. It is well established that electronically excited states of molecules of photochemical interest, such as the halogens themselves, interhalogens, hydrogen halides, and organic halides, correlate with $^2P_{1/2}$ halogen atoms. A dramatic effect demonstrating this has been the construction of the iodine atom photodissociation laser[3] in which the emitting state is the $5p^5$ $^2P_{1/2}$ state.

Many of the recent studies of electronically excited halogen atoms have exploited the long mean radiative life times associated with these states, resulting from the electric dipole forbidden transition to the $^2P_{3/2}$ state (hereafter referred to as the ground state). Absorption and emission spectroscopy on the $^2P_{1/2}$ atom, stimulated atomic emission, molecular halogen emission, and observation of the $^2P_{1/2}$ atoms by electron spin resonance have been the basis of many of the measurements. Experimental techniques have included flash photolysis, flow discharges, and shock tubes as well as classical photochemistry. The optical metastability of the $^2P_{1/2}$ atoms has resulted in this energized species being sufficiently long-lived over-all for chemical reaction by the excited atom to compete with spin orbit relaxation either by emission or collisional quenching. Thus $^2P_{1/2}$ atoms undergo both physical and chemical processes and these will be dealt with in detail. A relatively large amount of work has now been carried out on the spin orbit relaxation of $^2P_{1/2}$ halogen atoms and this has taken place at a time of increasing interest in energy transfer in general. It is with such processes that have been mentioned that this review is concerned.

II. CONSIDERATION OF ATOMIC STATES

A. Spin Orbit Coupling

The orbital motion of an electron about an atomic nucleus gives rise to a magnetic field which is proportional to the angular momentum of the electron. Associated with the intrinsic spin of the electron is a magnetic dipole moment in the direction of the axis of spin. The energy of this dipole in the magnetic field may therefore be calculated, with the appropriate relativistic correction, and this yields the spin orbit interaction operator which must then be included in the total Hamiltonian. It is this operator which gives rise to the states for the different J values, where J is the total angular momentum quantum number including spin. For heavy atoms, in which there are strong magnetic fields, the splitting between states characterized by different J values for the same electronic configuration, may be relatively large. Thus for the iodine atom in the ground state np^5 configuration, the multiplet splitting is nearly 1 eV. The details of the calculation of the spin orbit operator and the generation of the spin orbit states may be found in standard works on quantum mechanics.[4-8] The ^2P configuration is considered in detail by Eyring et al.[4] Inclusion of the spin orbit operator in the Hamiltonian leads to two states, $^2P_{J=3/2}$ and $^2P_{J=1/2}$, being characterized by four and two eigenfunctions, respectively. It may be shown that for an np configuration, the $J = 1/2$ state is of lower energy and while it can be readily demonstrated that an inversion of the J states for the np^q and np^{6-q} configurations takes place, it is particularly straightforward for the np and np^5 configurations (Landau et al.[8]). Thus for the halogen atoms with np^5 configurations, the J ground state is $^2P_{3/2}$ and the higher state $^2P_{1/2}$, the energies of which are shown in Table II.

TABLE II
Spin Orbit Energies of the $^2P_{1/2}$ States of the Halogen Atoms and Mean Radiative Lifetimes for Magnetic Dipole Emission (following Garstang[13])

Halogen atom	$E(^2P_{1/2}) - E(^2P_{3/2})$,[19] cm^{-1}	A_m, sec^{-1}	τ_e, sec	A_m, sec^{-1} (exptl)
F $2p^5$ $^2P_{1/2}$	404.0	0.0012	830	—
Cl $3p^5$ $^2P_{1/2}$	881	0.012	83	—
Br $4p^5$ $^2P_{1/2}$	3685	0.89	1.1	—
I $5p^5$ $^2P_{1/2}$	7603.15	7.8	0.13	~22[11]

B. Spontaneous Emission and the Mean Radiative Lifetimes of the Halogen Atoms in the np^5 $^2P_{1/2}$ States

Electronically excited halogen atoms in the np^5 $^2P_{1/2}$ state are optically metastable as the transition

$$np^5\ ^2P_{1/2} \longrightarrow np^5\ ^2P_{3/2} + h\nu \qquad (1)$$

is electric dipole forbidden. The transition is magnetic dipole and electric quadrupole allowed[9] and thus these states would be expected to be characterized by long mean radiative lifetimes (τ_e) for spontaneous emission. Experimental determination of mean radiative lifetimes of strongly forbidden atomic transitions in the laboratory as opposed to estimates obtained from astronomical observations are very few. Furthermore, these astronomical observations are essentially limited at present to relative values for different transitions.[10] To the best knowledge of the authors, the experimental determination of τ_e for $I(5p^5\ ^2P_{1/2})$ by Husain and Wiesenfeld[11] is the only absolute measurement for a strongly forbidden transition and its importance is that it can be used to test the general basis of theoretical calculations of this quantity. The experimental difficulty in measuring τ_e in the laboratory for strongly forbidden transitions for *transient* atoms lies in the production of sufficiently high concentrations of such energized atoms to yield adequate emission intensities. Further, the contribution of spontaneous emission as a rate coefficient [A_{nm} (sec^{-1}) = $1/\tau_e$ (sec)] to an over-all decay rate of the excited atoms is usually very small on account of more rapid removal processes. The experimental determination of τ_e by attempting to record the absolute absorption intensity for the transition $X(np^5\ ^2P_{1/2}) \leftarrow X(np^5\ ^2P_{3/2})$ will also be rendered difficult by the transient nature of these atoms. The basis of the quantification of the coefficients for spontaneous emission for the halogen atoms in the $^2P_{1/2}$ state, though such emission can be detected, is principally theoretical, the exception being $I(5p^5\ ^2P_{1/2})$ to be discussed later from the experimental viewpoint.

The calculation of the Einstein coefficient for magnetic dipole radiation in atoms has been considered by Shortley[12] and applied to a number of forbidden transitions by Garstang.[10,13] Spontaneous emission for the transition $^2P_{1/2} \rightarrow ^2P_{3/2}$ in the case of the halogen atoms is determined overwhelmingly by the magnetic dipole contribution. The magnetic dipole matrix element reduces to the form (Garstang[10])

$$\left\langle SLJ \left| \frac{eh}{4\pi m_e c}(\underline{L} + 2\underline{S}) \right| SLJ' \right\rangle \qquad (2)$$

where the symbols have their usual significance[10] and, in the form required, we calculate

$$S_m = \left| \left\langle SLJ \left| \frac{eh}{4\pi m_e c} (\underline{L} + 2\underline{S}) \right| SL(J+1) \right\rangle \right|^2 \quad (3)$$

which is a function of S, L, and J only for LS coupling as the radial part of the wave function in the matrix element does not appear in the final result of S_m. The result is given by Shortley (Ref. 12, Eq. 3)

$$S_m = [+]^2 \frac{(J - S + L + 1)(J + S - L + 1)(J + S + L + 2)(S + L - J)}{(4J + 1)} \quad (4)$$

in units of $(-eh/4\pi m_e c)^2$. For the np and np^5 configurations ($J = \frac{1}{2}$, $S = \frac{1}{2}$, and $L = 1$ for $^2P_{1/2} \to {}^2P_{3/2}$) this element $S_m = \frac{4}{3}$ in units of $(-eh/4\pi m_e c)^2$ (Garstang[13]). Hence the Einstein coefficient for magnetic dipole emission (A_m) between the $^2P_{1/2}$ and $^2P_{3/2}$ states is given by Shortley[12] as

$$A_m = \frac{1}{(2J+1)} \cdot \frac{64\pi^4}{3h} (-eh/4\pi m_e c)^2 S_m \omega^3 \quad (5)$$

where $S_m = \frac{4}{3}$. Thus, for these transitions, A_m depends only upon ω^3 ($\omega =$ energy separation between $^2P_{1/2}$ and $^2P_{3/2}$ levels in units of cm^{-1}) and the results are given in Table II following Garstang.[13] (The authors are deeply indebted to Professor Garstang for helpful discussion on this subject.) The result of this calculation should be satisfactory to within, say, 20–30% as the result is independent of the radial part of the wave function.

Garstang[14] has developed the equations for electric quadrupole emission. Here the transition probability does depend on the radial part of the wave function $P(np)$ in the form $\int_0^\infty r^2 P^2(np) \, dr$. This integral is solved generally for r^k by Condon and Shortley.[7] The resulting coefficient for electric quadrupole emission (A_q), however, is small compared with A_m and will be neglected here. The values of A_q for I($5p^5$ $^2P_{1/2}$) and Br($4p^5$ $^2P_{1/2}$) are given by Garstang[13] as 0.055 and 8.3 × 10^{-4} sec^{-1}, respectively, for example.

It can be seen from Table II that unless experimental conditions are carefully chosen, the contribution from emission to the over-all decay of X(np^5 $^2P_{1/2}$) can usually be neglected as more rapid kinetic processes result in the removal of the excited atoms. Further, as the mean radiative lifetime is long, it is possible to detect these atoms by absorption spectroscopy. (See Section VII.B, general kinetic equation for the removal of $^2P_{1/2}$ atoms.)

C. Atomic Transitions

The wavelengths associated with the $^2P_{1/2} \to {}^2P_{3/2}$ transitions of the halogen atoms are given in Table III. In the types of controlled experiments to be

TABLE III
Wavelength of emission $X[5p^5(^2P^0_{1/2})] \rightarrow X[5p^5(^2P^0_{3/2})] + h\nu$

X	I	Br	Cl	F
λ, μ^a	1.315	2.714	11.351	24.752

[a] $1\mu = 10^4$ Å.

described on the excited halogen atoms, only emission from iodine[3,11,15] and bromine[16,17] atoms in the $^2P_{1/2}$ state have been detected. Figure 1 illustrates those transitions that have been observed in absorption in the vacuum ultraviolet for $I(5^2P_{1/2})$ and $I(5^2P_{3/2})$. Table IV summarizes the transitions that have

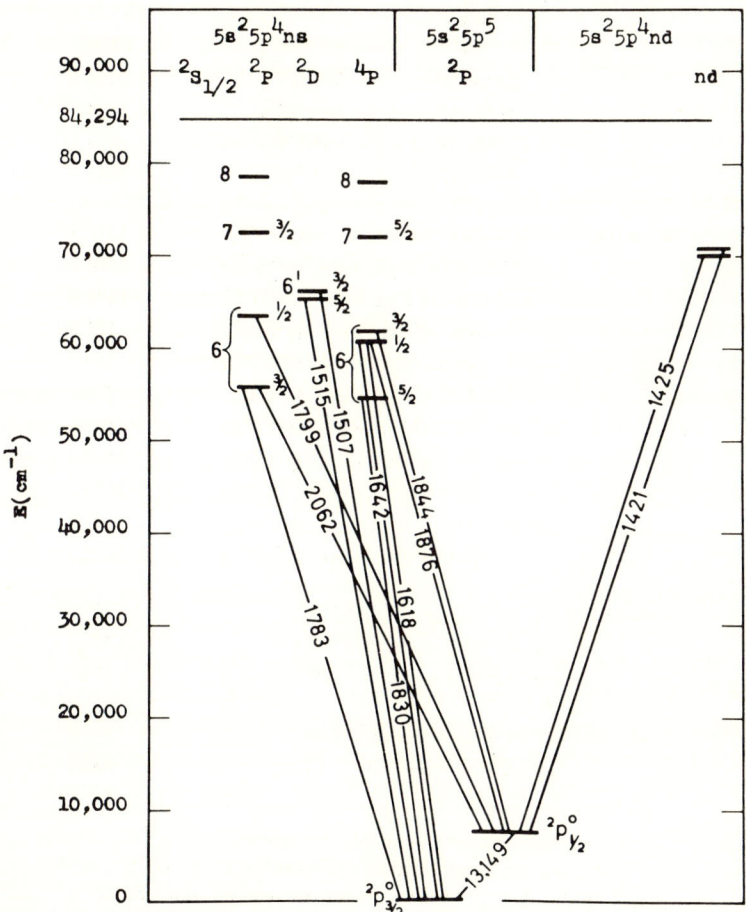

Fig. 1. Energy level diagram for atomic iodine transitions (Å) observed in absorption.

TABLE IV
Optical Transitions of the Halogen Atoms observed by Absorption Spectroscopy in the Vacuum Ultraviolet

		λ, Å
Iodine:[27,30,75]	$5p^4nd5.1_{3/2} \longleftarrow 5p^5(^2P^0_{3/2})$	1421.4
	$5p^4nd5_{5/2} \longleftarrow 5p^5(^2P^0_{3/2})$	1425.5
	$5p^46s'(^2D_{3/2}) \longleftarrow 5p^5(^2P^0_{3/2})$	1507
	$5p^46s'(^2D_{5/2}) \longleftarrow 5p^5(^2P^0_{3/2})$	1514.7
	$5p^46s(^4P_{3/2}) \longleftarrow 5p^5(^2P^0_{3/2})$	1617.6
	$5p^46s(^4P_{1/2}) \longleftarrow 5p^5(^2P^0_{3/2})$	1642.1
	$5p^46s(^2P_{3/2}) \longleftarrow 5p^5(^2P^0_{3/2})$	1782.8
	$5p^46s(^2P_{1/2}) \longleftarrow 5p^5(^2P^0_{1/2})$	1799.1
	$5p^46s(^4P_{5/2}) \longleftarrow 5p^5(^2P^0_{3/2})$	1830.4
	$5p^46s(^4P_{3/2}) \longleftarrow 5p^5(^2P^0_{1/2})$	1844.5
	$5p^46s(^4P_{1/2}) \longleftarrow 5p^5(^2P^0_{1/2})$	1876.4
	$5p^46s(^2P_{3/2}) \longleftarrow 5p^5(^2P^0_{1/2})$	2062.3
Bromine:[76]	$4p^4(^1D_2)5s(^2D_{5/2}) \longleftarrow 4p^5(^2P^0_{3/2})$	1317.7
	$4p^4(^1D_2)5s(^2D_{3/2}) \longleftarrow 4p^5(^2P^0_{1/2})$	1384.6
	$4p^45s(^2P_{1/2}) \longleftarrow 4p^5(^2P^0_{3/2})$	1449.9
	$4p^45s(^2P_{1/2}) \longleftarrow 4p^5(^2P^0_{1/2})$	1531.7
	$4p^45s(^4P_{3/2}) \longleftarrow 4p^5(^2P^0_{3/2})$	1540.7
	$4p^45s(^2P_{3/2}) \longleftarrow 4p^5(^2P^0_{1/2})$	1574.8
	$4p^45s(^4P_{5/2}) \longleftarrow 4p^5(^2P^0_{3/2})$	1576.4
	$4p^45s(^4P_{1/2}) \longleftarrow 4p^5(^2P^0_{1/2})$	1582.3
	$4p^45s(^4P_{3/2}) \longleftarrow 4p^5(^2P^0_{1/2})$	1633.4
	$4s^24p^4(^1D)5s(^2D_{3/2}) \longleftarrow 4p^5(^2P^0_{3/2})$[a]	1293.3
	$4p^45s(^2P_{3/2}) \longleftarrow 4p^5(^2P^0_{3/2})$[a]	1488.5
Chlorine:[29]	$3p^44s(^2P_{1/2}) \longleftarrow 3p^5(^2P^0_{3/2})$	1335.7
	$3p^44s(^2P_{3/2}) \longleftarrow 3p^5(^2P^0_{3/2})$	1347.2
	$3p^44s(^2P_{1/2}) \longleftarrow 3p^5(^2P^0_{1/2})$	1351.7
Fluorine:[b]	$2p^43s(^2P_{3/2}) \longleftarrow 2p^5(^2P^0_{3/2})$	954.8
	$2p^43s(^2P_{1/2}) \longleftarrow 2p^5(^2P^0_{1/2})$	955.5

[a] Transitions obscured by molecular spectra.
[b] These are the longest wavelength transitions of the $^2P_{3/2}$ and $^2P_{1/2}$ F atoms. They lie below the low wavelength limit cut off of pure LiF ($\lambda > 1050$ Å) and have not therefore been detected in this region.

been observed in absorption for all of the halogen atoms. Relative intensities are found to be in accord with the absolute f values calculated recently by Lawrence.[18] These transitions and others are well established in emission from discharges and an extensive set of term values are listed for the electronic states.[19,154,155,156]

III. EXPERIMENTAL METHODS FOR STUDYING ELECTRONICALLY EXCITED HALOGEN ATOMS

The experimental methods that have been employed to study the kinetics of electronically excited halogen atoms may be conveniently divided into two groups; namely, those in which the excited atoms may be monitored directly and those in which a molecular product of the reaction of a $^2P_{1/2}$ atom is followed. The majority of these investigations fall into the first category. Time-resolved studies in absorption of the $^2P_{1/2}$ atoms following flash photolytic dissociation of suitable halides may be employed on account of the long mean radiative lifetime in the excited state and the relatively long over-all lifetimes under normal experimental conditions. The atomic transitions of interest lie mainly in the vacuum ultraviolet with one exception in the far ultraviolet (Table IV). Time-resolved emission of the strongly forbidden transition $^2P_{1/2} \rightarrow {}^2P_{3/2} + h\nu$ in the infrared (Table III) has only been applied as yet to the iodine atom following flash photolytic dissociation of appropriate iodides. These transitions have been monitored in continuous photolysis and discharge experiments in flow systems for iodine and bromine atoms. Stimulated emission for the $^2P_{1/2} \rightarrow {}^2P_{3/2}$ transition has only been observed so far for the iodine atom at 1.315 μ (Table III) resulting from the population inversion following the flash photolysis of a number of iodides (see footnote, p. 15). Electronically excited halogen atoms in flow discharge systems have been detected by electron spin resonance although limited kinetic measurements have been made, the emphasis being on the spectroscopy of these excited states. The methods of monitoring the molecular products of the chemical reactions of the $^2P_{1/2}$ atoms include emission from excited electronic states of molecular halogens which correlate with these excited atoms, and those which monitor the products of atomic abstraction by the $^2P_{1/2}$ atoms with stable molecules. Detection of the molecular halogen emission has been employed in kinetic studies in flow discharge systems, shock tube studies, and following the flash photolysis of an iodide in a static system. The monitoring of the chemical products of the reactions of electronically excited halogen atoms include classical photochemical investigations and the observation of such products in flash photolysis experiments. These are the basic principles of the methods to be described in further detail.

A. Time-Resolved Absorption Spectroscopy in the Ultraviolet and Vacuum Ultraviolet Following Flash Photolytic Initiation

The method of flash photolysis and kinetic spectroscopy is well established.[20] The reaction system is subjected to an intense flash of light, obtained by discharging a bank of condensers through a quartz lamp filled with a noble gas. A pulse of energy in the region of 2000 joules and of duration 5–20 μsec results, giving continuous light over a convenient wavelength region. Following photochemical initiation by the high intensity flash, a second lamp of lower energy and shorter duration but of higher radiation temperature is triggered photoelectrically to take absorption spectra of the reaction system at predetermined time delays.[21] The spectrum may be monitored photographically at a given time delay over the whole wavelength region or continuously at a single wave band using photoelectric techniques.[22] In recent years, rapid scan spectrometers have been developed for the visible and infrared regions which will scan and generate photoelectric output over a large wavelength region in a single flash photolysis experiment.[23,24] The atomic transitions of the halogens lie in the vacuum ultraviolet and therefore it is necessary to evacuate the optical path in order to avoid absorption by atmospheric gases, the method of photolytic initiation remaining essentially the same. The flash photolysis apparatus first employing kinetic spectroscopy in the vacuum ultraviolet was constructed by Herzberg and Shoosmith.[25] Following this, Thrush[26] designed an apparatus of lower optical resolution but more suitable for recording the many spectrograms at different time delays required for kinetic investigation. The authors[27–29] have employed concave grating spectrographs of limited optical resolution to record in kinetic experiments the transient spectra of atomic iodine, bromine, and chlorine in the $^2P_{1/2}$ and $^2P_{3/2}$ states following flash photolysis.

Figure 2 shows a block diagram of the apparatus employed by the authors.

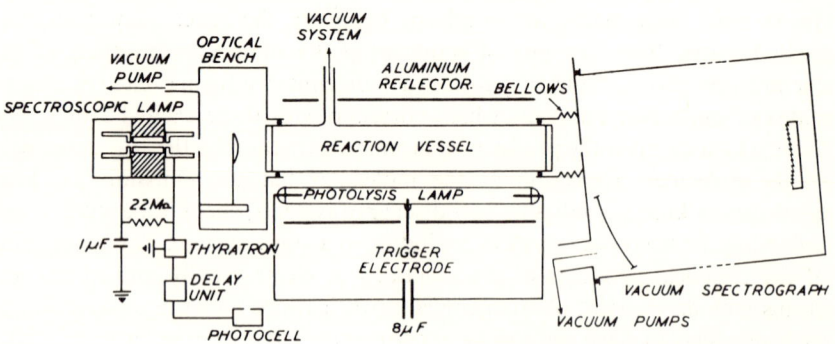

Fig. 2. Block diagram of apparatus for kinetic spectroscopy in the vacuum ultraviolet.

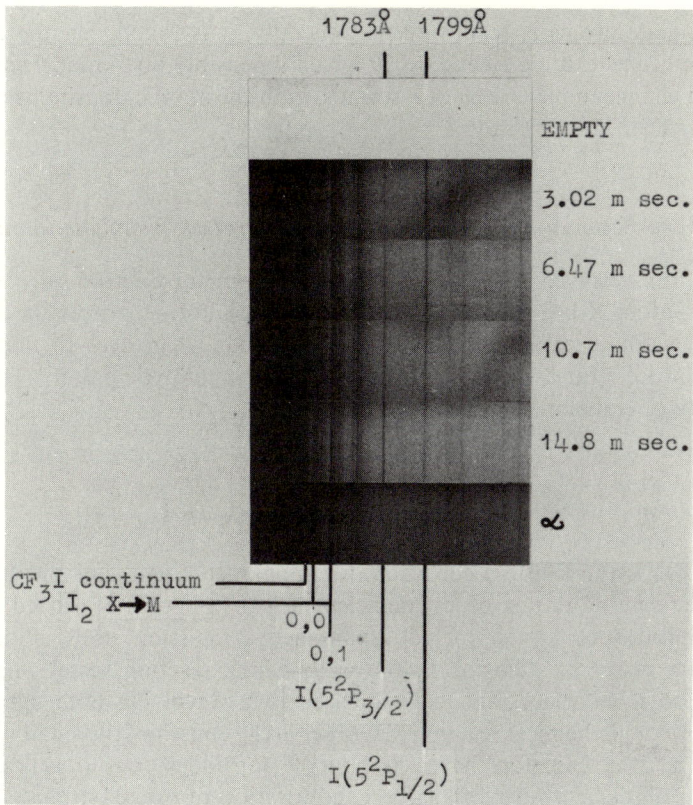

Fig. 3. Decay of I($5^2P_{1/2}$) in spectroscopically pure argon. $p_{CF_3I} = 0.5\tau$, $p_{Ar} = 50\tau$, flash energy = 1024 J.

The slit and plate holder of the vacuum spectrograph lie on the Rowlands circle. The resulting photographic plates are photometered in order to translate plate densities into both relative and absolute atomic concentrations. A transition of I($5p^5$ $^2P_{1/2}$) at 2062 Å (Table IV) can be readily observed and does not require the use of a vacuum spectrograph. Donovan and Husain[30] observed this transition in absorption with a 1-m vacuum spectrograph, and later Donovan, Hathorn, and Husain[31] employed a small, optically fast Hilger Model 301 quartz spectrograph to monitor this transition in kinetic experiments. Vacuum ultraviolet measurements enjoy the advantage that both the $^2P_{1/2}$ and $^2P_{3/2}$ atoms for iodine, bromine, and chlorine may be monitored so that differentiation may be made between relaxation and chemical reaction. Figure 3 shows the time-resolved absorption spectra of I($5^2P_{1/2}$) and I($5^2P_{3/2}$) in the vacuum ultraviolet following the flash photolysis of CF_3I in the presence of excess argon. The slow relaxation of the excited atoms yielding ground state atoms is readily seen. Generally, it is found that the excited halogen atoms exhibit kinetic decay which is sensibly first-order. Hence the relative change in plate density associated with the atomic transitions may be employed for kinetic analysis.

B. Time-Resolved Atomic Emission Following Flash Photolytic Initiation

The time-resolved spontaneous emission from an electronically excited halogen atom $X(np^5\ ^2P_{1/2})$ produced in a flash photolysis process in a static system is found experimentally to be described by an over-all first-order kinetic process for a long interval following the photolytic pulse.[11] Thus the intensity of emission (I_{emm}) is given by

$$I_{emm} = A_{nm}[X(^2P_{1/2})_t]$$
$$= A_{nm}[X(^2P_{1/2})_{t=0}] \exp(-kt) \qquad (6)$$

where A_{nm} is the Einstein coefficient for spontaneous emission and k represents the sum of all first-order coefficients for the removal of excited atoms. Contributions to k can include spontaneous emission itself, stimulated emission, decay by diffusion to the walls of the reaction vessel where deactivation takes place, and collisional removal involving both spin orbit relaxation and chemical reaction. To observe the emission, the small value of A_{nm} in a given case for the electric dipole forbidden transition (Table II) must be offset by a relatively large concentration of excited atoms. As yet, for static flash photolysis experiments, spontaneous emission has only been detected and monitored for iodine atoms where A_{nm} is largest for the halogens in the $^2P_{1/2}$ state and the populations of $I(5^2P_{1/2})$ produced on photolyzing CF_3I very large.[3,11,27] In the experiments of Husain and Wiesenfeld,[11] the emission from $I(5^2P_{1/2})$ produced on the flash photolysis of CF_3I in inert gases enters the slit of a high aperture Bausch and Lomb grating monochromator and the output from a lead sulfide photoconductive cell, situated at the exit slit, is displayed on the screen of an oscilloscope. Figure 4 shows typical emission traces which indicate the long lifetimes of $I(5^2P_{1/2})$ in the presence of argon, of the order of milliseconds. Here the decay is primarily controlled by the rate of diffusion to the walls of the reaction vessel followed by deactivation, as the 21.7 kcal of electronic energy cannot be readily transferred to translational energy in the inert gas. These traces may then be processed for kinetic data. The plot of log (I_{emm}) versus time is sensibly linear for approximately 10 msec. The slope k, which is the over-all first-order coefficient, may be broken down into the different contributions to the over-all decay by variation of the reactant conditions.

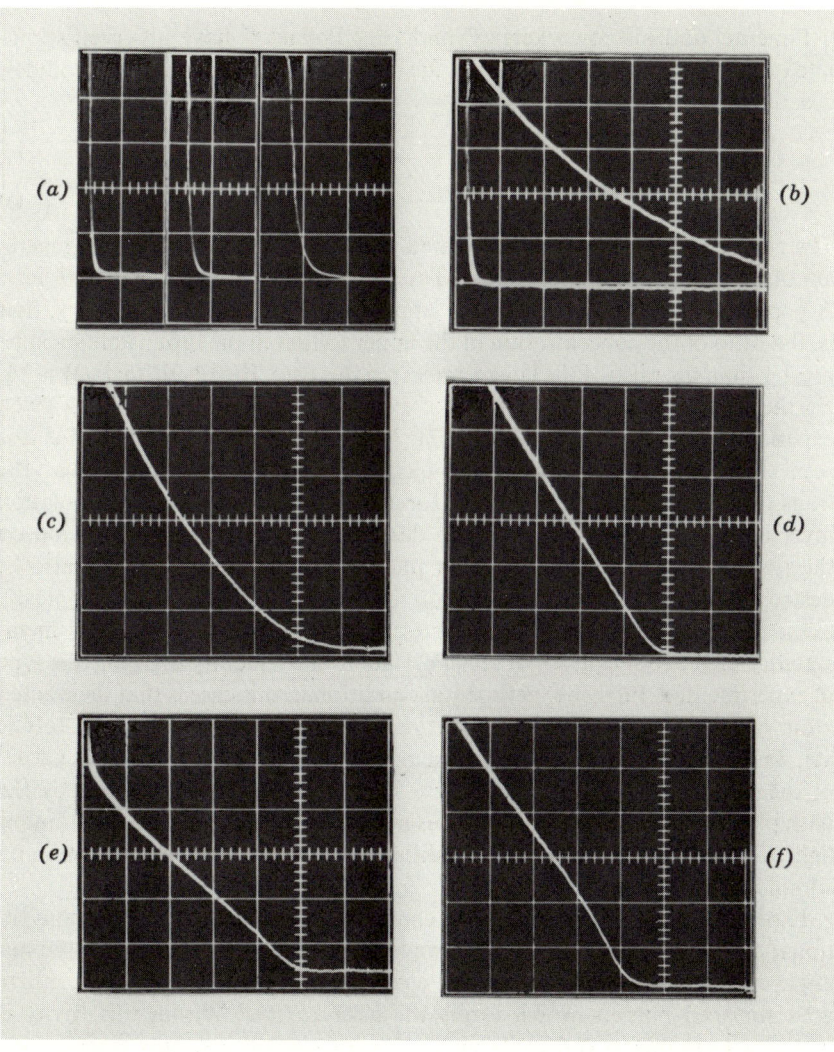

Fig. 4. Decay of the emission from $I(5^2P_{1/2})$ at different pressures of argon. $p_{CF_3I} = 1.00$ mm Hg, $E = 1767$ J. (a) Time response of apparatus. Photoflash on empty reaction vessel. Sensitivity, 10 mV/d. Time scales, left to right: 2, 1, and 0.5 msec/d. (b) $p_{Ar} = 10$ mm Hg, 2 mV/d, 2 msec/d. (c) $p_{Ar} = 20$ mm Hg, 2 mV/d, 5 msec/d. (d) $p_{Ar} = 40$ mm Hg, 2 mV/d, 5 msec/d. (e) $p_{Ar} = 50$ mm Hg, 2 mV/d, 5 msec/d. (f) $p_{Ar} = 60$ mm Hg, 2 mV/d, 5 msec/d.

C. Stimulated Emission from Excited Halogen Atoms Following Flash Photolysis

Pimentel and his coworkers[3,32] and later Pollack[33] have observed stimulated emission from $I(5^2P_{1/2})$ in the flash photolysis of a number of iodides, described by the processes

$$RI + h\nu(uv) \longrightarrow R + I(5^2P_{1/2}) \tag{7}$$

$$I(5^2P_{1/2}) + h\nu(1.315\ \mu) \longrightarrow I(5^2P_{3/2}) + 2h\nu(1.315\ \mu) \tag{8}$$

The conditions under which stimulated emission may be observed in general are discussed in detail elsewhere.[34] Three conditions will be mentioned here. A population inversion between the upper and lower states is necessary, that is, the ratio of the concentration in the upper to that in the lower state must be greater than the ratio of the degeneracies; in this case, $[I(5^2P_{1/2})]/[I(5^2P_{3/2})] > 2/4$. Further, a threshold concentration of excited atoms is necessary for a given experimental arrangement to overcome losses in the optical cavity. Thus it is seen that laser action from electronically excited states requires that the transition is sufficiently forbidden for the maintenance of the population inversion which would otherwise be destroyed by spontaneous emission on the time scale associated with flash photolysis. The inversion may also be destroyed by collisional deactivation of the species in the electronically excited state. This is discussed later (see Section VII). The optically metastability of the iodine atom in the $^2P_{1/2}$ state is conveniently high for this type of experiment. A further experimental condition in this case is that the excited atom must be produced in a population inversion in the primary photolytic act. This will depend on the photochemical excitation process and the nature of the appropriate molecular spectra. These conditions are fulfilled by the flash photolysis of a number of iodides in the quartz ultraviolet by continuous light. The details of the physical conditions will be discussed in the section on stimulated emission (Section VI).

Experimentally, the arrangement employed is similar to that for conventional flash photolysis studies. However, the reaction vessel has Brewster angle end plates and is situated in a confocal cavity formed by two concave gold surfaced mirrors which forms the laser cavity.[3] Within this cavity is placed a 1-cm quartz cell containing carbon tetrachloride, fully transmitting at 1.315 μ. If attenuation was required, the cell was filled with the appropriate mixture of water and acetone. A flat plate of quartz at 10° from the cavity axis deflected 6.4% of the light into an external optical system leading to a monochromator, the output of which was focused onto an indium antimonide photoelectric detector and displayed on the screen of an oscilloscope. The result is typically an envelope of stimulated emission spikes of over-all length of the order of 10 μsec., as opposed to the smooth spontaneous emission trace of some 10 msec and greater in the experiments of Husain and Wiesenfeld.[11] The stimulated emission profile does not lend itself readily to detailed kinetic analysis other than in the interpretation of the time delay between photochemical initiation and stimulated emission and the variation of the laser threshold with experimental conditions. The output as a function of experimental conditions has been studied by Pollack[33] and the resulting curves interpreted in terms of a simplified model. It should be emphasized that the absence of laser action in a given experiment does not necessarily indicate the absence of a population inversion since the condition of a suitable absolute concentration of excited atoms is required, over and above the condition of inversion. Nevertheless, within these conditions, Pimentel and his coworkers have greatly added to our understanding of primary photochemical and energy transfer processes. Stimulated emission has not been observed from the $^2P_{1/2}$ states of other halogen atoms.† Our knowledge of energy transfer processes in these systems is developed sufficiently for this to be interpreted with some confidence and will be given in Section VI on stimulated emission.

D. Atomic Emission in Flow Systems

Polanyi[35] and his coworkers have developed a number of methods for monitoring energized species produced in chemical reactions by following their infrared emission in appropriate cases. The majority of these studies have been concerned with emission from vibrationally excited diatomic molecules produced in metathetical reactions of atoms and diatomic molecules or in direct electronic energy transfer from an energized atom to a diatomic molecule. More recently, the infrared emissions from $I(5^2P_{1/2})$ and $Br(4^2P_{1/2})$ (Table III) have been used to monitor these excited atoms when they are produced in simple chemical and photochemical reactions.[15,17,34,36] This method measures the stationary concentrations of $I(5^2P_{1/2})$ and $Br(4^2P_{1/2})$ which will represent a balance between the production and removal processes of these energized atoms in given chemical systems. In the photolysis of HI by light of wavelength 2537 Å ($Hg\ 6^3P_1 \rightarrow Hg\ 6^1S_0 + h\nu$),[15,36] the observed emission intensity from $I(5^2P_{1/2})$ will express the stationary state concentration of the excited atoms in a mechanism including the processes:

$$HI + h\nu \longrightarrow H + I(5^2P_{1/2}), I(5^2P_{3/2}) \qquad (9)$$

$$H + HI \longrightarrow H_2 + I(5^2P_{1/2}), I(5^2P_{3/2}) \qquad (10)$$

$$I(5^2P_{1/2}) + M \longrightarrow I(5^2P_{3/2}) + M \qquad (11)$$

† Stimulated emission from $Br(4p^5\ ^2P_{1/2})$ following the flash photolysis of IBr has now been reported (C. R. Giuliano and L. D. Hess, J. Appl. Phys., **40**, 2428 (1969)).

and the removal of $I(5^2P_{1/2})$ by spontaneous emission and diffusion to the walls followed by deactivation. Reaction (10) has been studied independently by following the emission from $I(5^2P_{1/2})$ resulting from the reaction of H atoms, produced in a discharge of molecular H_2, with HI.[36] This mechanism has been confirmed by kinetic spectroscopic measurements in absorption in the vacuum ultraviolet.[37]

Experimentally, flow systems have been employed in which the emission from the excited atoms in a volume that can be regarded as the reaction vessel is integrated. This is in contrast to the study of the emission of light in systems involving, for example, H, N, and O atoms, where the emission is generally studied radially as a function of distance down the flow tube.[38] Thus $I(5^2P_{1/2})$ in the photolysis of HI has been studied by continuous photolysis in a low-pressure flow system[15,36] in which HI is pumped through a cylindrical, quartz reaction vessel and irradiated radially. The emission at 1.315 μ is observed axially by means of a cooled ($-80°C$) lead sulfide photoconductive cell at the output slit of a grating spectrometer, the stationary input radiation at this wavelength having been chopped mechanically at 480 Hz prior to entering the spectrometer. This is the same experimental arrangement that Polanyi et al.[39] have employed to study the infrared emission from vibrationally excited CO and NO formed from electronic energy transfer from $Hg(6^3P_{1,0})$. Alternatively, the integrated emission from $I(5^2P_{1/2})$ may be observed axially when H atoms from a Wood's discharge of H_2 or from a discharge obtained by passing the gas through a quartz tube in a microwave cavity are mixed with HI and excess argon in a rapid flow system. Weak emission at 1.315 μ was observed with this arrangement when Cl and Br atoms were added to HI but strong emission resulted from O + HI.[36]

Emission at 2.71 μ has been observed by Polanyi et al.[17] when H atoms from a Wood's discharge have been mixed with Br_2 and HBr in a fast flow system. Here, the integrated emission from inside a 12-liter spherical bulb, coated internally for reflection with a thin layer of gold or gold and magnesium fluoride, passed through a sapphire slot and was incident on a grating spectrometer coupled with a PbS photoconductive cell at $-80°C$ as previously described. The kinetics of the production of $Br(4^2P_{1/2})$ in these systems have not been as fully elucidated as the analogous systems for $I(5^2P_{1/2})$. Polanyi et al.[17] suggest that the reaction between H and HBr may give rise to these excited atoms. In earlier experiments involving reaction between H + Cl_2 and H + HCl by Cashion and Polanyi,[40] emission from $Cl(3^2P_{1/2})$ was not reported. An important aspect in these experiments that Polanyi and his coworkers have developed is the calibration of the photoelectric output from the photocell monochromator arrangement for absolute intensities. This is achieved absolutely by recording the thermal, equilibrium infrared emission from a diatomic molecule such as CO or HCl, thus enabling the

E. Electron Spin Resonance Studies of Electronically Excited Halogen Atoms

In recent years, there has been a considerable growth of output on esr studies of transient species in the gas phase, particularly of atoms and diatomic molecules of kinetic interest. Resonant absorption has been reported for the $^2P_{3/2}$ states of F, Cl, Br, and I[41] and, more recently, Carrington and his coworkers have studied the spectra of the $^2P_{1/2}$ states of the halogen atoms F and Cl in some detail.[42] Gases such as CF_4 and CF_3Cl were subjected to a microwave discharge to produce F and Cl atoms, respectively, and the products rapidly pumped into the cavity of an esr spectrometer. Non-Boltzmann distributions of F and Cl in the upper states were observed but the emphasis was kept to the spectroscopic analysis with only a preliminary measurement on the lifetime of the excited species with varying experimental conditions. The further study of the lifetime of $F\,^2P_{1/2}$ in a non-Boltzmann population (Table I) will be particularly useful as this data cannot be forthcoming from vacuum ultraviolet kinetic spectroscopic measurements. The atomic transitions of fluorine lie below the cutoff of lithium fluoride which is the optical material employed for vacuum ultraviolet studies (Table IV). Measurements on the other atoms will provide a useful confirmation of relaxation rates determined by different techniques.

F. Molecular Halogen Emission in Flow Discharge Systems

Molecular halogen emission in flow systems in which a halogen is partially dissociated into atoms by means of a discharge may be used to monitor the concentration of the recombining atoms along the flow. Since this review is concerned with electronically excited halogen atoms, our interest in this context is focused on molecular halogen emission from states that correlate with one or more excited atoms. The bound state correlating with $^2P_{1/2} + \,^2P_{3/2}$ atoms is the $^3\Pi_{0u^+}$ state and in atomic recombination processes, this can be populated via two alternative routes: the direct recombination of a $^2P_{1/2}$ and a $^2P_{3/2}$ atom, stabilized by an over-all three-body process, or the recombination of two ground state atoms to yield the molecular halogen in a repulsive state followed by an inverse predissociation into the $^3\Pi_{0u^+}$ state. The kinetic study of such processes requires a correlation between the molecular emission and the kinetics of the atoms. An important aspect of this type of investigation is a detailed understanding of the appropriate potential energy curves, particularly in the region of crossing between the repulsive state correlating with ground state atoms and the $^3\Pi_{0u^+}$ state. This is usually ascertained by observation of perturbations in the rotational structure when two curves approach closely or cross. Thus the observation of molecular emission from vibrational states in the $^3\Pi_{0u^+}$ state above that where crossing takes place supports direct recombination between $^2P_{1/2}$ and $^2P_{3/2}$ atoms, and conversely, the absence of such emission supports a mechanism involving inverse predissociation with the proviso that this could conceivably result from rapid vibrational relaxation following direct recombination. Further, the temperature dependence of the $^3\Pi_{0u^+}$ emission for an inverse predissociation mechanism should correspond to the energy barrier between the dissociated $^2P_{3/2}$ atoms and the point of crossing of the potential energy curves. This energy barrier is often small, though not for I_2 (see Section VIII.D), and does not always constitute a critical mechanistic test. The detailed mechanism in particular cases will be considered in the section on atomic recombination (Section VIII). Kinetic studies on the emission from the $^3\Pi_{0u^+}$ state in flow discharge systems have been made for Br_2[43,44], Cl_2,[45-48] and BrCl.[49] Only continuous emission has been observed for I_2 in a flow discharge system in an active nitrogen-iodine flame,[50] and it has been proposed by Phillips and Freeman that this results from the recombination of two ground state atoms. Certainly the Boltzmann population of $I(5^2P_{1/2})$ would be negligibly small (Table I). To the best knowledge of the authors, studies of this type for F_2 have not been reported. Orgyzlo[51] has reported this material to be too difficult to handle in the conventional flow discharge system. He has also reported difficulty in obtaining iodine atoms from an electrically heated discharge in a flow system.

Chlorine and bromine atoms have been monitored in flow discharge systems by measuring the heat liberated on a probe when a pair of atoms recombine on the surface.[43,51,52] Ogryzlo[51] has employed a moveable nickel calorimeter for both Cl and Br atoms. Chemical titration of the chlorine atoms may then be developed by comparing the effect of the addition of a material that rapidly removes the atoms observed with the calorimetric probe. Ogryzlo[51] has found that NOCl may be used for titration of the Cl atoms as the reaction

$$Cl + NOCl \longrightarrow NO + Cl_2 \quad (12)$$

is rapid and that reformation of NOCl via the reactions

$$Cl + NO + M \longrightarrow NOCl + M \quad (13)$$

$$Cl_2 + 2NO \longrightarrow 2NOCl \quad (14)$$

is comparatively slow. The end point of the titration is observed as the intercept at [Cl] = 0 of the plot of the probe response versus the flow rate of NOCl. It also appears that the reaction

$$Cl + ICl \longrightarrow Cl_2 + I \quad (15)$$

may be of use as a titrant process.[51] As yet, detailed chemical titration techniques for Br atoms have not been reported. Neither the probe nor the chemical titration methods differentiate between $^2P_{1/2}$ and $^2P_{3/2}$ atoms although, in principle, there are differences for the heats liberated on the probes for these atoms. The role of $^2P_{1/2}$ atoms in such systems is determined from the nature of the molecular halogen emission $^3\Pi_{0u^+} \to X^1\Sigma_g^+ + h\nu$, and its dependence on the over-all atomic concentrations. Hutton and Wright[47] have found that $I_{emm}(B-X)$ is proportional to $[Cl]^2$, the low wavelength limit corresponding to what appeared then to be the onset of predissociation at $v' = 13$. Bader and Ogryzlo[45] have found that the $B-X$ emission for Br_2 was proportional to $[Br]^2$; again levels above the onset of predissociation ($v' > 3$) were not observed. Clyne et al.,[49] however, have detected emission from $Cl_2(^3\Pi_{0u^+})$ involving levels up to $v' = 14$. Their recent work[48] has also shown that the dependence of the emission on $[Cl]^2$ is not rigorously held over the whole range of Cl atom concentration as could reasonably be expected on account of the simplifications in the analysis of the stationary state scheme leading to this relation.[47] Any major role played by $^2P_{1/2}$ atoms is thus considered in relation to the thermal (Boltzmann) population in this state at the temperature in the flow discharge system (near room temperature), the nature of the banded emission from the $^3\Pi_{0u^+}$ state particularly with regard to the onset of predissociation, and the importance of vibrational relaxation processes in the upper state. These will be discussed in detail in the section on atomic recombination (Section VIII).

G. Molecular Halogen Emission in Shock-Heated Gases

Emission from the $^3\Pi_{0u^+}$ state has been observed when both Br_2 and Cl_2 have been adiabatically shock heated[53-56] together with molecular emission from states correlating with two ground state atoms. Continuous emission from $Br_2(^3\Pi_{0u^+})$ was observed by Palmer and Hornig[2,53] from shock heated bromine and attributed to direct two-body recombination between $Br(^2P_{1/2})$ and $Br(^2P_{3/2})$. The emission from $Cl_2(^3\Pi_{0u^+})$ in shock tube studies[54,55] is observed to be both banded and continuous and results from an equilibrium concentration of $Cl_2(^3\Pi_{0u^+})$. Emission was not observed when fluorine gas was shocked[56] and has not been reported in studies of I_2.[57] Experimentally, the emission profiles from behind the shock front at different wavelengths can be studied by means of wave band filters together with the appropriate photoelectric recording. In the case of Cl_2,[55] it was possible to take an emission spectrogram on a photographic plate by superposition of the emission from repeatedly shocked samples of the same gaseous mixture. These emission studies have extended our understanding of the nature of the thermal dissociation of a diatomic molecule, of the role played by electronically excited halogen atoms, and of radiation in the recombination process. The function of the technique is not to provide an environment for studying the chemistry of excited halogen atoms in general, but to study the part they play in the dissociation and recombination processes of the molecular halogens, in particular.

H. Time-Resolved Molecular Halogen Emission Following Flash Photolysis

Time-resolved emission from $I_2(B^3\Pi_{0u^+})$ has been studied in a static system by Abrahamson, Husain, and Wiesenfeld[23] following the flash photolysis of CF_3I. The basis of the mechanism giving rise to the emission may be written:

$$CF_3I + h\nu \longrightarrow CF_3 + I(5^2P_{1/2}) \tag{16}$$

$$I(5^2P_{1/2}) \longrightarrow I(5^2P_{3/2}) \tag{17}$$

$$I(5^2P_{1/2}) + I(5^2P_{3/2}) + M \longrightarrow I_2(B^3\Pi_{0u^+}) + M \tag{18}$$

$$I_2(B^3\Pi_{0u^+}) \longrightarrow I_2(X^1\Sigma_g^+) + h\nu \tag{19}$$

Thus, following the flash photolysis of a CF_3I/inert gas mixture, the emission is incident upon a monochromator and the intensity recorded as a function of time. In the above mechanism, (17) represents the sum of the first-order coefficients of all the processes giving rise to the removal of $I(5^2P_{1/2})$; namely, spontaneous emission, diffusion to the walls of the reaction vessel followed by efficient deactivation, and collisional deactivation in the gas phase. Processes (18) and (19), representing part of the over-all atomic recombination mechanism, hardly affect the concentration of $I(5^2P_{1/2})$ over much of the period in which the $I_2(B-X)$ emission is observed where the decay of $I(5^2P_{1/2})$ is sensibly first-order, as seen in earlier studies (see Section III.B). Thus the nonsteady state differential equations for $I(5^2P_{1/2})$, $I(5^2P_{3/2})$, and $I_2(B^3\Pi_{0u^+})$ can be solved with a number of experimentally justifiable simplifying assumptions and the molecular emission can be used to study the kinetics of the excited atoms. The emission in a CF_3I/inert gas mixture is long lived, of the order of 10 msec, and comparable with the lifetime of the optically metastable $I(5^2P_{1/2})$ (Fig. 5). The mean radiative lifetime of $I_2(B^3\Pi_{0u^+})$ is 7.2×10^{-7} sec[58] and thus the partially forbidden $I_2(B-X)$ emission follows its long lived kinetic precursor, $I(5^2P_{1/2})$. The emission can either be studied at one wavelength as a function of time, or, as the emission is relatively long lived, a rapid scan spectrometer which scans across the whole wavelength region 4620–7570 Å every 1.25 msec can be employed.[23] The $I_2(B-X)$ emission thus starts at the onset of the $B-X$ continuum normally observed in absorption,

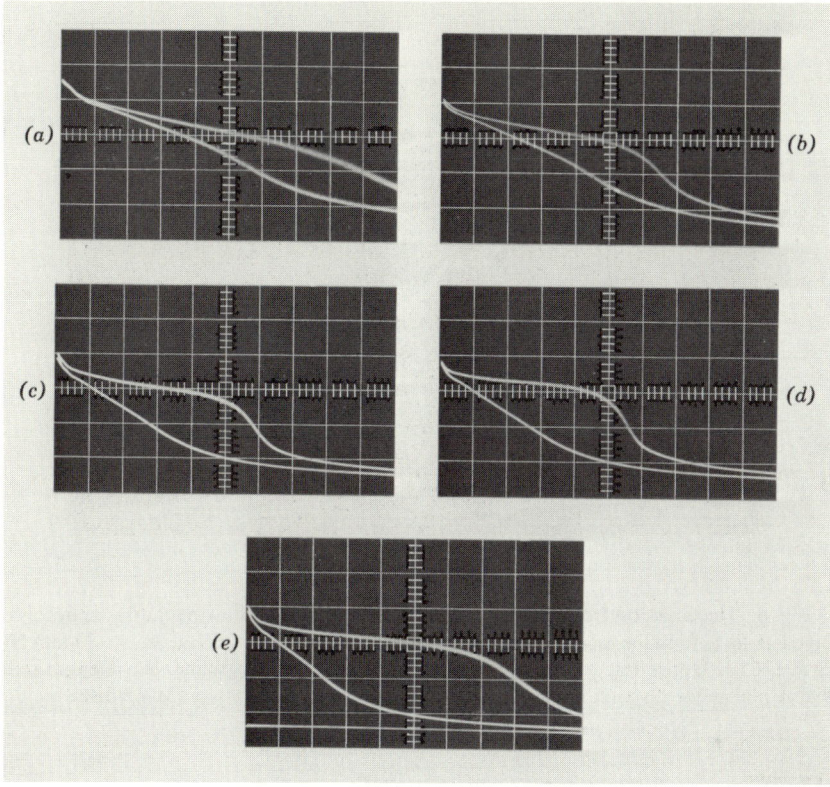

Fig. 5. Time-dependent emission of $I_2(B^3\Pi_{0u^+}) \rightarrow I_2(X^1\Sigma_g^+) + h\nu$ following the flash photolysis of CF_3I. Wavelength monitored at 5430 Å. Upper trace, $I_2(B - X)$ emission; lower trace, baseline. Horizontal scale, 1 msec/d; vertical scale 0.5 mV/d. (a) to (d), $p_{Ar} = 40$ mm Hg, (e) $p_{Ar} = 10$ mm Hg. p_{CF_3I}: (a) 1.0 mm Hg, (b) 2.0 mm Hg, (c) 3.0 mm Hg, (d) 4.0 mm Hg, (e) 4.0 mm Hg.

and extends into the banded region to longer wavelengths (Fig. 6). This clearly shows the production of $I_2(B^3\Pi_{0u^+})$ by direct three-body recombination via (18) as the emission is banded, and not by means of inverse predissociation. In any event, the energy barrier to inverse predissociation would be too high (16.5 kcal)[23] in this case. Similar emissions from the $^3\Pi_{0u^+}$ states on a time-resolved basis following the flash photolysis of other suitable halides have not been reported to date. The observation of $I_2(B-X)$ is facilitated by the very high concentrations of $I(5^2P_{1/2})$ that result on flash photolyzing CF_3I in the ultraviolet.[3,27] The $^3\Pi_{0u^+}$ states of Br_2 and Cl_2 have been detected by kinetic spectroscopy in absorption on flash photolyzing

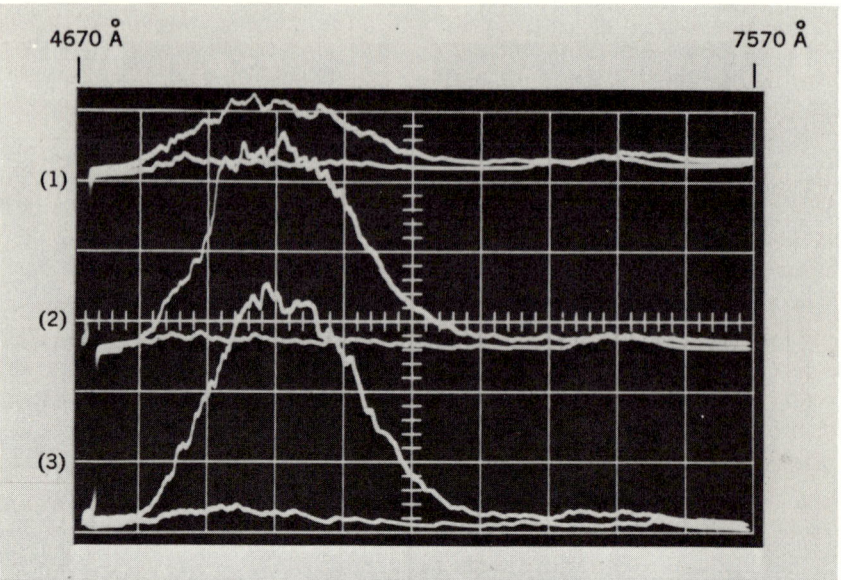

Fig. 6. The emission from $I_2(B^3\Pi_{0u^+}) \rightarrow I_2(X^1\Sigma_g^+) + h\nu$ following the flash photolysis of CF_3I as a function of wavelength and time. $p_{CF_3I} = 2.0$ mm Hg, $p_{Ar} = 80$ mm Hg, $E = 1767$ J. Upper trace, $I_2(B \rightarrow X)$ emission; lower trace, baseline. Wavelength scale 290 Å per major division. Delay times: (1) 2.0 msec, (2) 3.25 msec, (3) 4.5 msec.

these halogens.[59] The lifetimes of the excited molecular states were essentially comparable to the excitation flash and a detailed kinetic analysis of the mode of formation was not carried out. Briggs and Norrish[59] proposed that both mechanisms involving direct excitation by the photolysis flash raising the molecule in the $X^1\Sigma_g^+$ state to the $^3\Pi_{0u^+}$ state and also recombination of ground and excited atoms into this state could account for the observations.

I. Time-Resolved Mass Spectrometric Studies

The rapid formation of molecular iodine following the flash photolytic dissociation of CH_3I has been observed by time-resolved mass spectrometry.[60] This has been attributed to the reaction (20) rather than to slow termolecular recombination. The experimental difficulties associated with sampling by this technique have been discussed by Meyer.[61] This reaction is further discussed in Section IX.D on reaction of $I(5^2P_{1/2})$ with alkyl iodides.

$$I(5^2P_{1/2}) + CH_3I \longrightarrow I_2 + CH_3 \qquad \Delta H = -3.3 \text{ kcal} \qquad (20)$$

J. Classical Photochemical Studies

In classical photochemical studies, it has generally been appreciated that $^2P_{1/2}$ halogen atoms could be produced in the photolyses of halides.[62,63] However, often no differentiation was made in mechanistic descriptions as differences in reactions of halogen atoms in the $^2P_{1/2}$ and $^2P_{3/2}$ states could not be made. This is particularly true for chlorine atoms where the small difference in electronic energy (Table I) renders the atoms similar in chemical reactivity in most cases. In the case of iodine, however, the difference in electronic energy of 21.7 kcal causes a large difference in chemical reactivity which may be observed experimentally. This is particularly noticeable in metathetical atomic abstraction reactions of $I(5^2P_{1/2})$ and $I(5^2P_{3/2})$. This difference has been exploited in classical photochemical studies as well as in flash photolytic investigations. In straight photolysis experiments combined with conventional chemical analysis, this has been observed in H atom abstraction reactions for paraffins.[64-66] The thermochemistry of abstraction from ethane and propane

$$I(5^2P_{1/2}), I(5^2P_{3/2}) + C_2H_6 \longrightarrow HI + C_2H_5 \quad \Delta H = 4.9, 26.6 \text{ kcal mole}^{-1} \quad (21)$$

$$I(5^2P_{1/2}), I5^2P_{3/2}) + C_3H_8 \longrightarrow HI + C_3H_7 \quad \Delta H = 1.9, 23.6 \text{ kcal mole}^{-1} \quad (22)$$

clearly indicate the facility of reaction by $I(5^2P_{1/2})$ on an energy basis. Experimentally, iodine vapor is photolyzed in the transition $(B^3\Pi_{0u^+} \leftarrow X^1\Sigma_g^+)$ both in the continuum and in the banded region via a "collisional release" mechanism (see Section IV.B) to yield

$$I_2 + h\nu \longrightarrow I(5^2P_{1/2}) + I(5^2P_{3/2}) \quad (23)$$

in the presence of propane and other added inert gases.[65,66] The ethyl and propyl radicals resulting from the above reactions with the excited atoms then undergo further reaction with the I_2 present:

$$C_2H_5 + I_2 \longrightarrow C_2H_5I + I(5^2P_{3/2}) \quad (24)$$

$$C_3H_7 + I_2 \longrightarrow C_3H_7I + I(5^2P_{3/2}) \quad (25)$$

and the quantum yield for alkyl iodide formation is measured as a function of different reactant conditions. These two systems have been studied in detail by Callear and Wilson[65,66] and clarify the earlier observations of Filseth and Willard.[64] The application of stationary state kinetics and the final expression for the quantum yield describes the result in terms of the rate contributions to the collisional quenching of $I(5^2P_{1/2})$ by the different gases present *relative* to that of the abstraction reaction. Thus these studies can be put on an absolute basis using absolute rate measurements on the deactivation of these excited atoms determined by kinetic spectroscopic methods (see Sections III.A and III.B). The chemical reactions of $I(5^2P_{1/2})$, including these systems, are discussed in detail later (Section IX).

IV. PRIMARY PHOTOCHEMICAL PROCESSES RESULTING IN THE PRODUCTION OF ELECTRONICALLY EXCITED HALOGEN ATOMS

A. General Considerations

A consideration of the production of electronically excited halogen atoms in primary photochemical processes, particularly for diatomic molecules, may be based on detailed spectroscopic studies where these are available. Although in the case of the photolysis of the homonuclear halogens the atomic products of dissociation may be predicted with confidence, this is not so straightforward with the interhalogens as the appropriate potential energy diagrams, and the effects of "avoided crossing" are not so well understood. With polyatomic halides, the electronic states are generally not sufficiently defined to allow the prediction of the spin orbit state of a halogen atom resulting on photolysis. States resulting from continua, particularly those which are not derived from the extension of banded structure at longer wavelength, also may not be fully understood† (see footnote). Hence the prediction of the particular spin orbit state of an atom produced on photolysis may be very difficult when the splittings between J multiplets are small. Further, a continuum may be associated with the overlapping of transitions for a number of systems and the relative contributions from each unknown. Methods of direct observation for atoms resulting from photolysis will therefore confirm the details in some processes and provide the only information in others. An important aspect of methods of direct observation involves the time resolution of the experiment; namely, whether electronic energy transfer has altered significantly the distribution in electronic states before observations are made. This will differ for different experimental systems and for different chemical systems. For example, molecular I_2 deactivates $I(5^2P_{1/2})$ more efficiently than CF_3I by a factor of approximately 2×10^{4} [67] and thus for given experimental conditions, the distribution in the $^2P_{1/2}$ and $^2P_{3/2}$ states of the iodine atom from CF_3I will represent more closely the initial distribution in a kinetic spectroscopic observation. Similarly, the $^2P_{1/2}$ states of iodine, bromine, and chlorine relax at different rates. In many cases, although the observations may not be close to the initial distribution, they will still permit the elucidation of some important primary processes.

In one case the large difference in chemical reactivity between the excited and ground state iodine atoms has permitted classical investigations to be used in the elucidation of a primary photochemical process.[65,66] This method

† The technique of "Translational Spectroscopy", developed by Wilson and coworkers, now allows such continua to be assigned and the electronic states of the photolysis fragments to be determined. Further details are given in an addendum at the end of this review.

could, in principle, be extended to the other halogen atoms, and in particular, to the bromine atom (Section IX.B).

B. Primary Processes, the Halogens Cl_2, Br_2, and I_2

The photodissociation products of the homonuclear halogens in the visible and ultraviolet are now comparatively well established in view of the detailed spectroscopic studies that have been made. The strongest absorption system observed in this spectral region is associated with a transition to the $^3\Pi_{0u^+}$ state which correlates with $X(^2P_{3/2}) + X(^2P_{1/2})$. Thus photoexcitation to the continuum associated with this state leads directly to the formation of an excited atom, while excitation to the banded region followed by predissociation will lead only to ground state atoms.

Mathieson and Rees[68] have given the most detailed potential energy diagram of I_2. The strong absorption observed in the visible region is associated with a transition to the $I_2(B^3\Pi_{0u^+})$ state, which converges to a continuum at $\lambda = 4995$ Å, thus giving rise to excited atoms for irradiation at shorter wavelengths. The $B^3\Pi_{0u^+}$ state is predissociated at $v' = 26$ by $I_2(C^3\Sigma_u^+)$[68] and leads to ground state atoms when transfer to this state is induced by strong magnetic fields[69] or perturbation following collisions.[70] The presence of a further predissociating state which crosses the $B^3\Pi_{0u^+}$ state close to the continuum has recently been proposed.[71] The precise nature of this state has not been established; however, it appears to induce very efficient spontaneous dissociation. Inverse predissociation accompanied by emission following atomic recombination is a sensitive method of determining the vibrational level in the $^3\Pi_{0u^+}$ state at which predissociation occurs; however, this has not been possible in the case of I_2 in view of the large energy barrier involved[23] (see Section VIII.D). The $B^3\Pi_{0u^+}$ state of Br_2 is predissociated at $v' = 3$,[43,72,73] possibly by the $A^1\Pi_{1u}$ state. This is well supported by emission from $Br_2(^3\Pi_{0u^+})$ from $v' \leq 3$ resulting from recombining bromine atoms. The detection of emission from $Cl_2(A^3\Pi_{0u^+})$ ($v' \leq 14$) has been observed following the recombination of chlorine atoms by Clyne and Coxon.[49] Parts of the potential energy diagrams for Cl_2 are given in a number of papers.[47,59,74]

Transient atoms resulting from the isothermal photolytic dissociation of the halogens have been observed by kinetic spectroscopy in absorption in the vacuum ultraviolet. Both $I(5^2P_{1/2})$ and $I(5^2P_{3/2})$ have been detected from the photolysis of I_2; however, the absorption transition of the excited atom was observed only weakly as undissociated I_2 rapidly quenched a large fraction of the excited atoms, with an efficiency of 1 in 30 collisions, before the first observation was made.[75] Similarly, a weak absorption by $Br(4^2P_{1/2})$ and a strong one by $Br(4^2P_{3/2})$ was recorded following the photolysis of Br_2, again

Br($4^2P_{1/2}$) being rapidly quenched by any Br$_2$ present.[76] While Cl($3^2P_{1/2}$) would be expected from the photolysis of Cl$_2$, the strongest absorption transition of the excited atom at 1351.7 Å (Table IV) was obscured by the molecular spectrum of undissociated Cl$_2$ and only an absorption transition of the ground state atom at 1335.7 Å (Table IV) could be detected through a "window" in the vacuum ultraviolet molecular spectrum.[29]

Nonpredissociative photolytic fission by photoexcitation into the banded region close to the continuum has recently been observed in iodine and bromine molecules. Callear and Wilson[65,66] have observed photochemical reaction when molecular iodine is photolyzed at energies below that corresponding to the continuum of the system I$_2$($B^3\Pi_{0u^+} \leftarrow X^1\Sigma_g^+$), in the presence of propane, following an earlier preliminary observation by Filseth and Willard.[64] The production of I($5^2P_{1/2}$) was demonstrated for these conditions by its observation in absorption following flash photolysis.[77] The nature of the dissociation process has been postulated by Callear et al. as due to collisional release with species M, namely,

$$M + I_2(B^3\Pi_{0u^+}) \longrightarrow MI + I(5^2P_{1/2}) \qquad (26)$$

Tiffany[78,79] has employed a tuned, pulsed ruby laser to excite Br$_2$ to within 500–800 cm^{-1} below the dissociation continuum of the $^3\Pi_{1u}$ state (correlating with ground state atoms) and has observed the reaction of the bromine atoms resulting from the dissociation. By contrast with the collisional release mechanism, Tiffany has proposed a process in which the energy for dissociation for a small number of the Br$_2$($^3\Pi_{1u}$) molecules into ground state atoms is provided by collisions.

C. Primary Process, the Interhalogens

The electronic spectra of the interhalogen molecules IBr,[69,80–90] ICl,[80–82,84,85,90–92] and BrCl[49,82,84,85,90,93] have been studied in some detail in the visible, ultraviolet, and vacuum ultraviolet. The states of IBr and ICl, analogs to the $^3\Pi_{0u^+}$ and associated repulsive states of the homonuclear halogens, are characterized by a perturbation formally described in terms of a strong "avoided crossing" (Fig. 7). This arises from a loss of g-u symmetry in the interhalogens with consequent mixing of the $^3\Pi_{0^+}$ and O$^+$ states. However, discrete quantum states associated with the $^3\Pi_{0^+}$ state may be observed at wavelengths corresponding to energies above the potential maximum formed by this avoided crossing. This has been described by Van Vleck[69] as "the vestigial remains of the original $^3\Pi_{0^+}$ state." The atomic states resulting from excitation above the potential maximum are thus determined by factors governing radiationless transitions; namely, the energy of splitting between the two molecular states, the absolute difference in the slopes of the potential energy curves for the states where crossing takes place,

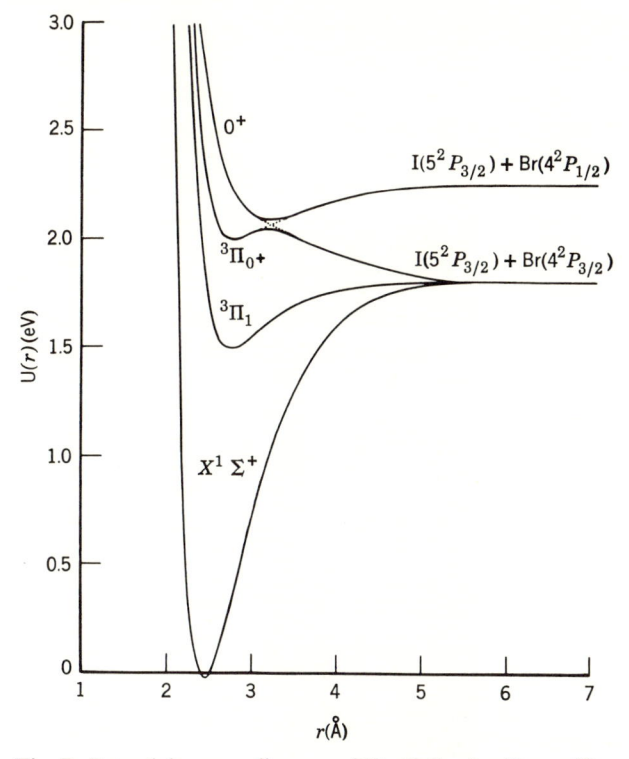

Fig. 7. Potential energy diagram of IBr (following Brown[83]).

and the rate of change of internuclear distance on passing through the region of intersection.[94–96] Donovan and Husain[90] have studied the atomic states directly by kinetic spectroscopy in absorption in the vacuum ultraviolet following the flash photolysis of the interhalogens. Photoexcitation of IBr to a repulsive region of the $^3\Pi_{0^+}$ state, well above the region of avoided crossing, gives rise to I($5^2P_{3/2}$) and Br($4^2P_{1/2}$) (see footnote p. 15):

$$\text{IBr}(X^1\Sigma) + h\nu \longrightarrow \text{I}(5^2P_{3/2}) + \text{Br}(4^2P_{1/2}) \tag{27}$$

Thus, under these conditions, the IBr molecule moves through the region of avoided crossing too rapidly for spin orbit mixing to take place and gives rise to products associated with the original $^3\Pi_{0^+}$ state, rather than the fully mixed final $^3\Pi_{0^+}$ state [correlating with I($5^2P_{3/2}$) + Br($4^2P_{3/2}$)] (Fig. 8 and 9). Br($4^2P_{3/2}$) was not detected due to (a) its lack of production in the primary photolytic process described and (b) rapid reaction of Br($4^2P_{1/2}$) with IBr (see Section IX.F) rather than spin orbit relaxation. Similar results were found for ICl although in this case only I($5^2P_{3/2}$) was observed. The absence of the excited chlorine atom can be attributed to its rapid reaction with ICl following its production on photolysis.[90] Although the system BrCl($^3\Pi_{0^+} - X^1\Sigma$) has

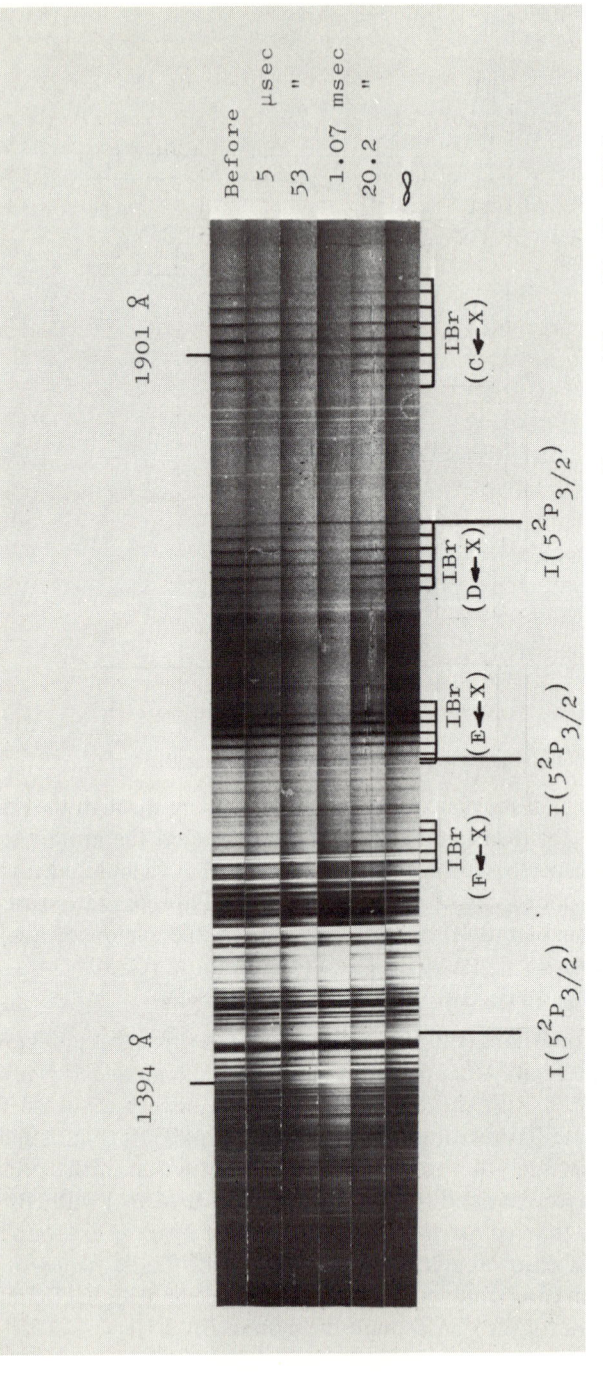

Fig. 8. Spectra resulting from the flash photolysis of IBr. $p_{IBr} = 0.07$ mm Hg, $p_{Ar} = 100$ mm Hg, $E = 780$ J.

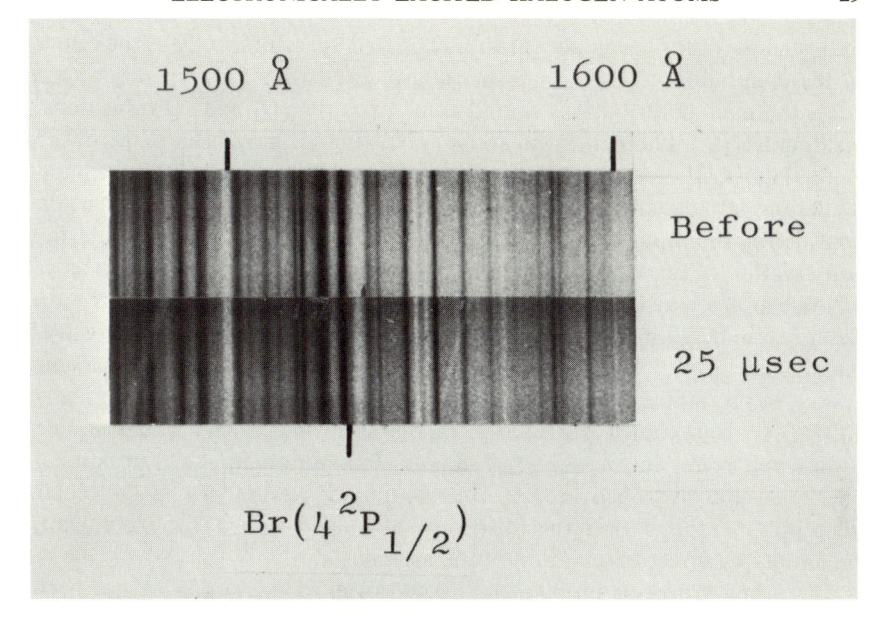

Fig. 9. Electronically excited bromine atoms, Br($4^2P_{1/2}$), resulting from the flash photolysis of IBr. $p_{IBr} = 0.03$ mm Hg, $p_{Ar} = 100$ mm Hg, $E = 780$ J.

now been observed in both emission [49] and absorption,[93] the details of the potential energy diagram of this molecule are not established, particularly that part due to the absorption continuum of this molecule in the quartz region. The photolysis of BrCl resulted only in the detection of Br($4^2P_{3/2}$) and the primary process may be written:[90]

$$\text{BrCl}(X^1\Sigma) + h\nu\ (\lambda > 2000\ \text{Å}) \longrightarrow \text{Br}(4^2P_{3/2}) + \text{Cl}(3^2P_{1/2, 3/2}) \tag{28}$$

D. Primary Processes, the Hydrogen Halides HCl, HBr, and HI

The absorption spectra of the hydrogen halides are continuous in the ultraviolet (HI,[97–99] HBr,[100] HCl[98,101]) and result from excitation to a number of repulsive states. HF shows no appreciable absorption in the ultraviolet.[102] Mulliken[103–105] has considered the electronic states of the hydrogen halides in some detail and has concluded that the observed continua result from three overlapping transitions from the ground state to three repulsive states, designated Q states for these particular molecules.[103] The $^3\Pi_{0^+}$ state correlates with electronically excited halogen atoms ($^2P_{1/2}$) whereas the other two, $^3\Pi_1$ and $^1\Pi$, correlate with ground state hydrogen and halogen atoms. The transitions have been designated $Q_0(^3\Pi_{0^+} \leftarrow X^1\Sigma^+)$, $^1Q(^1\Pi \leftarrow X^1\Sigma^+)$ and $^3Q_1(^3\Pi_1 \leftarrow X^1\Sigma^+)$. The HI molecule has been considered in some detail.[98,104] Recent Franck-Condon considerations of Mulliken's potential energy diagram[104] for HI by Martin and Willard[106] has indicated that the Q_0 component should contribute significantly at the center of the continuum at 2180 Å and hence give rise to $I(5^2P_{1/2})$, whereas at the edges the main contributions should come from the $^3Q_1$ and ^1Q transitions, yielding $I(5^2P_{3/2})$. Photochemical studies involving HI have been numerous.[62] Both HI and DI are convenient sources of translationally hot hydrogen and deuterium atoms on photolyzing these molecules in the ultraviolet. By considering the input of light energy above the dissociation energy and the conservation of linear momentum on photolytic fission, the detailed kinetics of these atoms may be studied as a function of kinetic energy. This has been described in detail most recently by Kuppermann.[107] Only in the long wavelength transition of the continuum ($^3Q_1$) is the distribution of the electronic energy in the iodine atoms known with any confidence, that is, exclusively $I(5^2P_{3/2})$.† An unknown distribution in the electronic energy of the iodine atoms will result in an unknown energy distribution in the translational energy of the hydrogen atoms. The continuous nature of the molecular absorption requires that this distribution be determined by experiment, preferably by direct observation of the atomic states.

The primary process on photolyzing hydrogen iodide at 2537 Å has been studied in detail by Polanyi et al.,[15,36] in particular, the relative yields of excited to ground state atoms generated at this wavelength. These are derived from a kinetic study of the spontaneous emission from $I(5^2P_{1/2})$ at 1.315 μ (Table III) in a flow system in which (a) H atoms are generated by the photolysis of HI and (b) atoms are generated in a Wood's discharge or microwave cavity. Following Polanyi,[15,36] we may write the relevant reactions as follows:

$$\text{HI} + h\nu(2537\text{ Å}) \begin{array}{c} \xrightarrow{k_4^*} \text{H} + I(5^2P_{1/2}) \\ \xrightarrow{k_5} \text{H} + I(5^2P_{3/2}) \end{array}$$

$$\text{H} + \text{HI} \begin{array}{c} \xrightarrow{k_6^*} \text{H}_2 + I(5^2P_{1/2}) \\ \xrightarrow{k_7} \text{H}_2 + I(5^2P_{3/2}) \end{array}$$

Polanyi's experiments yield the measurement of $k_4^*/k_5 + k_6^*/k_7$ for the condition (a), and k_6^*/k_7 for (b). Hence k_4^*/k_5 is determined by difference. The value of k_4^*/k_5 from Polanyi's measurements depends on the magnitude of

† The most recent study of the photolysis of HI at three wavelengths in a static system reports the following yields of $I(5^2P_{3/2})$:

93 ± 10% at 2537 Å
81 ± 10% at 2288 Å
100 ± 10% at 1850 Å

(L. E. Compton and R. M. Martin, J. Phys. Chem., **73**, 3474 (1969)).

TABLE V
Maxima and Extinction Coefficients of the Continuous
Absorption for the Hydrogen Halides $(Q \leftarrow N^1\Sigma^+)$[98]

	λ_{max}, Å	ϵ, M^{-1} cm^{-1}
HCl	1535	890
HBr	1785	525
HI	2150	175

the Einstein coefficient for spontaneous emission A_{nm} that is employed. This has been calculated by Garstang[13] to be 7.8 sec^{-1} and measured by Husain and Wiesenfeld[11] as 22 ± 6 sec^{-1}. Using the calculated value for A_{nm}, this yields $k_4^*/k_5 = 1.50 \pm 0.4$ whereas the measured value gives $k_4^*/k_5 = 0.55 \pm 0.25$ (neglecting the small ratio of k_6^*/k_7 measured as <0.02[36]). The smaller value is in accord with other recent work.[108] The significance of the difference between the two values for k_4^*/k_5 is principally that the higher value suggests that the production of an atomic iodine photodissociation laser from HI is a possibility whereas the smaller does not. The analogous ratios for HBr and HCl have not been determined by this method.

Kinetic spectroscopy in the vacuum ultraviolet has been employed to study the primary photolytic processes in the hydrogen halides. The nature of the method involves light absorption by the molecule over a range of wavelengths given out by the exciting photoflash rather than by monochromatic excitation. Romand[98] has reported the maxima of the $Q \leftarrow N$ absorption continua (Table V). Although the extinction coefficients increase in the order HI, HBr, HCl, the output of the photoflash lamp decreases at lower wavelength, the net effect being a reduction in the concentrations of atoms. Also, spin orbit relaxation of the excited halogen atoms in the presence of their respective hydrogen halides is rapid (see Section VII.G). Thus, kinetic spectroscopic measurements in the vacuum ultraviolet of the resulting transient atoms may be used to support Mulliken's predictions on the nature of the primary processes.

Flash photolysis of HI in a quartz system (lower wavelength cut off $\lambda > 2000$ Å) gives rise to approximately 20% I($5^2P_{1/2}$) and 80% I($5^2P_{3/2}$) at zero delay time, clearly involving excitation to the Q_0, $^3Q_1$, and ^1Q repulsive states in the continuum.[37] Kinetic spectroscopic observation following the flash photolysis of HBr with a Spectrosil (high purity quartz) system (λ air cutoff $\gtrsim 1850$ Å) gives rise to a weak absorption spectrum of Br($4^2P_{1/2}$) and a stronger one of Br($4^2P_{3/2}$).[76] Vacuum ultraviolet photolysis ($\lambda \gtrsim 1600$ Å) gave rise to a stronger spectrum of Br($4^2P_{1/2}$)[76] (Fig. 10), again supporting excitation via the transitions:

$$\text{HBr} + h\nu(Q_0) \longrightarrow \text{H} + \text{Br}(4^2P_{1/2}) \qquad (33)$$
$$\text{HBr} + h\nu(^3Q_1, {}^1Q) \longrightarrow \text{H} + \text{Br}(4^2P_{3/2}) \qquad (34)$$

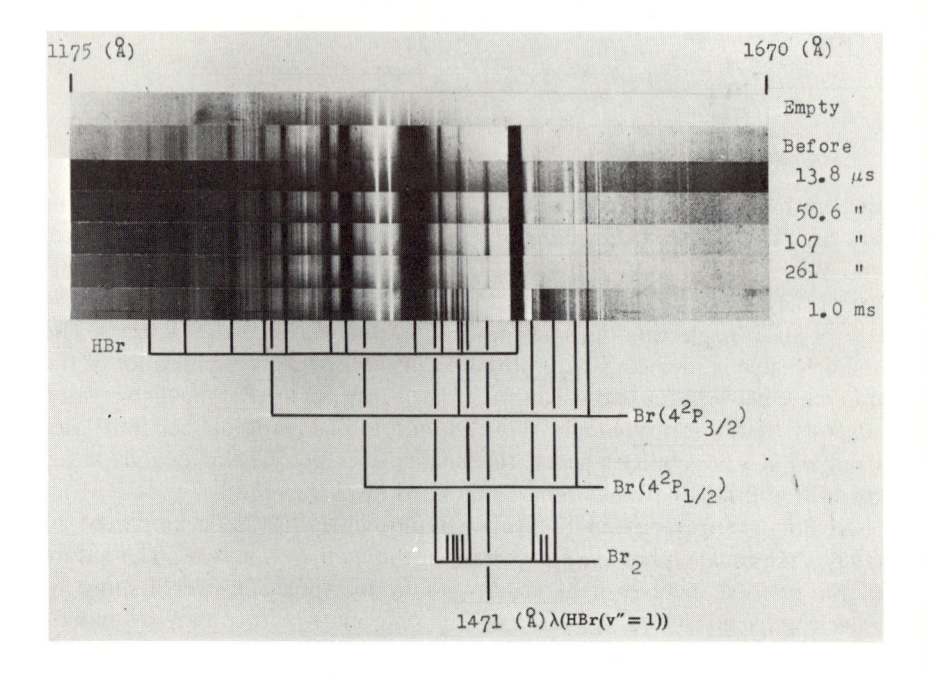

Fig. 10. Br($4^2P_{1/2}$) and Br($4^2P_{3/2}$) in the photolysis of HBr. $p_{HBr} = 0.5$ mm Hg, $p_{Ar} = 50$ mm Hg, $E = 576$ J, vacuum ultraviolet photolysis.

Excitation of HBr via low wavelength Rydberg transitions with the experimental arrangement employing Spectrosil was not significant. To achieve adequate excitation of the long wavelength continuum of HCl (Table V), it was necessary to photolyze the molecule with lithium fluoride optics ($\lambda > 1200$ Å), and under these conditions it was possible to detect both Cl($3^2P_{1/2}$) and Cl($3^2P_{3/2}$)[29] in accord with three $Q \leftarrow N$ transitions. Clearly Rydberg states may be excited by this method; however, the majority of the transitions are discrete,[109] and thus the atomic chlorine observed almost certainly resulted from excitation via the $Q \leftarrow N$ continuum.

E. Primary Processes, the Polyatomic Halides

The electronic spectra of large molecules, particularly when continuous in nature, do not always allow the confident prediction of spin orbit states

TABLE VI
Population Inversion between $I(5^2P_{1/2})$ and $I(5^2P_{3/2})$ Resulting from the Flash Photolysis of Polyatomic Iodides, Observed by the Stimulated Emission

$$I(5^2P_{1/2}) \xrightarrow{h\nu} I(5^2P_{3/2}) + 2h\nu$$

CH_3I [3,32,33]	CF_3I,[3,32,33] C_2F_5I,[32,33] $CF_3CF_2CF_2I$,[32,33]	
C_2H_5I,[32]	$CH_3CH_2CH_2I$,[32] $CH_3CH_2CH_2CH_2I$,[32] $(CH_3)_2CHCH_2I$,[32]	
[no stimulated emission from $(CH_3)_2CHI$, HI, and I_2]		

resulting from photolysis. However, Goodeve et al.[110,111] have treated the methyl halides on the basis of a diatomic model and have explained the ultraviolet continuum of CH_3I in terms of two overlapping systems involving transitions to repulsive states; one strong component giving rise to $I(5^2P_{1/2})$ and a weak one yielding $I(5^2P_{3/2})$. Mulliken[105] has interpreted the weak component as a $^3Q \leftarrow N$ transition, the strong component as $Q_0 \leftarrow N$, and concluded that according to molecular orbital theory, there must also be a third component $^1Q \leftarrow N$ which is presumably hidden by the strong $Q_0 \leftarrow N$ transition. Porret and Goodeve[111] have considered a range of alkyl bromides and iodides. The electronic spectrum of methyl iodide is dealt with by Walsh[112] and the electronic states of many of the large molecular halides to be considered here are included in the recent book by Herzberg.[113] However, even such detailed considerations of molecular spectra do not at present allow the prediction of the spin orbit states on photolysis. The general approach, therefore, must be experimental and involve the direct observation of the relevant atomic states. The methods that have been used most extensively to observe $^2P_{1/2}$ halogen atoms in primary photochemical processes with polyatomic halides are the study of stimulated emission in the infrared and time-resolved absorption spectroscopy in the vacuum ultraviolet.

Table VI lists polyatomic iodides for which atomic iodine photodissociation laser action has been observed, thus yielding direct observation of $I(5^2P_{1/2})$ in a population inversion with respect to the ground state atom. Stimulated emission has not yet been observed in any system from $Br(4^2P_{1/2})$ (see footnote, p. 15) or $Cl(3^2P_{1/2})$. Direct observation using kinetic absorption spectroscopy in the vacuum ultraviolet offers a further advantage over emission studies; namely, that both ground state and excited atoms may be observed. Thus the relative yields of the two states may be determined as the f values for the relevant transitions have been calculated.[18] Table VII records the results of positive experiments in which electronically excited halogen atoms have been observed by kinetic spectroscopy in absorption in the vacuum ultraviolet (Figs. 3, 8, 9, and 10). Generally, the $^2P_{3/2}$ atom was observed accompanying the $^2P_{1/2}$ atom unless the absorption transition of the ground state atom was

TABLE VII
Electronically Excited Halogen Atoms ($^2P_{1/2}$) Observed by Kinetic Spectroscopy in Absorption Following the Flash Photolysis of Polyatomic Halides

Iodides: CF$_3$I,[27,30] CH$_3$I,[30,114,115] C$_2$H$_5$I,[30] 1-C$_3$H$_7$I,[30,115] n-C$_4$H$_9$I,[30] 2-C$_3$H$_7$I [115]

Bromides: CF$_3$Br,[76] CHCl$_2$Br,[76] [only Br($4^2P_{3/2}$) from CH$_3$Br, C$_2$H$_5$Br, and CH$_2$Br$_2$][76]

Chlorides: CF$_3$Cl,[29] CCl$_4$,[29] [Cl($3^2P_{3/2}$) only from AsCl$_3$][29]

obscured by a molecular absorption. Spontaneous emission from I($5^2P_{1/2}$) following the flash photolysis of CF$_3$I has been observed by Husain and Wiesenfeld,[11] in accordance with the production of the excited atom in a primary photolytic step. Time-resolved emission from I$_2$($B^3\Pi_{0u^+}$) following the flash photolysis of CF$_3$I and resulting from the recombination of I($5^2P_{1/2}$) and I($5^2P_{3/2}$) in the presence of a third body has been studied by Abrahamson, Husain, and Wiessenfeld,[23] and also shows that the excited atoms are produced in the primary photolytic process (Fig. 5). However, the inability to detect excited halogen atoms in any given experiment does not necessarily eliminate its production in a primary photolytic process on account of the role played by spin orbit relaxation which may be rapid under particular conditions.

V. SECONDARY PROCESSES GIVING RISE TO THE PRODUCTION OF $^2P_{1/2}$ HALOGEN ATOMS VIA ATOMIC METATHETICAL REACTIONS

The production of an electronically excited atom following a metathetical reaction between a ground state atom and a diatomic molecule is a novel type of process and of theoretical interest.[35] Polanyi et al.[15,36] have studied this in detail for the reactions:

$$\text{H} + \text{HI} \xrightarrow{k_1^*} \text{H}_2 + \text{I}(5^2P_{1/2}) \quad (35)$$

$$\text{H} + \text{HI} \xrightarrow{k_2} \text{H}_2 + \text{I}(5^2P_{3/2}) \quad (36)$$

(Section IV) by monitoring the emission from I($5^2P_{1/2}$) when hydrogen atoms from a discharge are mixed with HI. These workers have found that $k_1^*/k_2 <$ 0.02. Donovan and Husain,[37] who monitored both I($5^2P_{1/2}$) and I($5^2P_{3/2}$) in absorption in the vacuum ultraviolet following the flash photolysis of HI, observed that I($5^2P_{1/2}$) maintained an approximately stationary state for about 300 μsec following photolysis, in which production by reaction (35) is balanced by spin orbit relaxation. Although these qualitative results indicate that $k_1^* \ll k_2$, the ratio k_1^*/k_2 from the absorption measurements does not appear to be quite as low as that reported by Polanyi.[15,36]

The weak emission from $I(5^2P_{1/2})$ observed when bromine and chlorine atoms are allowed to react with hydrogen iodide has been studied by Polanyi et al.[36] Although the relative contributions from the reactions:

$$Cl + HI \longrightarrow HCl + I(5^2P_{1/2}) \quad \Delta H = -34 \text{ kcal} \quad (37)$$

$$Br + HI \longrightarrow HBr + I(5^2P_{1/2}) \quad \Delta H = -16 \text{ kcal} \quad (38)$$

and secondary, energy transfer processes could not be separated, the over-all yield of excited atoms was shown to be very low. However, strong emission was observed when HI was mixed with discharged oxygen in a flow system.[36] Although the apparent fractional yield of $I(5^2P_{1/2})$ from the reaction

$$O + HI \longrightarrow OH + I(5^2P_{1/2}) \quad \Delta H = -31 \text{ kcal} \quad (39)$$

was high, the possibility of the participation of other species in the discharge in producing $I(5^2P_{1/2})$ was recognized. It is tempting to attribute this efficient production, at least in part, to the near resonant transfer from $O_2(^1\Delta_g)$:†

$$O_2(^1\Delta_g) + I(5^2P_{3/2}) \longrightarrow O_2(^3\Sigma_g^-) + I(5^2P_{1/2}) \quad \Delta H = -0.8 \text{ kcal} \quad (40)$$

in accord with a similar suggestion made by Ogryzlo[116] to account for emission from $I(5^2P_{1/2})$ when I_2 was mixed with atom free discharged oxygen.

The analogous reactions producing bromine atoms have not been studied in detail. Cashion and Polanyi[16] have reported the emission from $Br(4^2P_{1/2})$ in a flow system. A preliminary study on the kinetics of this emission when hydrogen atoms from a discharge react with Br_2 and HBr has shown that k^*/k for $H + Br_2$ is markedly smaller than k^*/k for $H + HBr$.[17] Vacuum ultraviolet measurements on $Br(4^2P_{1/2})$[76] following the flash photolysis of HBr appear to indicate that $[Br(4^2P_{1/2})]$ proceeds through a maximum at which production via chemical reaction between $H + HBr$ is balanced by electronic quenching.

Emission from $Cl(3^2P_{1/2})$ has not been detected in flow systems[40] although this excited atom has recently been observed in absorption following the vacuum ultraviolet flash photolysis of a number of chlorides, including HCl.[29] The over-all rate for the reaction

$$H + HCl \longrightarrow H_2 + Cl(^2P_{1/2}, ^2P_{3/2}) \quad (41)$$

is sufficiently slow at room temperature[117] that the production of $Cl(3^2P_{1/2})$ in this process is hidden by rapid spin orbit relaxation. Similar considerations will apply to flow systems and accounts for the absence of emission in such

† See also the recent work of Thrush et al. (R. G. Derwent, D. R. Kearns and B. A. Thrush, Chem. Phys. Lett., 6, 115 (1970)).

experiments.[40] The treatment of this general problem by a potential energy surface approach is very difficult on account of the crossing of surfaces which would clearly be involved in a process giving rise to excited atoms from ground state reactants.

VI. STIMULATED EMISSION

A. Population Inversion and the Atomic Photodissociation Laser

The optical and collisional metastability of the $np^5\ ^2P_{1/2}$ configuration for the halogen atoms provides a particularly convenient means for light amplification by stimulated emission via the transition $^2P_{1/2} \to\ ^2P_{3/2}$, provided that a population inversion (Section III.C) can be achieved and that certain threshold requirements can be met. Photodissociation lasers offer an efficient means of achieving population inversion by providing this directly in the initial photochemical process, thus avoiding the difficulties associated with optical "pumping" into upper states by excitation of ground state atoms. Pimentel et al.[3] first demonstrated that the flash photodissociation of the lower alkyl and fluorinated alkyl iodides gave rise to the required conditions and that laser action could be achieved with the appropriate optical cavity. This system provides one of the highest optical gains for any gas laser constructed, and indeed, it was a direct result of this high efficiency that led to its accidental discovery in an apparatus designed primarily for the detection of transient species in the infra red.[118] The high gains observed are such that power in the killowatt range may be achieved. One practical limitation with these alkyl iodide systems is the requirement that a fresh charge of the iodide be introduced into the optical cavity after relatively few flashes, owing to the depletion of the parent molecule and the production of species that "quench" laser action.

The possibility of constructing a photodissociation laser involving I_2 must at first have appeared extremely attractive as such a system could, in principle, be used indefinitely without the necessity of changing the gaseous mixture in the cavity. Photodissociation of I_2 in the $(B \leftarrow X)$ continuum ($\lambda < 4995$ Å) gives rise to equal concentrations of $I(5^2P_{1/2})$ and $I(5^2P_{3/2})$ which constitutes a population inversion. Unfortunately, however, the iodine molecule shows a remarkable efficiency for inducing spin orbit relaxation of the $^2P_{1/2}$ state, requiring ca. 30 collisions.[75] Thus for a given degree of dissociation, an optimum concentration will exist for which the rate at which the threshold concentration of $I(5^2P_{1/2})$ can be achieved will be just greater than the rate at which the inversion is lost through fast relaxation processes. In principle therefore, a photodissociation laser may indeed be constructed using molecular iodine, provided that a sufficiently high degree of photolysis can be obtained in an interval of ca. 5 μsec; this might well be feasible today using the recently described Z-pinch technique.[119,157]

A third approach to the construction of an iodine atom photodissociation laser has employed hydrogen iodide as the initial reagent. However, although the primary photolytic process may give rise to a population inversion at the particular wavelength used ($\lambda = 2537$ Å),[15,36] the fast secondary reaction

$$H + HI \longrightarrow H_2 + I \tag{42}$$

is known to yield almost exclusively ground state atoms[15,36] (Section V). Thus unless the primary step yields $>67\%$ of the atoms in the $^2P_{1/2}$ configuration, neglecting any reduction in the radiation density arising from optical losses or spin orbit relaxation, there will be no possibility of achieving amplification. The further possibility of efficiently scavenging the hydrogen atoms without simultaneously quenching the excited iodine atoms appears remote.

B. Quenching of Stimulated Emission

"Quenching" of stimulated emission may arise from three processes: physical relaxation of the $^2P_{1/2}$ state to the ground state, chemical reaction favoring the removal of the excited state, and the direct production of ground state atoms either by secondary chemical or physical processes. All these effects have now been observed although it has not always been clear from the laser experiments alone which process is involved. However, once the mechanism has been established, a detailed study of relative efficiencies for spin orbit relaxation (Section VII) may, in principle, be carried out using a laser. An approximate model has been given for the power output of a number of alkyl iodide photodissociation lasers which incorporates a simplified quenching mechanism.[33]

The initial observations on the alkyl iodide photodissociation laser clearly indicated that CF_3I was more efficient than CH_3I in producing stimulated emission,[3] and further investigation using kinetic spectroscopic observation in the vacuum ultraviolet has now established that the fluorinated compound is several orders of magnitude less efficient than CH_3I in causing spin orbit relaxation.[27,114] Furthermore, investigations with the latter technique have established that reactions of the type

$$I(^2P_{1/2}) + RI \longrightarrow I_2 + R\cdot \tag{43}$$

are of minor importance at room temperature[114] although they may be significant at higher temperatures.[60] It should be noted, however, that the temperature coefficient for spin orbit relaxation is positive for all cases so far investigated,[66] and it is by no means clear which process will dominate as the temperature is raised. Thus the observed "quenching" of laser action when high flash intensities are employed in the absence of excess inert gas[3] ("adiabatic flash heating") must at present be attributed to a combination of these two effects. At sufficiently high temperatures direct pyrolysis of the iodides may occur and the inversion may be lost following the production of ground state atoms.

The addition of small partial pressures of molecular iodine to a gaseous mixture of CF_3I and inert gas in the laser cavity leads to progressively longer delays before the onset of stimulated emission, and also reduces the over-all intensity, until finally at ca. 1.5τ of I_2 complete "quenching" occurs.[32] It was correctly inferred from these results that the production of ground state atoms (which would destroy the inversion) was of secondary importance relative to the efficient quenching of excited iodine atoms by I_2, a point that has been independently established using kinetic absorption spectroscopy.[75] By contrast, it was found that relatively high pressures of $2\text{-}C_3H_7I$, which does not itself give rise to stimulated emission on photolysis, could be added to a given CF_3I mixture before "quenching" occurred, indicating that in this case the direct production of ground state atoms from the photolysis of $2\text{-}C_3H_7I$ is the dominant effect.[32] These observations have been further extended to other members of the homologous series, and when taken with the data obtained by kinetic spectroscopy clearly demonstrate that the proportion of excited atoms produced in the initial photolytic process falls progressively with molecular complexity.[32,114] One further striking effect is the difference in behavior for $2\text{-}C_3H_7I$ and $1\text{-}C_3H_7I$. Although both iodides exhibit similar efficiencies for spin orbit relaxation, the preferential yield of ground state atoms from the secondary iodide[115] inhibits stimulated emission.

The quenching of iodine atom laser action by Br_2,[31] which is almost as efficient as I_2, is now known to occur via the fast reaction

$$I(^2P_{1/2}) + Br_2 \longrightarrow IBr + Br(^2P_{3/2}) \qquad (44)$$

The analogous reaction is also established for Cl_2, but is somewhat slower.[31] Thus the close similarity between the "quenching" efficiencies for the molecular halogens strongly suggests that quenching by I_2 proceeds by means of a very fast exchange reaction analogous to (44) (Section IX.E).

C. Future Possibilities, the Bromine and Chlorine Atom Lasers

Although a population inversion for the two states of the bromine atom has been observed following the vacuum ultraviolet photolysis of CF_3Br,[76] it has not yet proved possible to meet the threshold conditions for stimulated emission. The work reported in Ref. 76 showed that the intensity of the

1531.9 Å transition for $Br(^2P_{1/2})$ in absorption was greater than that for $Br(^2P_{3/2})$ at 1540.8 Å, and on the then available data relating to emission intensities it was inferred that this constituted a population inversion. Recent calculations by Lawrence[18] give the ratio of the Einstein coefficients for spontaneous emission as 1531.9 Å/1540.8 Å = 1.5. The ratio of the respective intensities observed in absorption for the excited to the ground state transitions is $>1.6 \pm 0.2$. This means that the population of the two states is, in fact, equal within experimental error and that inversion is achieved. It may be expected that the threshold conditions for a bromine atom laser will require significantly higher concentrations of $Br(^2P_{1/2})$ relative to that needed for amplification with $I(^2P_{1/2})$ in view of the longer mean radiative lifetime. Attempts to increase further the population of $Br(^2P_{1/2})$ using mixtures of CF_3I and CF_3Br, suggest that the transfer of energy by the process

$$I(^2P_{1/2}) + Br(^2P_{3/2}) \longrightarrow Br(^2P_{1/2}) + I(^2P_{3/2}) \tag{45}$$

is inefficient (see footnote, p. 15).

It has now been shown that a population inversion is not achieved between the $^2P_{1/2}$ and $^2P_{3/2}$ states of the chlorine atom in the vacuum ultraviolet flash photolysis of CF_3Cl,[29] and thus the construction of a chlorine atom photodissociation laser appears remote at present.

VII. SPIN ORBIT RELAXATION

A. General Considerations

Spin orbit relaxation may occur either by emission of a photon or by collisions of the second kind. Radiative decay for the forbidden transition $(^2P_{1/2} \to {}^2P_{3/2})$ will be a relatively slow process (Section II.B). Collisions of the second kind, whereby the electronic energy of the atom is transferred to one of the degrees of freedom associated with the colliding species, generally show a considerable range of efficiency, depending on the nature of the collision partner and the energy to be transferred. Thus electronically excited halogen atoms provide a particularly useful basis for discussing relaxation in view of the variation of spin orbit energies involved (Table I).

Although various theoretical treatments for this type of process have been proposed, they do not as yet provide an absolute basis for calculation.† Further, approximations are generally involved which require certain limiting conditions, particularly with respect to the nature of the interaction potential.

---

† Andreev and Nikitin have recently reported a detailed calculation for the collisional relaxation of $I(5^2P_{1/2})$ and $Br(4^2P_{1/2})$ by N_2 and CO (E. A. Andreev and E. E. Nikitin, Theoret. Chim. Acta (Berl.), **17**, 171 (1970)).

A full review of the theory is not possible here. However, an outline of the more important factors that influence relaxation will facilitate a discussion of the data. It is convenient to recognize three cases for spin orbit relaxation: first, relaxation between potential surfaces that are essentially parallel in the region accessible for thermal velocities; second, potential surfaces that converge or cross due to a strong coupling between electronic and nuclear motion; and third, potential surfaces that "cross" due to a coincidental electronic degeneracy at a given point on the multidimensional potential energy surface. For weak spin orbit coupling, it may be expected that the states of a given multiplet will exhibit similar chemical behavior, and thus that the potential surfaces on which collisions may be considered to occur should be essentially parallel for low relative kinetic energies. Clearly, this may not be the case where strong chemical interaction is involved, as atomic states, having different symmetries, may form molecular orbitals of significantly different symmetry and therefore different energy (e.g., $X(^2P_{3/2})$ + $X(^2P_{1/2})$ and $X(^2P_{3/2})$ + $X(^2P_{3/2})$, where X = Cl, Br, I).[68] The latter case will be considered later and we here deal with those collisions for which the potential surfaces are essentially parallel, the simplest example from the theoretical point of view being a collision between the excited atom and an inert gas atom. Transition from one surface to another must occur by tunneling and may be treated in terms of the "distorted wave approximation" first given by Zener[120] and which is well established as a basis for the direct calculation of vibration-vibration and vibration-translation relaxation probabilities.[121,122] The collision is represented by an incoming translational wave which increases both in amplitude and wavelength at the classical turning point; for an elastically scattered particle this wave also describes the recoil. Following inelastic scattering the recoil is represented by a wave originating on the lower potential surface, and thus *one* of the critical parameters for relaxation will be the rate of tunneling between the two surfaces. The rate of tunneling will be given by the "overlap" between the translational wave functions, and will clearly be greatest when the two surfaces are close (i.e., the energy to be converted to translation is small) and when the waves are incident on a steeply repulsive potential. Furthermore, tunneling should increase for light particles, and it is significant that for the cases so far investigated, helium shows such a trend in relation to the inert gases.[123] For a more complete discussion of these points, the reader is referred to a recent article by Callear.[124] In the case for spin orbit relaxation one further term will be important in determining the rate of transfer and is connected with the efficiency of coupling between the degrees of freedom involved. For collisions between atoms, the coupling arises from the interaction between the magnetic moment associated with the rotating nuclei and that due to the electronic angular momentum, together with a contribution from the direct coupling between the angular momenta. For collisions with diatomic species, a coupling may result from the interaction between the in-plane motion of the nuclei and the electronic energy of the quasitriatomic molecule (vibronic coupling). This may clearly be extended to polyatomic molecules. It may be anticipated that direct calculation of the coupling term will be difficult, although such calculations are established for cases where the potential surfaces are known to "cross" (see later, this Section).

The above considerations should be applicable to collisions between an excited atom and chemically inert species. However, in those cases where significant chemical interaction occurs, it is necessary to consider the near convergence or crossing of potential surfaces induced by nuclear motion. Thus for collisions between atoms, the actual "coupling case," which describes the effective coupling between the electronic and rotational angular momenta in the quasimolecule, may change as the internuclear separation changes. Further, increasing rotational energy may lead to a change of coupling for a given internuclear separation. These effects, which are well established for stable diatomic molecules,[125] may be expected to be of greatest importance for collisions involving strongly attractive potentials. However, it may be noted that no clear distinction exists between the case for weak and strong interaction, other than that the crossing of potential surfaces may occur for relatively low thermal energies in the latter case. Similar considerations will apply to collisions with diatomic and polyatomic species when attractive interactions are involved, and have been discussed[126] for the process:

$$\text{Hg}(^3P_1) + \text{N}_2(v'' = 0) \longrightarrow \text{Hg}(^3P_0) + \text{N}_2(v'' = 1) \tag{46}$$

The final case which may be relevant to a discussion of spin orbit relaxation is that for which the surfaces representing the electronic potential energy of the system "cross" or lead to "avoided crossing" due to a local degeneracy of the electronic wave functions. This has been applied in a particularly illuminating way by Nikitin[127] to the quenching of sodium resonance fluorescence. Here again, the transition from one surface to another may be facilitated by rotation of the nuclei.[94–96,127]

B. General Equation for the Removal of Electronically Excited Halogen Atoms

Before discussing individual results for spin orbit relaxation, a general framework for the kinetic treatment will be given. The rate of relaxation in a system involving only two states,

$$\text{X}^* \underset{k_2}{\overset{k_1}{\rightleftarrows}} \text{X} \qquad (-\Delta E) \tag{47}$$

will be given by the expression:

$$\frac{-d[X^*]}{dt} = k_1[X^*] - k_2[X] = k_1\left\{[X^*] - [X]\frac{g_1}{g_2}\exp(-\Delta E/RT)\right\} \quad (48)$$

The value of k_2 is determined according to the principle of detailed balancing. For thermal energies less than or equal to 300°K, the last term in Eq. (48) may be neglected for the initial decay, provided a large excess of ground state atoms is not present (Table I). Thus the measured decay coefficients and data derived from these relate to k_1.

The over-all first-order decay coefficient (k_1) may be broken down into a number of terms:

$$\frac{-d[X^*]}{dt} = \left(\sum_q k_q[Q] + \beta + A_{nm} + \rho A_{nm}\right)[X^*] \quad (49)$$

where $\sum_q k_q[Q]$ is the sum of all first-order homogeneous collisional deactivation rate coefficients by species Q, β is the rate coefficient for diffusion to the vessel wall, A_{nm} is the Einstein coefficient for spontaneous emission, and ρA_{nm} represents decay by stimulated emission of radiation. In addition to these contributions, others representing chemical removal may be relevant and under certain conditions will be similar in form to the first term in Eq. (49), but may be omitted for the present considerations (see however, Section IX). Furthermore, although stimulated emission may occur when high partial pressures of CF_3I are flash photolyzed,[3] even without mirrors in the optical cavity, calculations show the threshold is several orders of magnitude too high for the conditions pertaining to the results presented here.

C. Diffusion of $I(5P_{1/2})$ in Inert Gases

The homogeneous quenching of $I(5^2P_{1/2})$ generated in the photolysis of CF_3I is sufficiently slow so that the dominant term in Eq. (49) is apparently that due to diffusion to the walls of the reaction vessel (followed by efficient deactivation). Thus it is observed that as the total pressure of inert gas present is lowered, the rate of relaxation increases, showing an inverse first power relation[27] (Fig. 11). The rate of diffusion to the vessel walls is given by the diffusion equation:

$$\nabla^2[I(5^2P_{1/2})] = \frac{1}{D}\cdot\partial[I(5^2P_{1/2})]/\partial t \quad (50)$$

where D is the diffusion coefficient at a given pressure (being inversely proportional to the pressure). For a cylindrical vessel of length l and radius r, the "long time" solution to Eq. (50) takes the form:[128]

$$[I(5^2P_{1/2})]_t = [I(5^2P_{1/2})]_{t=0}\exp(-\beta t) \quad (51)$$

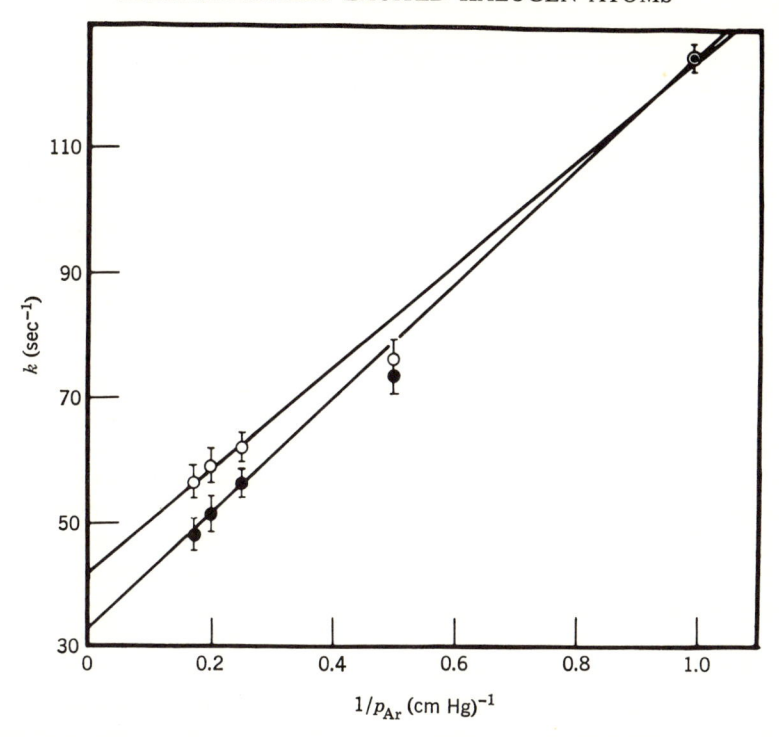

Fig. 11. First-order decay coefficient (k) for the emission from I($5^2P_{1/2}$) as a function of pressure of Ar. $p_{CF_3I} = 1.0$ mm Hg, $E = 1767$ J. (○) k observed. (●) k corrected for quenching by 0.4 ppm O_2 in Ar.

where

$$\beta = D\left(\frac{\pi^2}{l^2} + \frac{5.81}{r^2}\right) \qquad (52)$$

which is consistent with the observed relationship. However, a comparison of the diffusion coefficients determined by vacuum ultraviolet kinetic spectroscopy,[27,37,67] time-resolved emission,[11] flow tube studies,[15] and kinetic mass spectrometry[60] indicates that the observed values are considerably higher than those computed using the first approximation Chapman-Enskog equation[129] for the diffusion of Xe in the relevant inert gas (Table VIII). This implies that either one or both of the force constants (σ_{12}, ϵ_{12}) are significantly different for I($5^2P_{1/2}$) and Xe in a given inert gas. The most detailed study for diffusion in the inert gas argon has been carried out by time-resolved emission.[11] Here the various contributions to the over-all decay have been carefully defined and that due to diffusion determined to

TABLE VIII

Diffusion Coefficients (cm² sec⁻¹) for $I(5^2P_{1/2})$ in the Noble Gases (at 1 atm pressure)

	D, vacuum uv [27,37,67]	D, ir emission [11] static system	D, flow tube [15,36]	D, calcd [129]
He	1.1	—	—	0.55 (Xe in He)
Ar	0.27	0.41 ± 0.05	0.60	0.10 (Xe in Ar)
Xe	0.25	—	—	0.057 (Xe in Xe)

within ca. 10%. More extensive results were obtained using kinetic spectroscopy in the vacuum ultraviolet[27,37,67]; however, the limited accuracy achieved in these experiments does not allow a detailed comparison. Thus, as the various contributions due to spontaneous emission, quenching by CF_3I, and diffusion to the vessel wall could not be quantitatively determined for the latter experiments, the results should be regarded as correct to within a factor of 2. The study of the diffusion of $I(5^2P_{1/2})$ imposes severe restrictions on the purity of the inert gases employed, particularly on account of the extremely efficient collisional deactivation by traces of O_2. Thus Husain and Wiesenfeld[11] found it necessary to correct their data for the final trace (0.4 ppm) of O_2 present in the inert gas (Fig. 11). The diversity of experimental conditions, including pressure range and vessel dimensions which lead to the same value for the diffusion coefficient, is compelling evidence for the proposed mode of decay. However, a detailed understanding of the relevant effects leading to the anomalously high coefficient will require further work.

The cross section for the quenching of $Br(4^2P_{1/2})$ by CF_3Br[76] is considerably greater than that for $I(5^2P_{1/2})$ by CF_3I, thus diffusion of $Br(4^2P_{1/2})$ in the inert gas is dominated by collisional quenching and has not been examined. It appears unlikely that comparably detailed studies for the lighter halogen atoms will be possible.

D. Experimental Determination of the Mean Radiative Lifetime of $I(5^2P_{1/2})$

The term in Eq. (49) which describes radiative decay is relatively small under conditions where stimulated emission does not occur. Thus, only when all other contributions to the over-all decay are accurately determined will this term be accessible. At present only time-resolved atomic emission studies provide the required precision.[11] The mean radiative lifetime has been obtained by observing the variation in the first-order decay coefficient for this emission as a function of inert gas pressure (Fig. 11). The slope of the graph so obtained yields the diffusion coefficient, while the intercept represents the sum of the contributions from quenching by CF_3I and radiative decay. Independent experiments were used to establish the quenching term for CF_3I, the difference between this and the observed intercept yielding a value of $\tau_{e(obs)} = 0.045$ sec[11] for the mean radiative lifetime. This compares very favorably with the value calculated by Garstang[13] ($\tau_{e(calc)} = 1/A_{nm} = 0.13$ sec).

The calculated mean radiative lifetime for $Br(4^2P_{1/2})$ is almost an order of magnitude longer[13] (Table II) than that for $I(5^2P_{1/2})$, thus although this forbidden transition has been observed in emission by Polanyi et al.,[17] it appears unlikely that the contribution from this emission can be made kinetically significant. The experimental determination of τ_e for the lighter halogens, using the techniques described for $I(5^2P_{1/2})$, thus appears remote at present.

E. Collisional Quenching of $X(^2P_{1/2})$ by X_2 where X = Halogen

The efficiencies for spin orbit relaxation of electronically excited halogen atoms for all collisional partners investigated to date are given in Table IX.

TABLE IX
Comparison of the Probabilities (P)[a] for Spin Orbit Relaxation of $X(^2P_{1/2})$ on collision with various species M

M	$P_{I(5^2P_{1/2})}$[b]	$P_{Br(4^2P_{1/2})}$[c]	$P_{Cl(3^2P_{1/2})}$[d]
He	$< 10^{-8}$	—	—
Ar	$< 10^{-8}$	$< 10^{-6}$	$< 10^{-6}$
Xe	$< 10^{-8}$	—	—
$I(5^2P_{3/2})$	$< 10^{-4}$	—	—
$Cl(3^2P_{3/2})$	—	—	—
$H(1^2S_{1/2})$	—	—	~ 1
N_2	8.0×10^{-7}	1.1×10^{-5}	—
CO	4.8×10^{-6}	3.2×10^{-5}	—
CO_2	5.7×10^{-7}	—	—
SF_6	1.1×10^{-7}[e]	—	—
CF_4	2.2×10^{-6}[e]	9.5×10^{-4}	—
N_2O	4.8×10^{-6}	—	—
CF_3I	1.6×10^{-6}	—	—
CF_3Br	—	2.6×10^{-3}	—
CF_3Cl	—	—	7.8×10^{-4}
CCl_4	—	—	0.15
CH_3I	1.0×10^{-2}	—	—
C_2H_5I	1.0×10^{-3}	—	—
n-C_3H_7I	0.9×10^{-3}	—	—
i-C_3H_7I	0.9×10^{-3}	—	—

continued

TABLE IX continued

M	$P_{I(5^2P_{1/2})}$ [b]	$P_{Br(4^2P_{1/2})}$ [c]	$P_{Cl(3^2P_{1/2})}$ [d]
n-C_4H_9I	1.4×10^{-3}	—	—
i-C_4H_9I	1.3×10^{-3}	—	—
t-C_4H_9I	1.7×10^{-3}	—	—
HI	8.0×10^{-4}	—	—
DI	7.0×10^{-4}	—	—
HBr	—	6.8×10^{-3}	—
HCl	—	—	3×10^{-2}
D_2	4.5×10^{-6} [e]	1.8×10^{-2}	—
H_2	5.8×10^{-4} [e]	7.8×10^{-3}	8.0×10^{-3}
CH_4	1.8×10^{-4}	1.4×10^{-2}	—
CF_3H	2.0×10^{-4}	—	—
C_3H_8	1.9×10^{-4}	—	—
$CH_2=CH-CH_2Cl$	5.6×10^{-4}	—	—
$CH_2=CH-CH_2Br$	1.2×10^{-3}	—	—
$CH_2=CH-CH_2I$	2.2×10^{-3}	—	—
$CH_2=CH_2$	2.6×10^{-4}	—	—
$CF_2=CFH$	2.6×10^{-4}	—	—
$CF_2=CF_2$	2.0×10^{-5}	—	—
$CH_3CH=CH_2$	5.0×10^{-4}	—	—
$CH_3-CH_2-CH=CH_2$	4.9×10^{-4}	—	—
$trans$-$CH_3CH=CHCH_3$	8.4×10^{-4}	—	—
cis-$CH_3CH=CHCH_3$	7.5×10^{-4}	—	—
$(CH_3)_2C=CH_2$	7.8×10^{-4}	—	—
D_2O	2.8×10^{-4}	4.9×10^{-2}	—
H_2O	4.3×10^{-3}	0.16	—
ICN	2.4×10^{-4}	—	—
I_2	2.8×10^{-2} (Reaction ?)	—	—
Br_2	Reaction	9.5×10^{-2} (Reaction ?)	—
Cl_2	Reaction	—	—
NOCl	Reaction	—	—
NOBr	Reaction	—	—
O_2	4.9×10^{-2}	0.16	—
NO	5.2×10^{-2}	0.22	—

[a] The probabilities given depend on the values taken for "collision cross sections"; the values used here are taken from one source.[129] Where values are not known, those for the most similar species are taken. Some of the more important values employed are: $I(5^2P_{1/2})$, $I(5^2P_{3/2})$ = Xe = 4.06 Å; $Br(4^2P_{1/2})$ = Kr = 3.61 Å; $Cl(3^2P_{1/2})$, $Cl(3^2P_{3/2})$ = Ar = 3.44 Å; $H(1^2S_{1/2})$ = He = 2.6 Å.

[b] References 11, 27, 28, 31, 37, 67, 75, and 115.

[c] References 76 and 131.

[d] Reference 29.

[e] Values given in Ref. 11, 27, 37, and 67 differ significantly in these cases, only the preferred values are given here.

Although the first kinetic measurements involving the direct observation of $I(5^2P_{1/2})$ in the vacuum ultraviolet[75] indicated that molecular iodine was remarkably efficient at inducing spin orbit relaxation on collision, the details of the mechanism involved have not as yet been unambiguously determined. Some insight is provided by the rapid chemical reactions of $I(5^2P_{1/2})$ with both Br_2 and Cl_2 to form the interhalogen compounds IBr and ICl, respectively.[31] Thus it is observed that the second-order rate constants for these reactions are close to that observed for "quenching" by I_2, indicating that the latter case may also involve the rapid exchange reaction:

$$I(5^2P_{1/2})^{(1)} + I_2^{(2)(2)} \longrightarrow I(5^2P_{3/2})^{(2)} + I_2^{(1)(2)} \qquad (53)$$

Clearly isotopic labeling experiments could be used to resolve this point.

The analogous cases for $Br(4^2P_{1/2})$ and $Cl(3^2P_{1/2})$ are also likely to involve chemical reaction and further discussion is therefore deferred to Section IX.

F. Collisional Quenching by Atoms

Spin orbit relaxation during collision with inert gas atoms appears to be extremely inefficient for the cases so far investigated, requiring $> 10^8$ collisions for $I(5^2P_{1/2})$ with He, Ar, and Xe[27,37,67] (Table IX). However, from the foregoing discussion (Section VII.A) it seems likely that the inert gases will be of importance only in those cases where spin orbit coupling in the atom is weak [i.e., $Cl(3^2P_{1/2})$ and $F(2^2P_{1/2})$]. Unfortunately, this is the area for which data is sparse, and the only available work at present is that for $Cl(3^2P_{1/2})$ with argon,[29] where no appreciable quenching could be observed for pressures of argon up to 1 atm, thus establishing a lower limit of 10^6 collisions for transfer. Further experiments with helium are clearly desirable.

Relaxation on collision with ground state halogen atoms might be expected to be a relatively efficient process as strongly attractive potential curves exist in all cases. However, direct investigations present considerable experimental difficulties, and only upper limits for the efficiencies have been established. Thus, following the flash photolysis of CF_3I, the relaxation of $I(5^2P_{1/2})$ shows no significant deviation from a monotonous first-order decay, despite the observed steady increase of $I(5^2P_{3/2})$, until appreciable concentrations of I_2 result from atomic recombination.[11,27,37,67] This implies that relaxation of $I(5^2P_{1/2})$ by $I(5^2P_{3/2})$ involves $> 10^4$ collisions, which is somewhat surprising as predissociation by the $C^3\Sigma_u^+$ state, following collision along the $B^3\Pi_{0u^+}$ curve, could provide an efficient path. It appears that in this case, the rate at which the atoms move through the region of crossing is too great for any appreciable mixing to occur. The presence of magnetic fields[69] or collisional perturbations[70] may facilitate mixing, and thus enhance the efficiency for this process. Recent experiments by Wasserman et al.[71]

using esr techniques to detect $I(5^2P_{3/2})$ atoms suggest that an efficient path exists for spontaneous predissociation resulting from transfer from high vibrational levels of the $B^3\Pi_{0u^+}$ state to a hitherto unknown repulsive state. However, studies involving excitation to a known group or even individual quantum states, will be required before these experiments can be used to elucidate the mechanism of spin orbit relaxation by ground state atoms.

The analogous process for bromine atoms has not been examined in such detail and even an estimate of an upper limit for the collision efficiency is not possible. However, relaxation of $Cl(3^2P_{1/2})$ following the vacuum ultraviolet flash photolysis of CCl_4 is independent of the radical concentration[29] which indicates a lower limit of 10 collisions for relaxation by $Cl(3^2P_{3/2})$. The high efficiency for quenching by CCl_4 does not at present allow a more refined estimate; however, further experiments with CF_3Cl may do so in view of the lower quenching efficiency of this molecule (Section VIII). By contrast the relaxation of $Cl(3^2P_{1/2})$ by $H(1^2S_{1/2})$ atoms appears to occur with close to unit efficiency. Photolysis of HCl in the vacuum ultraviolet gives rise to sufficiently high concentrations of $Cl(3^2P_{1/2})$ to allow kinetic studies,[29] and it has been shown that the rate of decay is dependent on the degree of photolysis. Under the conditions used, the hydrogen atoms formed in the initial photochemical process are removed very slowly, relative to the rate of spin orbit relaxation. Thus in view of the limit established for quenching by $Cl(3^2P_{3/2})$, it is inferred that the important radical quenching process involves hydrogen atoms. Consideration of the concentrations of the various species present ([H] ~ 10^{-2}[HCl]) indicates that the hydrogen atom is extremely efficient, and requires approximately only one collision. It has been suggested therefore that the crossing between the $^3\Pi_{0^+}$ and $^3\Pi_1$ or $^1\Pi$ states[104] of HCl occurs in a region accessible to most thermal collisions, and that relaxation proceeds by such radiationless transitions.

G. Collision with Diatomic Molecules

Even for those diatomic species that are chemically inert to $I(5^2P_{1/2})$, relaxation efficiencies are at least two orders of magnitude greater than those for the noble gas atoms. This suggests that the efficiency for coupling between electronic and nuclear angular momentum in a quasitriatomic molecule, through vibronic interaction,† is significantly greater than that following rotational motion in quasidiatomic molecules. The results for relaxation of $I(5^2P_{1/2})$ by nitrogen and carbon monoxide, determined by both time-resolved emission[11] in the infrared and kinetic absorption spectroscopy in the vacuum

---

† The quenching of $Br(4^2P_{1/2})$ by HBr has now been shown to give rise to HBr (v″ = 1) (R. J. Donovan, D. Husain and C. D. Stevenson, Trans. Faraday Soc., **66**, 2148 (1970)).

ultraviolet,[37,67] are in very close agreement and must be considered as the most reliable determinations made to date. Similar results have been obtained for Br($4^2P_{1/2}$), the relaxation efficiencies being approximately five times greater than for I($5^2P_{1/2}$) with both N_2 and CO. This must result from the increased overlap of the translational wave functions, following the reduced separation of the potential surfaces for bromine atoms. These observations are in accord with the predictions of Linnett et al.[130] concerning the low efficiency for transfer involving a multivibrational transition in the diatomic partner.

For collisions with X($^2P_{1/2}$) and H_2, D_2, and HX, strongly attractive potential surfaces may be expected. The enhanced relaxation rates, being typically two orders of magnitude greater than those for the inert molecules N_2 and CO, support this view. The discrepancies observed between the relaxation rates determined by time-resolved emission,[11] flow tube[15] and vacuum ultraviolet studies[37,67] are not yet understood. However, we now feel that the isotope effect observed when the efficiencies of H_2 and D_2 are compared by time-resolved emission[11] is the most reliable,† which suggests that either near resonance with vibrational quantum states in the diatomic collision partner or some more general reduced mass effect plays a significant role.

The diatomic molecules NO and O_2 may also be expected to exhibit a strong chemical affinity for X($^2P_{1/2}$); however, the exceptionally high relaxation efficiencies observed for these species implies that some other specific effect is involved. As both molecules are paramagnetic, it has been suggested that the inhomogeneous magnetic field present during collision leads to an increased transition probability.[37,67,131] Following the theory developed by Wigner[132] for ortho-para conversion of H_2 catalyzed by paramagnetic species, the efficiency for relaxation should depend on the square of the magnetic moment. This appears to be the case for both I($5^2P_{1/2}$) and Br($4^2P_{1/2}$).

H. Quenching by Polyatomic Molecules

The lower quenching efficiencies observed for I($5^2P_{1/2}$) with CF_4 and SF_6 using time-resolved emission techniques[11] are preferred here on the grounds that the presence of trace impurities may make very significant contributions to the observed decay rates. Thus, in view of the difficulties associated with removing impurities in the 1 ppm range, a lower value is generally to be preferred, as this must surely indicate that a purer sample of gas was used. However, whichever value is taken, it is clear that quenching by the spherically

† Examination of the mass spectral analyses suggests that the data from Ref. 11 is most reliable (unpublished results).

symmetric and chemically inert molecules CF_4 and SF_6 is not enhanced relative to inert diatomic molecules by increased molecular complexity. Further, for those polyatomic species which may be expected to show marked chemical affinity, no pronounced trend in efficiency with molecular complexity has been observed. Thus H_2, CH_4, and C_3H_8 all show closely similar efficiencies. The closely similar efficiencies observed for CH_4, CF_3H, and the alkyl iodides as a group suggests that where strongly attractive potentials are involved, the collision is effectively localized. Again the striking isotope effect observed for H_2O and D_2O may be noted. It has been suggested that this could be attributed to the near resonant transfer involving combination frequencies in vibration. However, this may be fortuitous and a more general reduced mass effect, applicable to all systems involving these atoms, may be responsible. The data for the olefins satisfies the same correlation that has been reported for the addition of electrophilic reagents to these molecules.[133] The second-order rate constants for relaxation increase with decreasing molecular excitation energy[28] (Fig. 12), indicating that electronic rearrangement of the type leading to interaction with the π system influences the interaction potential associated with the collision.

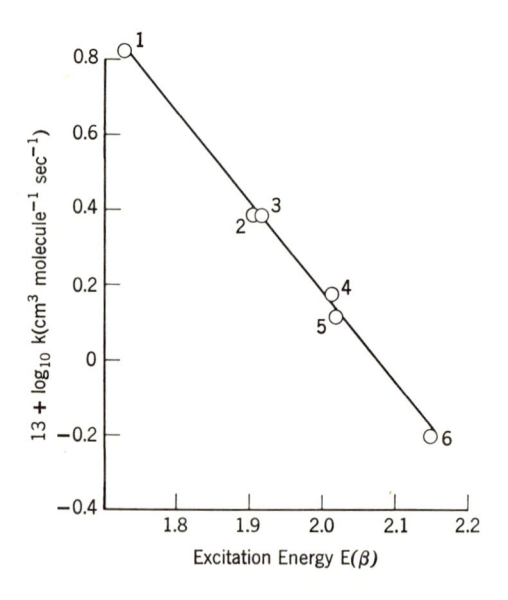

Fig. 12. Correlation of the rate constant (k) for the spin orbit relaxation of $I(5^2P_{1/2})$ with the excitation energy [$E(\beta)$] of the colliding olefin. (1) Tetramethylethylene. (2) cis- and trans-2-Butene. (3) Isobutene. (4) 1-Butene. (5) Propylene. (6) Ethylene. $E(\beta) = E$ (units of a resonance integral β for a standard C=C bond).[133]

VIII. X($^2P_{1/2}$) IN ATOMIC RECOMBINATION

A. Molecular Emission Accompanying Atomic Recombination

The production of $X_2(^3\Pi_{0u^+})$ which correlates with $X(^2P_{1/2}) + X(^2P_{3/2})$ can result from either direct recombination or inverse predissociation. The emission $^3\Pi_{0u^+} \rightarrow X^1\Sigma_g^+$ resulting from direct recombination can either be continuous or banded depending upon whether the recombination is a two-body or three-body process. If an inverse predissociation takes place, the emission is banded and the short wavelength limit corresponds to transitions from the vibrational level v' in the $^3\Pi_{0u^+}$ state at the region of crossing with the repulsive, predissociating state. Which of these two mechanisms operates will depend upon the appropriate potential energy diagram, the energy barrier to inverse predissociation, the efficiency of transfer from one potential energy curve to another, and the temperature which will determine the Boltzmann distribution in the $^2P_{1/2}$ state. Thus differences will be expected between the mechanism taking place in a flow discharge tube at room temperature and that in shock heated gases at high temperatures in which the atoms are generated by thermal dissociation. A mechanistic description in terms of inverse predissociation is a statement that the emitting molecular state correlates with excited as well as ground state atoms, and that the excited atom need play no role in the recombination process.

B. Halogen Atom Recombination in Flow Discharges

Emission from the $^3\Pi_{0u^+}$ state to the ground state during atomic recombination in flow discharges has been observed for Cl_2, Br_2, and BrCl ($^3\Pi_{0^+}$), that for Cl_2 having been studied in most detail.[45–49,134–136] Bader and Ogryzlo[45] and Hutton and Wright[47] have found that the emission intensity $I \propto [Cl]^2$ but more recent very detailed work by Clyne and Stedman[48,134–136] has shown that the squared dependence is only observed in the high energy tail of the emission, and that the power in $[Cl]^n$ varies with wavelength. Thus for 9600 Å $> \lambda >$ 5000 Å, $1 < n < 2$ approximately. To account for this, Clyne and Stedman included the quenching step:

$$Cl(3^2P_{3/2}) + Cl_2(^3\Pi_{0u^+}) \longrightarrow Cl_2 \text{ (nonradiative)} + Cl(3^2P_{3/2}) \qquad (54)$$

in their mechanism. Bader and Ogryzlo[45] proposed an inverse predissociation mechanism in accord with their spectroscopic observation that no emission occurred from $v' > 13$; Clyne and Coxon[49] have detected emission from $v' = 14$ but this simply requires a detailed reconsideration of the region of crossing of potential energy curves where predissociation takes place. On the other hand, although Hutton and Wright[47] also observe transitions from $v' \le 13$, these workers propose a direct recombination mechanism between Cl($3^2P_{3/2}$) and the small Boltzmann population of Cl($3^2P_{1/2}$), attributing the lack of emission from higher levels than $v' = 13$ to rapid vibrational relaxation. Clyne and Stedman[48] have found that the activation energy for the emission from Cl$_2$($^3\Pi_{0u^+}$) in the flow discharge system ($E = -2.0$ kcal mole^{-1}) was virtually equal to that for the total removal of Cl atoms. The 2.5 kcal of electronic energy in Cl($3^2P_{1/2}$) would be expected to give rise to a significant difference between these activation energies had Cl$_2$($^3\Pi_{0u^+}$) been formed by direct recombination.

Emission from Br$_2$($B^3\Pi_{0u^+} \to X^1\Sigma_g^+$) has been observed by Gibbs and Ogryzlo[43] who have shown, from measurement of the concentration of bromine atoms with a moveable nickel wire calorimeter, that the emission intensity $I \propto [\text{Br}]^2$. Transitions were reported for $v' \le 3$, and a consideration of the vibrational constant B as a function of vibrational quantum number in the $B^3\Pi_{0u^+}$ state indicates either the close proximity of a weakly stable O_u^- state or a predissociating state. Certainly, the Boltzmann population of Br($4^2P_{1/2}$) at room temperature (Table I) is too low to account for direct recombination. Clyne and Coxon[48] have also studied the emission from molecular bromine in detail with a low dispersion, high aperture spectrograph, and have shown that the major emitting species is Br$_2$($^3\Pi_{1u}$) [correlating with Br($4^2P_{3/2}$) atoms] whereas emission from Br$_2$($B^3\Pi_{0u^+}$) is weak. This is attributed to the energy barrier of 350 cm^{-1} for formation of Br$_2$($B^3\Pi_{0u^+}, v' = 3$) following inverse predissociation and thus the majority of the recombining atoms lead to the $^3\Pi_{1u}$ state. This is in contrast to Cl$_2$ where no activation barrier existed for the formation of Cl$_2$($A^3\Pi_{0u^+}$) from two $^2P_{3/2}$ atoms.[45,49] Strong emission from BrCl($^3\Pi_{0^+} \to X^1\Sigma^+$), resulting from the recombination of bromine and chlorine atoms in a discharge flow system, has been observed by Clyne and Coxon[49] who have recently observed this system in absorption.[93] These authors favor direct population from Br($4^2P_{3/2}$) + Cl($3^2P_{3/2}$) to the $^3\Pi_{0^+}$ state. In this case, the predissociation discussed hitherto is replaced by an "avoided crossing" yielding a potential maximum in the $^3\Pi_{0^+}$ state[69] (see Section IV.C). There should be no appreciable energy barrier to this maximum[43] for BrCl.

C. Halogen Atom Recombination in the Shock Tube

The emission from molecular halogens in the $^3\Pi_{0u^+}$ state resulting from shock heating could, in principle, arise from both inverse predissociation and direct recombination on account of the large thermal populations of the excited atoms at these high temperatures. Emission from this state has been observed hitherto at high temperatures from iodine, bromine, and chlorine in static systems.[137,138] Continuous emission resulting from two-body recombination of

$$Br(4^2P_{1/2}) + Br(4^2P_{3/2}) \longrightarrow Br_2(B^3\Pi_{0u^+}) \longrightarrow Br_2(X^1\Sigma_g^+) + h\nu \quad (55)$$

following shock heating of Br_2 has been reported,[2,53] arising from the Boltzmann population of $Br(4^2P_{1/2})$ at high temperatures. Banded emission from $v' > 3$, the predissociation level in $Br_2(^3\Pi_{0u^+})$, would have been expected from an over-all three-body recombination process:

$$Br(4^2P_{1/2}) + Br(4^2P_{3/2}) + M \longrightarrow Br_2(^3\Pi_{0u^+}) + M \quad (56)$$

$$Br_2(^3\Pi_{0u^+}) \longrightarrow Br_2(X^1\Sigma_g^+) + h\nu \quad (57)$$

and from $v' \le 3$ for an inverse predissociation mechanism. The emission from $Cl_2(A^3\Pi_{0u^+})$ in the shock tube is both continuous and banded and results from the equilibrium population of $Cl_2(A^3\Pi_{0u^+})$ at, typically, 2200°K.[54,55] Thus no conclusions may be drawn regarding the kinetics of the chlorine atoms leading to this state. No emission has been observed from shocking fluorine,[56] neither has any been reported in shock tube studies of I_2.[57]

D. Halogen Atom Recombination in Flash Photolysis Experiments

There have been many studies of the recombination of iodine[139-141] and bromine[141,142] atoms following flash photolysis. Both Christie et al.[140] and Strong and coworkers[141] have found no effect on the recombination rate following photolytic dissociation with light of different wavelengths, which would generate different ratios of $I(5^2P_{1/2})/I(5^2P_{3/2})$ in the primary step. This can now be considered following the measurement by Donovan and Husain[75] of the rapid relaxation of $I(5^2P_{1/2})$ by I_2 which proceeds at 1 collision in 30 (Section VII). Typical pressures of I_2 in flash photolysis experiments for recombination studies are 0.13-0.18 mm Hg[140,141]; at a constant pressure of $I_2 = 0.15$ mm Hg, noting that $[I_2]$ changes slowly by termolecular recombination of atoms, the halflife of $I(5^2P_{1/2})$ is 27 μsec.[75] Recombination measurements commence typically around 1 msec or greater following the photolytic flash, by which time spin orbit relaxation is essentially complete. In the presence of third bodies which may also give rise to efficient electronic quenching, the halflife is much smaller (Section VII). Rapid relaxation of $Br(4^2P_{1/2})$ by Br_2 is, similarly, very rapid.[76] Thus excited atoms play no major role in the over-all recombination of iodine and bromine atoms, following partial photolytic dissociation of the molecular halogen at room temperature. The $^3\Pi_{0u^+}$ states of Br_2 and Cl_2 have been observed in absorption following the flash photolysis of Br_2 and Cl_2, respectively[59]; Briggs and Norrish[59] have attributed the formation of these electronically excited molecules to both direct molecular photoexcitation from the ground state and also recombination of $^2P_{1/2}$ and $^2P_{3/2}$ atoms. The recombination of chlorine atoms following flash photolysis has not been studied. The rate of dissociation in shock tubes at high temperatures has been investigated both spectroscopically [143] and mass spectrometrically.[144]

Time-resolved emission from $I_2(B^3\Pi_{0u^+})$ following the flash photolysis of CF_3I in the presence of an inert gas (see Experimental Methods, Section III.H) observed by Abrahamson, Husain, and Wiesenfeld,[23] illustrates that direct recombination between $I(5^2P_{1/2})$ and $I(5^2P_{3/2})$ atoms may readily take place provided electronic quenching of the excited atom is slow. The short wavelength limit of the $I_2(B-X)$ emission observed in these experiments corresponds to the onset of the long wavelength limit of the $B-X$ continuum observed in absorption (Fig. 6). Thus the banded emission results from three body recombination. Rewriting the equations as in Experimental Methods (III.H):

$$CF_3I + h\nu \xrightarrow{k_5} CF_3 + I(5^2P_{1/2}) \quad (58)$$

$$I(5^2P_{1/2}) \xrightarrow{k_6} I(5^2P_{3/2}) \quad (59)$$

$$I(5^2P_{1/2}) + I(5^2P_{3/2}) + M \xrightarrow{k_7} I_2(B^3\Pi_{0u^+}) + M \quad (60)$$

$$I_2(B^3\Pi_{0u^+}) \xrightarrow{k_8} I_2(X^1\Sigma_g^+) + h\nu \quad (61)$$

and following the solution of Abrahamson et al.[23] yields:

$$I_{emm}(B\text{-}X) = k_7[M][I(5^2P_{1/2})]_{t=0}^2 [\exp(-k_6 t) - \exp(-2k_6 t)] \quad (62)$$

The total absolute emission intensity was not measured, only the form of the emission which went through a maximum as a function of time, yielding k_6 (Fig. 13). Thus the value of k_7 remains unknown. The determination of this quantity would allow an interesting comparison with the rate constant for the recombination of two ground state atoms. k_6 is the over-all first-order coefficient for the decay of $I(5^2P_{1/2})$ to which there will be contributions from spontaneous emission, diffusion to the walls following deactivation, and collisionally induced spin orbit relaxation. These contributions can be measured by separate experiments on $I(5^2P_{1/2})$ and are described in Section VII. The agreement between the value of k_6 derived from the form of the $I_2(B-X)$ emission as a function of time and the value of k_6 calculated for the different contributions to the over-all decay is satisfactory within the experimentally justifiable, simplifying assumptions that were employed to yield a tractable and explicit solution of the time-dependent differential equation for $I_{emm}(B-X)$.[23] It must be emphasized that in these experiments, the formation of $I_2(B^3\Pi_{0u^+})$ represents a small "bleed off" of $I(5^2P_{1/2})$ which decay principally by an over-all first-order process, observed experimentally. The low

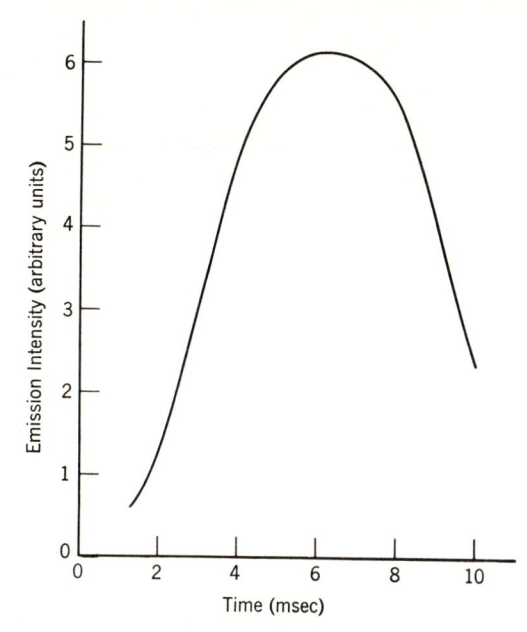

Fig. 13. Time-dependent emission from $I_2(B^3\Pi_{0u^+}) \rightarrow I_2(X^1\Sigma_g^+) + h\nu$ following the flash photolysis of CF_3I. $p_{CF_3I} = 4.0$ mm Hg, $p_{Ar} = 10.0$ mm Hg, $E = 1767$ J.

wavelength onset of the *B-X* emission and the kinetics of $I(5^2P_{1/2})$, $I(5^2P_{3/2})$, and $I_2(B^3\Pi_{0u^+})$ clearly indicate that direct atomic recombination between ground and excited atoms into $I_2(B^3\Pi_{0u^+})$ takes place, rather than inverse predissociation for which the energy barrier of 16.5 kcal would be prohibitive at room temperature. The ability to observe this over-all effect is simply the fortunate result of the large populations of $I(5^2P_{1/2})$ that result on flash photolyzing CF_3I.

IX. CHEMICAL REACTIONS OF ELECTRONICALLY EXCITED HALOGEN ATOMS

A. *The Thermochemistry of Some Reactions of* $X(^2P_{1/2})$

The difference in the spin orbit energies of the $^2P_{1/2} - {}^2P_{3/2}$ states of the halogen atoms is relatively large (Table I) and thus significant differences may be expected in the rates at which these atoms undergo chemical reaction. A comparison of the reaction rates of the atom in these two states is particularly interesting as the chemical interaction on collision with a given reactant would be similar as only spin orbit atomic energies are involved. Any

TABLE X
Thermochemistry of Some Reactions of $X(^2P_{1/2,3/2})$, $\Delta H_{298°K}$, kcal mole^{-1} [19,145]

	X	Cl	Br	I
$H_2 + X(^2P_{3/2}) \longrightarrow HX + H$		+1.1	+19.6	+32.9
$H_2 + X(^2P_{1/2}) \longrightarrow HX + H$		−1.4	+9.1	+11.2
$Br_2 + X(^2P_{3/2}) \longrightarrow XBr + Br(4^2P_{3/2})$		+4.4	0	+3.6
$Br_2 + X(^2P_{1/2}) \longrightarrow XBr + Br(4^2P_{3/2})$		+1.9	−10.5	−18.1
$i\text{-}C_3H_7\text{—}H + X(^2P_{3/2}) \longrightarrow i\text{-}C_3H_7 + HX$		−9.1	+6.4	+22.7
$i\text{-}C_3H_7\text{—}H + X(^2P_{1/2}) \longrightarrow i\text{-}C_3H_7 + HX$		−11.6	−4.1	+1.0
$CH_3X + X(^2P_{3/2}) \longrightarrow CH_3 + X_2$		+22.6	+20.9	+16.5
$CH_3X + X(^2P_{1/2}) \longrightarrow CH_3 + X_2$		+20.1	+10.4	−5.2

difference in reactivity would thus result entirely from the electronic energy of the atom. These differences will be most important in photochemical rather than thermal systems for which the Boltzmann population of excited atoms will be low (Table I). Although differences in rate for the reactions of the atoms in the two states may always be expected, the most striking should occur when a change in the sign of the thermochemistry is involved. Examples are given in Table X of some of the types of abstraction reactions of interest, a number of which have been studied and will be considered further in this section. The reactions of excited halogen atoms with H_2 have not yet been investigated. In terms of the rate constant for the metathetical atomic abstraction reaction of $Cl(3^2P_{1/2,3/2})$ with H_2, a factor of approximately 10 would be expected for the ratio of the rate constants at room temperature if the activation energy for reaction involving $Cl(3^2P_{1/2})$ were reduced by an amount equal to the electronic energy of the atom. This process could play an important role in the photochemical reaction between H_2 and Cl_2.

B. Spin Orbit Relaxation

To observe the kinetic effect of the chemical reaction of an electronically excited atom, the rate must be significantly different from that of the ground state atom, as spin orbit relaxation may compete with reaction. The separation of these processes is inherent in studies of the reactions of excited halogen atoms. Clearly, where a strong attractive potential facilitates chemical reaction, this chemical interaction will also aid spin orbit relaxation itself (Section VII). Thus relatively strongly endothermic, slow, chemical reaction is difficult to observe in competition with rapid electronic quenching which reduces the concentration of excited atom and cuts down the yield of products from chemical reaction. Thus for the H atoms abstraction in the secondary position in C_3H_8 by $I(5^2P_{1/2})$, which is close to thermoneutral ($\Delta H = +1$ kcal mole^{-1}),[19,145] Callear and Wilson[65,66] report that spin orbit relaxation is 6.38×10^3 times faster than the reaction at 30°C, and decreases to 3.05×10^3 at 90°C. The relative contributions of reaction and spin orbit relaxation when atomic abstraction is an exothermic process cannot be readily predicted theoretically. Unfortunately, a potential energy surface for the reaction and relaxation of an excited atom cannot be plotted with any confidence principally on account of the possible breakdown of the adiabatic approximation and the detailed consideration of the hypersurface in the regions of crossing. At present, recourse must be made to experiment and will be dealt with in a number of cases in the following Sections. Generally speaking, it is found that, in the presence of a diatomic molecule, exothermic atomic metathetical reaction predominates over relaxation whereas, for polyatomic molecules, quenching is more rapid than exothermic abstraction reaction.

C. H *Atom Abstraction from Paraffins by* $I(5^2P_{1/2})$ *by Classical Photochemical Investigation*

The abstraction of a hydrogen atom from both propane and ethane by $I(5^2P_{1/2})$ according to

$$I(5^2P_{1/2}) + C_3H_8 \xrightarrow{k_1} HI + C_3H_7 \quad \Delta H = +1.0 \text{ kcal mole}^{-1} \quad (63)$$

$$I(5^2P_{1/2}) + C_2H_6 \xrightarrow{k_2} HI + C_2H_5 \quad \Delta H = +4.1 \text{ kcal mole}^{-1} \quad (64)$$

has been studied by Callear and Wilson[65,66] by continuous photolysis of I_2 both in the continuum of the $B^3\Pi_{0u^+} \leftarrow X^1\Sigma_g^+$ system and by a proposed collisional release mechanism in the banded region (Section III.J). A stationary state kinetic treatment yields the rates of the above reactions relative to the electronic quenching of $I(5^2P_{1/2})$ by the species present. Callear and Wilson[65,66] take as their absolute standard the quenching efficiency of $I(5^2P_{1/2})$ by propane determined by Donovan and Husain,[27] yielding a quenching efficiency of $I(5^2P_{1/2})$ in these experiments by I_2 corresponding to approximately 1 in 6 collisions. This is approximately five times faster than the rate observed by vacuum ultraviolet absorption spectroscopy on the excited atom (Table IX), and very rapid compared to the corresponding processes with the other halogens and interhalogens[31] (see later, this Section). Thus, assuming that the activation energy for quenching of $I(5^2P_{1/2})$ by I_2 to be zero, Callear and Wilson report the Arrhenius parameters for reactions (63) and (64) as:

k_1: $\log_{10} A$ (cm³ molecule⁻¹ sec⁻¹) = −13.53; $\quad E = 5.0$ kcal mole⁻¹

k_2: $\log_{10} A$ (cm³ molecule⁻¹ sec⁻¹) = −13.36; $\quad E = 7.0$ kcal mole⁻¹

The most striking aspect of these results is the low Arrhenius A factors corresponding to p factors of the order of 10^{-3}. Callear and Wilson attribute this to a lack of equilibrium involving the transition state caused by relaxation to the lower surface. It is clearly possible in principle to extend this type of measurement to other hydrocarbons. Its extension to the reactions of $Br(4^2P_{1/2})$, although thermochemically favorable (Table X), would be very difficult experimentally on account of more rapid relaxation of $Br(4^2P_{1/2})$ (Table IX) yielding low stationary concentrations of the excited atoms, and possibly even lower yields of the products than with $I(5^2P_{1/2})$, which, in the case of propane,[65,66] involved measurement of the order of 10^{-9} moles in a given experiment.[65]

D. Reactions of $I(5^2P_{1/2})$ with Alkyl Iodides

Iodine atom abstraction from an alkyl iodide by $I(5^2P_{1/2})$ is an exothermic process and thus may compete with spin orbit relaxation:

$$I(5^2P_{1/2}) + RI \xrightarrow{k_3} I(5^2P_{3/2}) + RI \qquad (65)$$

$$I(5^2P_{1/2}) + RI \xrightarrow{k_4} I_2 + R\cdot \qquad (66)$$

The thermochemistry for abstraction reactions (66) for processes which have been investigated are given in Table XI. The corresponding reactions of $I(5^2P_{3/2})$ are strongly endothermic, and thus very much slower,[146] and will not be considered here. The determination of the relative and absolute magnitudes of k_3 and k_4 constitute our interest in this context. Although there have been many investigations into the photolyses of the alkyl iodides,[62,63,147–149] it is only recently that experimental differentiation has been attempted between the reactions of the $^2P_{1/2}$ and $^2P_{3/2}$ states of the iodine atom.

TABLE XI
Thermochemistry [19,145] of the Abstraction Reaction $I(5^2P_{1/2}) + RI \longrightarrow I_2 + R\cdot$

RI	CH_3I	C_2H_5I	n-C_3H_7I	i-C_3H_7I	n-C_4H_9I	i-C_4H_9I	t-C_4H_9I
ΔH, kcal mole⁻¹	−3.3	−6.3	−7.3	−11.3	−8.3	−9	−10.3

In a recent study of the decay of excited atoms in the presence of alkyl iodides, Donovan, Hathorn, and Husain[114,115] have determined the sum $k_3 + k_4$ for seven of these molecules by monitoring the excited iodine atom in the ultraviolet and vacuum ultraviolet using kinetic spectroscopy (Table XII). The rate of removal of the excited atoms is clearly very rapid. The separation of k_3 and k_4 has only been carried out for methyl iodide itself, principally because the transitions of the ground state iodine atom are masked by the molecular absorption of the alkyl iodides.[115] Kinetic spectroscopic investigation in the vacuum ultraviolet using low pressures of methyl iodide in order to penetrate the molecular spectrum[114,115] permits observation of the atomic transitions for both states and also the intense molecular system $I_2(M \leftarrow X)$. No significant yield of molecular iodine could be detected simultaneous with the relaxation of $I(5^2P_{1/2})$ but only following slow, termolecular atomic recombination (Fig. 14). This enables an upper limit to be placed on the rate of the reaction, yielding $k_4/k_3(300°K) < 1/10$. This procedure has not been repeated for higher iodides where the vacuum ultraviolet molecular absorption was too strong. Meyer[60,61] has attributed the rapid formation of I_2 detected by time-resolved mass spectrometry following the flash photolysis of CH_3I, to the abstraction (66) from CH_3I, and reports the Arrhenius parameters as $\log_{10} A$ (cm^3 molecule^{-1} sec^{-1}) = -14.52 and $E = 1.9$ kcal mole^{-1} which yields the value of $k_{4\,CH_3I}$ of 1.23×10^{-14} cm^3 molecule^{-1} sec^{-1} at 300°K. This therefore represents about 1% of the over-all decay of $I(5^2P_{1/2})$ compared with $k_3 + k_4$ measured spectroscopically (Table XII). A further estimate for $k_{4\,CH_3I}$ may be made by considering the upper limit for the quantum yield for the photochemical exchange of I_2,[131] photolyzed in the

TABLE XII
Data for Reactions of $I(5^2P_{1/2})$ + Alkyl Iodides RI

$$I(5^2P_{1/2}) + RI \xrightarrow{k_3} I(5^2P_{3/2}) + RI$$
$$I(5^2P_{1/2}) + RI \xrightarrow{k_4} I_2 + R\cdot$$

RI	$10^{13}(k_3 + k_4)$, cm^3 molecule^{-1} sec^{-1}	Number of collisions $Z/(k_3 + k_4)$	Collision cross section $\sigma_3^2 + \sigma_4^2$, Å^2
CH_3I	17	97	0.18
C_2H_5I	1.9 ± 0.2	974	2.0×10^{-2}
n-C_3H_7I	2.0 ± 0.2	1080	2.1×10^{-2}
i-C_3H_7I	2.0 ± 0.2	1054	2.2×10^{-2}
n-C_4H_9I	2.9 ± 0.2	715	3.2×10^{-2}
i-C_4H_9I	2.9 ± 0	783	3.1×10^{-2}
t-C_4H_9I	3.8 ± 1.3	591	4.2×10^{-2}

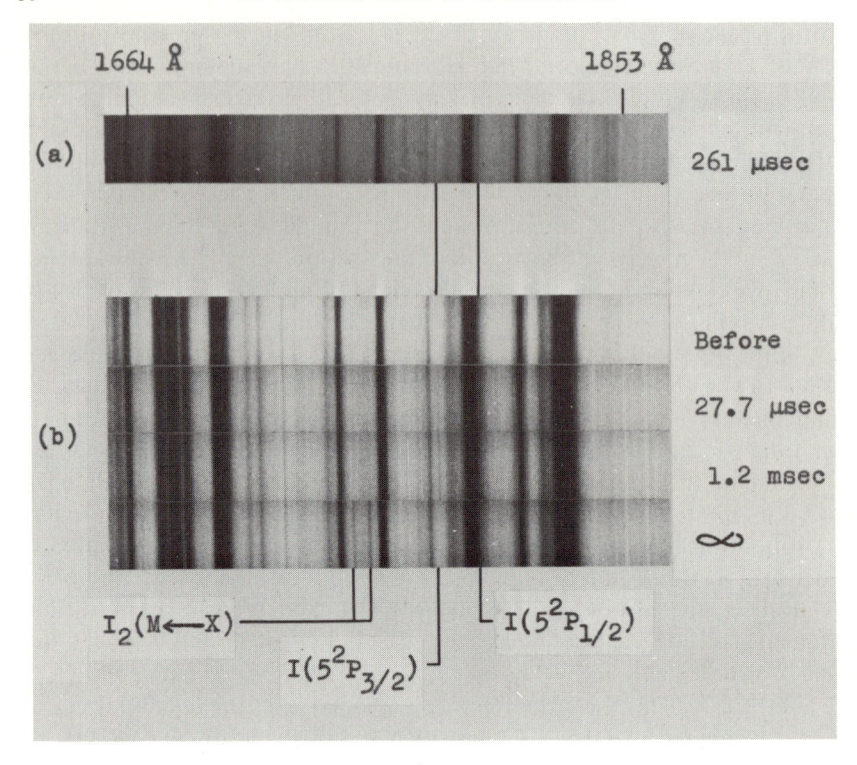

Fig. 14. Spin orbit relaxation of $I(5^2P_{1/2})$ in the photolysis of methyl iodide. (a) $p_{CH_3I} = 0.025$ mm Hg, $p_{Ar} = 50$ mm Hg, $E = 1024$ J. (b) $p_{CH_3I} = 0.20$ mm Hg, $p_{Ar} = 50$ mm Hg, $E = 1024$ J.

$I_2(B\text{-}X)$ continuum in the presence of CH_3I.[150] The processes giving rise to photochemical exchange may be written:

$$I_2^{131} + h\nu \longrightarrow I(5^2P_{1/2})^{131} + I(5^2P_{3/2})^{131} \tag{67}$$

$$I(5^2P_{1/2})^{131} + CH_3I \xrightarrow{k_6} I(5^2P_{3/2}) + CH_3I \tag{68}$$

$$I(5^2P_{1/2})^{131} + CH_3I \xrightarrow{k_7} I_2 + CH_3 \tag{69}$$

$$I(5^2P_{1/2})^{131} + I_2^{131} \xrightarrow{k_8} I(5^2P_{3/2}) + I_2^{131} \tag{70}$$

$$CH_3 + I_2^{131} \xrightarrow{k_9} CH_3I^{131} + I^{131} \tag{71}$$

Under the conditions employed by Aditya and Willard,[150] the CH_3 radical will be consumed by reaction (71) and hence the expression for the quantum yield becomes:

$$\phi = k_7[CH_3I]/\{(k_6 + k_7)[CH_3I] + k_8[I_2]\} \tag{72}$$

$k_6 + k_7$ is taken from Table XII, and k_8 may be taken from the data for the quenching of $I(5^2P_{1/2})$ by I_2 (Table IX). Thus the measured upper limit for the above quantum yield measured by Aditya and Willard[150] yields the value of $k_7 < 1.7 \times 10^{-15}$ cm^3 molecule^{-1} sec^{-1} at 300°K which is an order of magnitude lower than Meyer's value of 1.23×10^{-14} at this temperature.[60,61] Thus for CH$_3$I, any chemical reaction of $I(5^2P_{1/2})$ represents a very minor process in the over-all decay of $I(5^2P_{1/2})$ which is controlled by rapid spin orbit relaxation. This would appear to be the case for other alkyl iodides as the measured rates for the over-all decay fit in well within the body of data for energy transfer (Table IX). The low p factor of $\sim 10^{-4}$ from Meyer's value of k_7[60] appears to compare reasonably with the p factors for H atom abstraction from ethane and propane by $I(5^2P_{1/2})$ which Callear and Wilson[66] found to be $\sim 10^{-3}$.

E. Reactions of $I(5^2P_{1/2})$ and $I(5^2P_{3/2})$ with Halogens and Interhalogens

The electronic energy of the iodine atom in the $5^2P_{1/2}$ state renders a number of atomic reactions with halogen and interhalogen molecules exothermic which would be endothermic for the analogous reactions of the ground state. A number of these reactions, namely, (74), (76), (78), and (80) in the following set, have been studied by kinetic spectroscopy in absorption:[31]

$$I(5^2P_{1/2}) + Cl_2 \xrightarrow{k_{11}} I(5^2P_{3/2}) + Cl_2 \qquad \Delta H = -21.7 \text{ kcal mole}^{-1} \qquad (73)$$

$$I(5^2P_{1/2}) + Cl_2 \xrightarrow{k_{12}} ICl + Cl(3^2P_{3/2}) \qquad \Delta H = -14.3 \text{ kcal mole}^{-1} \qquad (74)$$

$$I(5^2P_{1/2}) + ICl \xrightarrow{k_{13}} I(5^2P_{3/2}) + ICl \qquad \Delta H = -21.7 \text{ kcal mole}^{-1} \qquad (75)$$

$$I(5^2P_{1/2}) + ICl \xrightarrow{k_{14}} I_2 + Cl(3^2P_{3/2}) \qquad \Delta H = -7.6 \text{ kcal mole}^{-1} \qquad (76)$$

$$I(5^2P_{1/2}) + Br_2 \xrightarrow{k_{15}} I(5^2P_{3/2}) + Br_2 \qquad \Delta H = -21.7 \text{ kcal mole}^{-1} \qquad (77)$$

$$I(5^2P_{1/2}) + Br_2 \xrightarrow{k_{16}} IBr + Br(4^2P_{3/2}) \qquad \Delta H = -18.2 \text{ kcal mole}^{-1} \qquad (78)$$

$$I(5^2P_{1/2}) + IBr \xrightarrow{k_{17}} I(5^2P_{3/2}) + IBr \qquad \Delta H = -21.7 \text{ kcal mole}^{-1} \qquad (79)$$

$$I(5^2P_{1/2}) + IBr \xrightarrow{k_{18}} I_2 + Br(4^2P_{3/2}) \qquad \Delta H = -15.4 \text{ kcal mole}^{-1} \qquad (80)$$

(All molecular states are ground states)

The decay of $I(5^2P_{1/2})$ in the presence of, for example, ICl yields the sum $k_{13} + k_{14}$, and inspection of the yields of $I(5^2P_{3/2})$ and I_2 by vacuum ultraviolet spectroscopy indicates that $k_{14} > k_{13}$ (Fig. 15c). Indeed, the small yield of $I(5^2P_{3/2})$ could be the result of two chemical processes; namely, reaction (76) and the exothermic reaction

$$Cl(3^2P_{3/2}) + I_2 \xrightarrow{k_{19}} ICl + I(5^2P_{3/2}) \qquad (81)$$

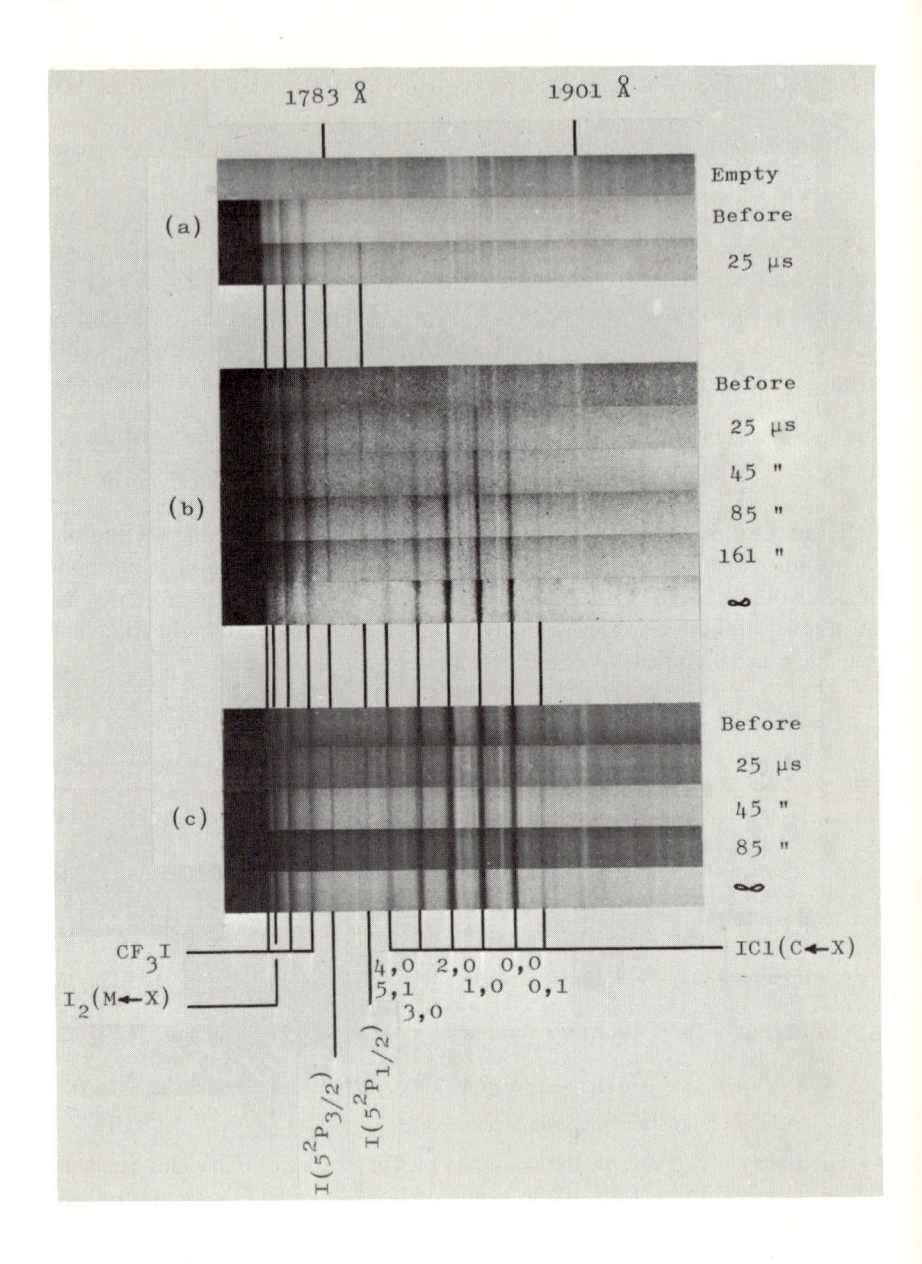

Fig. 15. Time-resolved vacuum ultraviolet spectra resulting from the reactions of $I(5^2P_{1/2})$ with Cl_2 and ICl. $p_{CF_3I} = 0.6$ mm Hg, $p_{Ar} = 50$ mm Hg, $E = 784$ J. (a) CF_3I/Ar alone. (b) $p_{Cl_2} = 0.5$ mm Hg, $p_{ICl} = 0.06$ mm Hg.

TABLE XIII
Comparison of Rate Constants for Reactions of $I(5^2P_{1/2})$ and $I(5^2P_{3/2})$

Reaction	$k_{300°K}$, cm^3 molecule^{-1} sec^{-1}	$N = \dfrac{Z}{k_{300°K}}$	E, kcal[a]	ΔH, kcal
$I(5^2P_{1/2}) + Cl_2 \longrightarrow ICl + Cl$	2.1×10^{-13}	930	2.7	-14.3
$I(5^2P_{1/2}) + Br_2 \longrightarrow IBr + Br$	1.5×10^{-12}	107	1.4	-18.2
$I(5^2P_{1/2}) + ICl \longrightarrow I_2 + Cl$	3.4×10^{-12}	51	1.0	-7.7
$I(5^2P_{1/2}) + IBr \longrightarrow I_2 + Br$	4.3×10^{-12}	39	0.8	-15.4
$I(5^2P_{1/2}) + I_2 \longrightarrow I_2 + I$	5.0×10^{-12}	30	0.7	-21.7
$I(5^2P_{3/2}) + Cl_2 \longrightarrow ICl + Cl$	6.1×10^{-21}	3.2×10^{10}	12.0	$+7.5$
$I(5^2P_{3/2}) + Br_2 \longrightarrow IBr + Br$	(2.7×10^{-14})	(5.9×10^3)	(3.8)	$+3.5$
$I(5^2P_{3/2}) + IBr \longrightarrow I_2 + Br$	$(\sim 10^{-17})$	$(\sim 10^7)$	—	$+6.3$

[a] Calculated assuming a steric factor = 0.1 with the exception of the reaction between $I(5^2P_{3/2}) + Cl_2$.[151]

rather than electronic quenching via reaction (75). Similar considerations apply to the other reactions studied. Further, the measurement of the decay rate for $I(5^2P_{1/2})$ in the presence of Cl_2, corrected for the removal of the excited atom by ICl produced in reaction (74) (using the measured value of $k_{13} + k_{14}$) yields $k_{11} + k_{12}$. Further, measurements in the vacuum ultraviolet show that $k_{12} > k_{11}$ (Fig. 15b). k_{12} can also be measured by integration of the rate equation for the production of ICl, and again, this indicates that the decay of $I(5^2P_{1/2})$ in the presence of Cl_2 is dominated by the chemical reaction (74). Thus the observed decay of the excited atom is attributed to chemical reaction in each case. By this means, the rate constants for the chemical reactions (74), (76), (78), and (80) have been measured at 300°K [31] (Table XIII).

The rates of reaction of the electronically excited iodine atoms have been compared directly with a number of analogous reactions of the ground state atoms in the following manner. The equilibrium constants K_{20}, K_{21}, and K_{22} for ground state reactants and products (since the small Boltzmann populations of excited states may be neglected) in the following processes

$$Cl + ICl \underset{(k-20)}{\rightleftharpoons} Cl_2 + I \quad k_{20} \quad (82)$$

$$Br + IBr \underset{(k-21)}{\rightleftharpoons} Br_2 + I \quad k_{21} \quad (83)$$

$$Br + I_2 \underset{(k-22)}{\rightleftharpoons} IBr + I \quad k_{22} \quad (84)$$

were calculated by statistical mechanics.[31] k_{-20} has been characterized following the classical photochemical study of Thrush et al.,[151] and the reactions

(83, k_{21}) and (84, k_{22}) may be estimated from kinetic spectroscopic measurements.[31] In this way, the rates of (82), (83), and (84) for the ground state atoms may be compared with those of the excited atoms (74), (76), and (78). The results are summarized in the Table XIII.

It is seen that the spin orbit energy of $I(5^2P_{1/2})$ may be efficiently utilized in reactions with the halogens and interhalogens. The analogous endothermic reactions of $I(5^2P_{3/2})$ are very much slower. The rate constant hitherto attributed to the relaxation of $I(5^2P_{1/2})$ on collision with I_2[75] is included in Table XIII, and is seen to be of the expected order of magnitude required for chemical reaction by comparison with the other halogens, namely,

$$I(5^2P_{1/2})^{(1)} + I_2^{(2)(2)} \longrightarrow I_2^{(1)(2)} + I(5^2P_{3/2})^{(2)} \tag{85}$$

This clearly will only be resolved by an isotopic labeling experiment in the gas phase. These results, which show that $I(5^2P_{1/2})$ undergoes rapid, exothermic chemical reaction with the halogens and the interhalogens, are in strong contrast to those for the paraffins, and particularly for the alkyl iodides, where spin orbit relaxation is dominant. Similarly, with the other polyatomic molecules studied, where atomic abstraction is an exothermic process, it is generally found that electronic quenching dominates chemical reaction (Section IX.G).

F. Some Reactions of $Br(4^2P_{1/2})$ and $Cl(3^2P_{1/2})$

The evidence for chemical reactions of $Br(4^2P_{1/2})$ and $Cl(3^2P_{1/2})$ results from recent work on the photochemistry of IBr and ICl studied by kinetic spectroscopy in the vacuum ultraviolet[90] (see Section IV.C). $Br(4^2P_{1/2})$ was observed at short delay in the photolysis of IBr (Fig. 9), in the absence of $Br(4^2P_{3/2})$, suggesting the reaction

$$Br(4^2P_{1/2}) + IBr \xrightarrow{k_{24}} Br_2 + I(5^2P_{3/2}) \qquad \Delta H = -14.0 \text{ kcal} \tag{86}$$

This is further supported by the observed decay of IBr and the simultaneous growth of $I(5^2P_{3/2})$ (Fig. 8). Similarly, although $Cl(3^2P_{1/2})$ could not be observed in the photolysis of ICl, the decay of ICl itself following the flash and the absence of $Cl(3^2P_{3/2})$ are attributed to the reaction

$$Cl(3^2P_{1/2}) + ICl \xrightarrow{k_{25}} Cl_2 + I(5^2P_{3/2}) \qquad \Delta H = -10.0 \text{ kcal} \tag{87}$$

Rate constants for reactions (86) and (87) have been reported as $k_{24}(300°K) \sim 3 \times 10^{-12}$ and $k_{25} \sim 3 \times 10^{-11}$ cm^3 molecule^{-1} sec^{-1}. k_{24} is probably correct to within a factor of 2 and k_{25} to within an order of magnitude. These represent high efficiencies corresponding to about 1 in 40 and 1 in 10 collisions for effective reaction. Thus we may conclude from Sections IX.E and IX.F that the collision between an excited halogen atom (Cl, Br, I $^2P_{1/2}$) with a halogen or interhalogen molecule leads predominantly to chemical reaction, exhibiting rates close to the collisional number, rather than spin orbit relaxation.

An interesting consideration may be put forward for the reaction

$$\text{Cl}(3^2P_{1/2}) + H_2 \xrightarrow{k_{26}} \text{HCl} + \text{Cl} \qquad (88)$$

The reaction

$$\text{H} + \text{HCl} \xrightarrow{k_{27}} H_2 + \text{Cl} \qquad (89)$$

has recently been studied in detail by Clyne and Stedman.[117] If the rate constant for reaction (88) is taken as that for the reverse of (89), calculated by means of the appropriate equilibrium constant,[117] with an activation energy reduced by the electronic energy of the chlorine atom in the $^2P_{1/2}$ state, this yields $k_{26}(300°K) = 2.0 \times 10^{-12}$ cm³ molecule⁻¹ sec⁻¹. This is comparable to the observed rate of relaxation of Cl($3^2P_{1/2}$) by H_2 (Table IX) and thus abstraction may dominate electronic quenching as for the reactions in Sections IX.E and IX.F. Although the Arrhenius p factors for the reactions of I($5^2P_{1/2}$) with polyatomic molecules are generally small and attributed to a lack of equilibrium involving the transition state (Section IX.C and IX.D), the corresponding parameters for reactions with diatomic molecules appear to be of normal magnitude. Hence the effect of the reaction (88) in the over-all photochemical reaction between H_2 and Cl_2 remains open, particularly as the propagation reaction

$$\text{H} + Cl_2 \longrightarrow \text{Cl}(3^2P_{1/2}) + \text{HCl} \qquad (90)$$

has not been elucidated experimentally (Section V).

G. I($5^2P_{1/2}$) with Polyatomic Molecules

A number of further systems in which atomic metathetical reaction by I($5^2P_{1/2}$) with polyatomic molecules is exothermic have been studied recently.[28]

1. Nitrosyl Halides.[28] The exothermic reactions of I($5^2P_{1/2}$) with NOCl and NOBr [19,145]

$$\text{I}(5^2P_{1/2}) + \text{NOCl} \xrightarrow{k_{29}} \text{ICl} + \text{NO} \quad \Delta H = -35.2 \text{ kcal mole}^{-1} \qquad (91)$$

$$\text{I}(5^2P_{3/2}) + \text{NOCl} \xrightarrow{k_{30}} \text{ICl} + \text{NO} \quad \Delta H = -13.5 \text{ kcal mole}^{-1} \qquad (92)$$

$$\text{I}(5^2P_{1/2}) + \text{NOBr} \xrightarrow{k_{31}} \text{IBr} + \text{NO} \quad \Delta H = -35.6 \text{ kcal mole}^{-1} \qquad (93)$$

$$\text{I}(5^2P_{3/2}) + \text{NOBr} \xrightarrow{k_{32}} \text{IBr} + \text{NO} \quad \Delta H = -13.9 \text{ kcal mole}^{-1} \qquad (94)$$

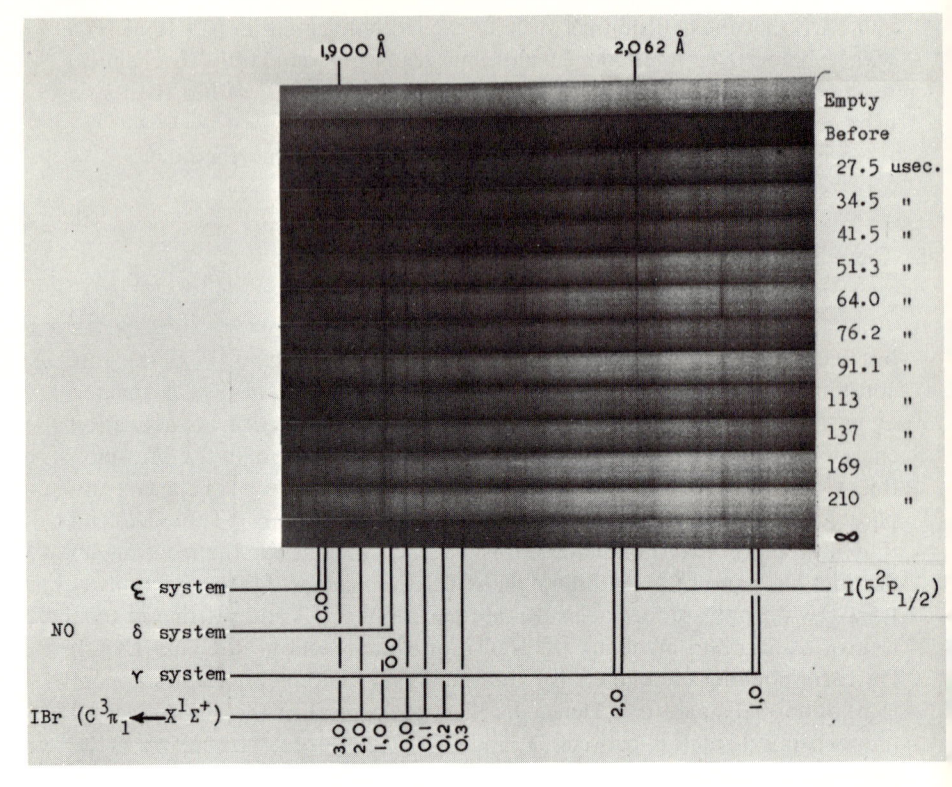

Fig. 16. The decay of $I(5^2P_{1/2})$ and the growth of IBr and NO in the photolysis of CF_3I in the presence of NOBr. $p_{CF_3I} = 4.0$ mm Hg, $p_{NOBr} = 0.06$ mm Hg, $p_{Ar} = 100$ mm Hg, $E = 2112$ J.

appear to dominate electronic quenching. Table XIV indicates that the overall rate of decay of $I(5^2P_{1/2})$ in the presence of these molecules is very rapid. The products of reaction may be observed spectroscopically (ICl, IBr $C^3\Pi_1 \leftarrow X^1\Sigma^+$; NO γ, δ, and ε bands, Fig. 16), but it is difficult to estimate the separate contributions from k_{33} and k_{34} to the sum $k_{33} + k_{34}$ (Table XIV) on account of the strong continuous spectra associated with NOX in the far ultraviolet. In the case of NOBr, it has been estimated that reaction accounts for over 50% of the over-all decay of $I(5^2P_{1/2})$.[28] A more precise estimate is not possible in view of the thermochemically favorable reactions involving ground state atoms.

2. Nitrous Oxide, Ozone, and Nitrogen Dioxide.[28] $I(5^2P_{1/2})$ undergoes slow, spin orbit relaxation in the presence of nitrous oxide (Table IX) rather than chemical reaction according to

$$I(5^2P_{1/2}) + N_2O \longrightarrow IO + N_2 \quad \Delta H = -25 \text{ kcal mole}^{-1} \quad (97)$$

TABLE XIV
Rate Data for the Decay of $I(5^2P_{1/2})$ in the Presence of Nitrosyl Halides[28]

$$I(5^2P_{1/2}) + NOX \xrightarrow{k_{33}} IX + NO \qquad (95)$$

$$I(5^2P_{1/2}) + NOX \xrightarrow{k_{34}} I(5^2P_{3/2}) + NOX \qquad (96)$$

X	$k_{33} + k_{34}$, cm³ molecule⁻¹ sec⁻¹	$P = \dfrac{k_{33} + k_{34}}{Z}$	ΔH, kcal, for reaction (95)
Cl	6.2×10^{-12}	3.3×10^{-2}	-35.2
Br	9.6×10^{-12}	6.0×10^{-2}	-35.6

The strong visible system of $IO(A^2\Pi \leftarrow X^2\Sigma)$[28,152] was not detected when $I(5^2P_{1/2})$ decayed in the presence of N_2O.

Although IO with up to four vibrational quanta in the ground electronic state can be observed in photolyzing CF_3I in the presence of ozone, there is no strong evidence for the exothermic reaction

$$I(5^2P_{1/2}) + O_3 \longrightarrow IO + O_2 \qquad \Delta H = -25.1 \text{ kcal} \qquad (98)$$

The absorption maxima in the ultraviolet of both CF_3I, the source of $I(5^2P_{1/2})$, and O_3 almost coincide,[153,62] and thus photolysis of both of these materials take place, giving rise to the reaction

$$O(^1D) + CF_3I \longrightarrow IO + CF_3 \qquad \Delta H = -32.0 \text{ kcal} \qquad (99)$$

The reaction of O^1D above can be shown to be important as the ultraviolet photolysis of NO_2, which gives rise to O^1D,[62] results in the production of IO in the presence of CF_3I.[28] The abstraction of atomic oxygen from NO_2 by $I(5^2P_{1/2})$ would be an endothermic process. Thus although reaction (99) appears to be established, its contribution following the photolysis of CF_3/O_3 mixtures may not be quantitatively assessed. The importance of reaction (98) therefore remains in doubt.

3. Olefins.[28] Spin orbit relaxation dominates the decay of $I(5^2P_{1/2})$ in the presence of olefins,[28] the data for which is given in Table IX. Since measurements on these systems are usually carried out in times short compared with those required for significant contributions from slow, termolecular atomic recombination, the evidence for the decay by a physical process can be seen from the reasonable balance maintained for $[I(5^2P_{1/2})] + [I(5^2P_{3/2})]$[28] over the main part of the decay. There is no evidence for any significant role by the reaction

$$I(5^2P_{1/2}) + CH_3-CH=CH_2 \longrightarrow HI + \cdot CH_2-CH=CH_2 \qquad \Delta H = -16 \text{ kcal} \qquad (100)$$

as the intense transitions of HI in the vacuum ultraviolet were not detected.[28]

4. Allyl Halides.[28] Although metathetical abstractions of a halogen atom from the allyl halides CH_2=CH—CH_2X, where X = Cl, Br, and I by $I(5^2P_{1/2})$, is exothermic to the extent of 11.4, 15.9, and 23.7 kcal mole^{-1}, respectively,[19,45] the decay of the electronically excited halogen atom is controlled by spin orbit relaxation as given in Table IX. The quenching efficiency increases in the order propylene, allyl chloride, bromide, and iodide, presumably on account of stronger chemical interaction, a trend observed in general for spin orbit relaxation with systems containing H, Cl, Br, and I (Table IX). The strong ultraviolet systems of ICl and IBr were not detected when $I(5^2P_{1/2})$ decayed in the presence of allyl chloride and bromide.

X. GENERAL CONCLUSIONS

Direct studies on electronically excited halogen atoms can be considered as having commenced with the observations on stimulated emission from $I(5^2P_{1/2})$ by Kasper and Pimentel in 1964. Since that date, many aspects of the fundamental processes that halogen atoms in the $np^5\ ^2P_{1/2}$ state undergo have been established by a variety of methods, particularly direct spectroscopic methods. Theoretical and experimental considerations of spontaneous emission, stimulated emission, gas phase diffusion of the excited atoms, spin orbit relaxation, and chemical reaction of these atoms have been the objects of recent investigations. Not all the techniques employed have been applied to the study of all the electronically excited halogen atoms, and, in general, the facility with which these atoms can at present be studied is in the order I > Br > Cl. The large amount of work that has been carried out on $I(5^2P_{1/2})$ is principally on account of the large populations that result on the photodissociation of various iodides, particularly the molecule CF_3I, the high optical as well as collisional metastability of the atom, and the subsequent ease with which the atom can be studied by spectroscopic methods. Although $Br(4^2P_{1/2})$ and $Cl(3^2P_{1/2})$ are more optically metastable than $I(5^2P_{1/2})$, many limitations on experimental methods for studying these lighter atoms result from their collisional reactivity. Kinetic studies of $F(2p^2\ ^2P_{1/2})$ can only conveniently be made at this present stage by electron spin resonance spectroscopy as the electronic transitions of the atom lie at too low a wavelength, and it is to be hoped that this work will be carried out in the near future, particularly the spin orbit relaxation of this atom following the production of a non-Boltzmann distribution in the excited state in a discharge. The chemistry of both $Br(4^2P_{1/2})$ and $Cl(3^2P_{1/2})$ has yet to be established in any detail, and careful use of existing methods should enable this to be done, if only for the bromine atom where a large difference in chemical reactivity for $Br(4^2P_{1/2})$ and $Br(4^2P_{3/2})$ is expected in some cases. The limiting factor in such investigations will again be the relative rates of chemical reaction and spin orbit relaxation.

A theoretical treatment of the factors governing the relative contributions of chemical and spin orbit relaxation when both of these processes are energetically favorable would be desirable. As this problem is hindered by the difficulty in constructing a suitable potential energy surface for the reaction of an excited atom, mainly owing to the uncertainty of the nature of the surface in the regions of crossing and the possible breakdown of the adiabatic approximation, the construction of parts of the appropriate surface from experimental results would seem to be the most fruitful direction at present. This will require relatively detailed determinations of the initial energy product distributions for both chemical reaction and electronic quenching, as well as measurement of the over-all rates involving each of these processes in a particular case. An obvious route toward this is the determination of the Arrhenius parameters for many of the rate processes that have been described here for single temperature measurements. Further, the developments of theories of spin orbit relaxation in forms that allow a clear comparison with experiment will facilitate this general approach as well as being of value per se. At present, the data on collisional relaxation can be usefully considered within the context of existing theories that have been briefly mentioned. The existing data is highly relevant to many fields including flow discharges, shock tubes, laser studies, energy transfer from atoms in general, and the photochemistry of electronically excited atoms. The many classical investigations on the photochemistry of halides may now be considered within the context of the data presented here.

Since the preparation of this Chapter, the main additional development in the field of electronically excited halogen atoms has been the extent to which the new technique of "Translational Spectroscopy" or "Photodissociation Recoil Spectroscopy" has involved these excited states in particular systems. In this method, developed by Wilson and his coworkers, a molecular beam of molecules is subjected to a short pulse of radiation and the distribution of translational energy of the photodissociated fragments is analysed with a quadrupole mass spectrometer.[i,ii] The resulting record, combined with the energetics of the dissociation of a halide molecule and of the light input, permits the inference of the presence or absence of $^2P_{1/2}$ halogen atoms in a given case. The *angular* distribution of products following the photolysis of molecular bromine and iodine was first reported in detail by Solomon.[iii] A number of photolytic sources have been employed by Wilson and his coworkers including the second harmonics of the ruby and neodymium doped glass lasers at $\lambda = 3471$ and 5310 Å, respectively,[i,ii] radiation from a "theta pinch" lamp[iv] and a tunable dye laser.[iv,v] Such experiments are essentially concerned with the details of molecular spectroscopy rather than the kinetics of any resulting electronically excited halogen atoms. The halide molecules whose photolytic dissociation by translational spectroscopy has been investigated include Cl_2,[vi–ix] Br_2,[ix–xi] I_2,[viii–x,xii,xiii] IBr,[viii,xiii,xiv] ICN,[i,ii,xv] NOCl[ix,x] and alkyl iodides.[i,ii] Clearly, $^2P_{1/2}$ halogen atoms will not be expected or observed in all cases. The wavelength of photolysis and the spectroscopic molecular states involved require consideration, and indeed, with laser initiation, two photon processes may also take place.[xiii] As examples, $Br(4^2P_{1/2})$, $Br(4^2P_{3/2})$ and $I(5^2P_{3/2})$ were observed following the photolysis of IBr at 5310 Å.[viii,xiii,xiv] The photolysis of I_2 at 5310 Å yielded two $I(5^2P_{3/2})$ atoms as expected from the known spectroscopic states; it further appears that two $I(5^2P_{1/2})$ atoms resulted from a two photon process at this wavelength involving the sequence $I_2(X^1\Sigma_g^+) \to {}^3\Pi_{0u^+} \to O_g^+ \to 2I(5^2P_{1/2})$.[viii–x,xii,xiii] The reader is referred to the papers of Wilson and his coworkers for the details of the photolytic fission of those molecules mentioned above.

i. K. R. Wilson, *Disc. Faraday Soc.*, **44**, 234 (1967).
ii. K. R. Wilson, *Bull Am. Phys. Soc.*, **13**, 494 (1968).
iii. J. Solomon, *J. Chem. Phys.*, **47**, 889 (1967).
iv. G. E. Busch, J. F. Cornelius, R. T. Mahoney, R. I. Morse, D. W. Schlosser and K. R. Wilson, *Rev. Sci. Instr.*, **41**, 1066 (1970).
v. G. E. Busch, R. T. Mahoney and K. R. Wilson, IEEE. *J. Quantum Electronics*, Vol. **QE6**, 171 (1970).
vi. R. W. Diesen, J. C. Wahr and S. E. Adler, *J. Chem. Phys.*, **50**, 3635 (1969).
vii. G. E. Busch, R. T. Mahoney, R. I. Morse and K. R. Wilson, *J. Chem. Phys.*, **51**, 449 (1969).
viii. R. T. Mahoney, R. J. Oldman, R. K. Sander and K. R. Wilson, 6th. International Conference on the Physics of Electronic and Atomic Collisions, M.I.T., Cambridge Mass., July, (1969), p. 538.
ix. K. R. Wilson, "Photofragment Spectroscopy of Dissociative Excited States" in "Excited State Symposium", J. N. Pitts, Ed., Environmental Resources Inc., Riverside, California (1970), Vol. 2, Wiley-Interscience, New York.
x. R. K. Sander, R. J. Oldman, R. T. Mahoney and K. R. Wilson, 24th. Symposium on Molecular Structure and Spectroscopy, Ohio State University, p. 32 (1969).
xi. R. T. Mahoney, R. J. Oldman, R. K. Sander and K. R. Wilson, *Bull. Am. Phys. Soc.*, **14**, 850 (1969).
xii. R. T. Mahoney and K. R. Wilson, *Bull. Am. Phys. Soc.*, **14**, 849 (1969).
xiii. G. E. Busch, R. T. Mahoney, R. I. Morse and K. R. Wilson, *J. Chem. Phys.*, **51**, 837 (1969).
xiv. M. L. Blethen, R. T. Mahoney, A. F. Tuck and K. R. Wilson, *Bull. Am. Phys. Soc.*, **14**, 821 (1969).
xv. K. E. Holdy, L. C. Klotz and K. R. Wilson, *J. Chem. Phys.*, **52**, 4588 (1970).

REFERENCES

1. G. C. Fettis and J. H. Knox, *Progress in Reaction Kinetics*, Vol. 2, Pergamon, London, 1964, p. 1.
2. H. B. Palmer, *J. Chem. Phys.*, **26**, 648 (1957).

3. J. V. V. Kasper and G. C. Pimentel, *Appl. Phys. Lett.*, **5**, 231 (1964).
4. H. Eyring, J. Walter, and G. E. Kemball, *Quantum Chemistry*, Wiley, New York, 1960, pp. 151–155.
5. L. I. Schiff, *Quantum Mechanics*, McGraw-Hill, New York, and Kogashuka, Tokyo, 1955, p. 292.
6. E. C. Kemble, *The Fundamental Principles of Quantum Mechanics with Elementary Applications*, Dover, New York, 1958, p. 501.
7. E. U. Condon and G. H. Shortley, *The Theory of Atomic Spectra*, Cambridge Univ. Press, London, 1963, p. 120.
8. L. D. Landau and E. M. Lifshitz, *Quantum Mechanics*, Pergamon, London, 1958, p. 230.
9. G. Herzberg, *Atomic Spectra and Atomic Structure*, Dover, New York, 1948.
10. R. H. Garstang, in *Forbidden Transitions in Atomic and Molecular Processes*, D. R. Bates, Ed. Academic Press, New York, 1962, p. 1.
11. D. Husain and J. R. Wiesenfeld, *Nature*, **213**, 1227 (1967); *Trans. Faraday Soc.*, **63**, 1349 (1967).
12. G. H. Shortley, *Phys. Rev.*, **57**, 225 (1940).
13. R. H. Garstang, *J. Res. Nat. Bur. Stand.*, *A*, **68**, 61 (1964).
14. R. H. Garstang, *Proc. Cambridge Phil. Soc.*, **53**, 214 (1957).
15. P. Cadman, J. C. Polanyi, and I. W. M. Smith, *J. Chim. Phys.*, **64** (1), 111 (1967).
16. J. K. Cashion and J. C. Polanyi, *Proc. Roy. Soc., Ser. A*, **258**, 570 (1960).
17. J. R. Airey, P. D. Pacey, and J. C. Polanyi, *11th International Symposium on Combustion*, Combustion Institute, Pittsburgh, 1967, p. 85.
18. G. M. Lawrence, *Astrophys. J,*, **48**, 261 (1967).
19. C. E. Moore, Ed., *Atomic Energy Levels*, National Bureau of Standards Circular 467, Washington, D.C., 1958.
20. R. G. W. Norrish and G. Porter, *Nature*, **164**, 658 (1949); G. Porter, *Proc. Roy. Soc., Ser. A*, **200**, 284 (1950); R. G. W. Norrish and G. Porter, *Quart. Rev.*, (London), **10**, 149, (1956); G. Porter, in *Photochemistry and Reaction Kinetics*, P. G. Ashmore, F. S. Dainton, and T. M. Sugden, Eds., Cambridge Univ. Press, London, 1967, p. 93; B. A. Thrush, in *Photochemistry and Reaction Kinetics*, P. G. Ashmore, F. S. Dainton, and T. M. Sugden, Eds., Cambridge Univ. Press, London, 1967, p. 112.
21. R. G. W. Norrish, G. Porter, and B. A. Thrush, *Proc. Roy. Soc., Ser. A*, **216**, 165 (1953).
22. R. G. W. Norrish, G. Porter, and B. A. Thrush, *5th International Symposium on Combustion*, Reinhold, New York, 1955, p. 651.
23. E. W. Abrahamson, D. Husain, and J. R. Wiesenfeld, *Trans. Faraday Soc.*, **64**, 833 (1968).
24. K. C. Herr and G. C. Pimentel, *Appl. Opt.*, **4**, 25 (1965); G. A. Carlson and G. C. Pimentel, *J. Chem. Phys.*, **44**, 4053 (1966).
25. G. Herzberg and J. Shoosmith, *Can. J. Phys.*, **34**, 523 (1956).
26. B. A. Thrush, *Proc. Roy. Soc., Ser. A*, **243**, 555 (1958).
27. R. J. Donovan and D. Husain, *Trans. Faraday Soc.*, **62**, 11 (1966).
28. F. G. M. Hathorn and D. Husain, *Trans. Faraday Soc.*, **65**, 2687 (1969).
29. R. J. Donovan, D. Husain, W. Braun, A. M. Bass, and D. D. Davis, *J. Chem. Phys.*, **50**, 4115, (1969).
30. R. J. Donovan and D. Husain, *Nature*, **209**, 609 (1966).
31. R. J. Donovan, F. G. M. Hathorn, and D. Husain, *Trans. Faraday Soc.*, **64**, 1228 (1968).
32. J. V. V. Kasper, J. H. Parker, and G. C. Pimentel, *J. Chem. Phys.*, **43**, 1827 (1965).

33. M. A. Pollack, *Appl. Phys. Lett.*, **8**, 36 (1966).
34. *Appl. Opt., Suppl. 2*, "Chemical Lasers," 1965.
35. J. C. Polanyi, *Chem. Britain*, **2**, 151 (1966).
36. P. Cadman and J. C. Polanyi, *J. Phys. Chem.*, **72**, 3715 (1968).
37. R. J. Donovan and D. Husain, *Trans. Faraday Soc.*, **62**, 1050 (1966).
38. B. A. Thrush, in *Progress in Reaction Kinetics*, Vol. 3, Pergamon, London, 1965, p. 63; B. Brocklehurst and K. R. Jennings, in *Progress in Reaction Kinetics*, Vol. 4, Pergamon, London, 1967, p. 1; F. Kaufman, in *Progress in Reaction Kinetics*, Vol. 1, Pergamon, London, 1961, p. 3.
39. G. Karl, P. Kruss, and J. C. Polanyi, *J. Chem. Phys.*, **46**, 224 (1967); G. Karl and J. C. Polanyi, *J. Chem. Phys.*, **38**, 271 (1963); G. Karl, P. Kruss, J. C. Polanyi, and I. W. M. Smith, *J. Chem. Phys.*, **46**, 244 (1967).
40. J. K. Cashion and J. C. Polanyi, *Proc. Roy. Soc., Ser. A*, **258**, 529 (1960).
41. H. E. Radford, V. W. Hughes, and V. Beltram-Lopez, *Phys. Rev.*, **123**, 153 (1961); V. Beltram-Lopez and H. G. Robinson, *Phys. Rev.*, **123**, 161 (1961); J. S. M. Harvey, R. A. Kamper, and K. R. Lea, *Proc. Phys. Soc.*, **B76**, 979 (1960); K. D. Bowers, R. A. Kamper, and C. D. Lustig, *Proc. Phys. Soc.*, **B70**, 1176 (1957).
42. A. Carrington, D. H. Levy, and T. A. Miller, *J. Chem. Phys.*, **45**, 4093 (1966).
43. D. E. Gibbs and E. A. Ogryzlo, *Can. J. Chem.*, **43**, 1905 (1965).
44. M. A. A. Clyne and J. A. Coxon, *J. Mol. Spectrosc.*, **23**, 258 (1967).
45. D. W. Bader and E. A. Ogryzlo, *J. Chem. Phys.*, **41**, 2926 (1964).
46. M. A. A. Clyne and J. A. Coxon, *Trans. Faraday Soc.*, **62**, 2175 (1966).
47. E. Hutton and M. Wright, *Trans. Faraday Soc.*, **61**, 78 (1965).
48. M. A. A. Clyne and D. H. Stedman, *Trans. Faraday Soc.*, **64**, 1816 (1968).
49. M. A. A. Clyne and J. A. Coxon, *Proc. Roy. Soc., Ser. A*, **298**, 424 (1967).
50. L. F. Phillips, *Can. J. Chem.*, **43**, 369 (1965); C. G. Freeman and L. F. Phillips, *J. Phys. Chem.*, **68**, 362 (1964).
51. E. A. Ogryzlo, *Can. J. Chem.*, **39**, 2556 (1961).
52. M. H. Booth and J. W. Linnett, *Nature*, **199**, 1181 (1963).
53. H. B. Palmer and D. F. Hornig, *J. Chem. Phys.*, **26**, 98 (1957).
54. D. Britton, D. J. Seery, and M. Van Thiele, *J. Phys. Chem.*, **69**, 834 (1965); D. Britton and M. Van Thiele, 18th *International Congress of Pure and Applied Chemistry*, Montreal, Butterworth, London, 1961, p. 6.
55. R. A. Carabetta and H. B. Palmer, *J. Chem. Phys.*, **46**, 1325 (1967); **49**, 2466 (1968).
56. C. D. Johnson and D. Britton, *J. Phys. Chem.*, **68**, 3032 (1964).
57. D. Britton, N. Davidson, W. Gehman, and G. Schott, *J. Chem. Phys.*, **25**, 804 (1956); D. Britton, N. Davidson, and G. Schott, *Discussions Faraday Soc.*, **17**, 58 (1954).
58. J. I. Steinfeld and W. Klemperer, *J. Chem. Phys.*, **42**, 3475 (1965).
59. A. G. Briggs and R. G. W. Norrish, *Proc. Roy. Soc., Ser. A*, **276**, 51 (1963).
60. R. T. Meyer, *J. Chem. Phys.*, **46**, 4146 (1967).
61. R. T. Meyer, *J. Phys. Chem.*, **72**, 1583 (1968).
62. J. G. Calvert and J. N. Pitts, *Photochemistry*, Wiley, New York, 1966.
63. W. A. Noyes and P. A. Leighton, *Photochemistry of Gases*, Reinhold, New York, 1941.
64. S. V. Filseth and J. E. Willard, *J. Amer. Chem. Soc.*, **84**, 3806 (1962).
65. A. B. Callear and J. H. Wilson, *Trans. Faraday Soc.*, **63**, 1358 (1967); A. B. Callear and J. H. Wilson, *Nature*, **211**, 517 (1966).
66. A. B. Callear and J. H. Wilson, *Trans. Faraday Soc.*, **63**, 1983 (1967).

67. R. J. Donovan and D. Husain, *Trans. Faraday Soc.*, **62**, 2023 (1966).
68. L. Mathieson and A. L. G. Rees, *J. Chem. Phys.*, **25**, 753 (1956).
69. J. H. Van Vleck, *Phys. Rev.*, **40**, 544 (1932).
70. E. Kondratjew and L. Polak, *Z. Sowjetunion*, **4**, 764 (1933).
71. E. Wasserman, W. E. Falconer, and W. A. Yager, *J. Chem. Phys.*, **49**, 1971 (1968).
72. A. L. G. Rees, *Proc. Phys. Soc.*, **59**, 998, 1008 (1947).
73. N. S. Bayliss and A. L. G. Rees, *J. Chem. Phys.*, **7**, 854 (1939).
74. R. S. Mulliken, *Rev. Mod. Phys.*, **4**, 1 (1932).
75. R. J. Donovan and D. Husain, *Nature*, **206**, 171 (1965).
76. R. J. Donovan and D. Husain, *Trans. Faraday Soc.*, **62**, 2643 (1966).
77. T. W. Broadbent, A. B. Callear, and H. K. Lee, *Trans. Faraday Soc.*, **64**, 2320 (1968).
78. W. B. Tiffany, *Appl. Opt.*, **7**, 67 (1968).
79. W. B. Tiffany, *J. Chem. Phys.*, **48**, 3019 (1968).
80. H. Cordes and H. Sponer, *Z. Phys.*, **63**, 334 (1930).
81. H. Cordes, *Z. Phys.*, **74**, 34 (1932).
82. H. Cordes and H. Sponer, *Z. Phys.*, **79**, 170 (1932).
83. W. Brown, *Phys. Rev.*, **42**, 355 (1932).
84. R. S. Mulliken, *Phys. Rev.*, **46**, 549 (1934).
85. D. J. Seery and D. Britton, *J. Phys. Chem.*, **68**, 2263 (1968).
86. L. E. Selin, *Ark. Fysik*, **21**, 479 (1962).
87. L. E. Selin and B. Söderborg, *Ark. Fysik*, **21**, 515 (1962).
88. P. V. B. Haranath and P. J. Rao, *Indian J. Phys.*, **31**, 156 (1957).
89. P. V. B. Haranath and P. J. Rao, *Indian J. Phys.*, **31**, 368 (1957).
90. R. J. Donovan and D. Husain, *Trans. Faraday Soc.*, **64**, 2325 (1968).
91. W. G. Brown and G. E. Gibson, *Phys. Rev.*, **40**, 529 (1932).
92. E. Hulthen, N. Järlsater, and L. Koffman, *Ark. Fysik*, **18**, 479 (1960).
93. M. A. A. Clyne and J. A. Coxon, *Nature*, **217**, 448 (1968).
94. L. Landau, *Phys. Z. Sowjetunion*, **2**, 46 (1932).
95. C. Zener, *Proc. Roy. Soc., Ser. A*, **137**, 696 (1932).
96. C. Zener, *Proc. Roy. Soc., Ser. A*, **140**, 660 (1933).
97. C. F. Goodeve and A. W. C. Taylor, *Proc. Roy. Soc., Ser. A*, **154**, 181 (1936).
98. J. Romand, *Ann. Phys.*, **4**, 527 (1949).
99. J. R. Bates, J. O. Halford, and L. C. Anderson, *J. Chem. Phys.*, **3**, 415 (1935).
100. C. F. Goodeve and A. W. C. Taylor, *Proc. Roy. Soc., Ser. A*, **152**, 221 (1935); J. R. Bates, J. O. Halford, and L. C. Anderson, *J. Chem. Phys.*, **3**, 531 (1935).
101. J. H. Baley, F. F. Rust, and W. E. Vaughan, *J. Amer. Chem. Soc.*, **70**, 2767 (1948).
102. H. J. Plumley, *Phys. Rev.*, **49**, 405 (1936); K. Siga and H. J. Plumley, *Phys. Rev.*, **48**, 105 (1935).
103. R. S. Mulliken, *Phys. Rev.*, **50**, 1017 (1936).
104. R. S. Mulliken, *Phys. Rev.*, **51**, 310 (1937).
105. R. S. Mulliken, *J. Chem. Phys.*, **8**, 382 (1940).
106. R. M. Martin and J. E. Willard, *J. Chem. Phys.*, **40**, 2999 (1964).
107. A. Kupperman, "Fast Reactions and Primary Processes in Chemical Kinetics," in *Proceedings of the Fifth Nobel Symposium, September, 1967, Stockholm*, S. Claesson, Ed., Wiley-Interscience, New York, and Almqvist and Wiksell, Stockholm, 1967.
108. D. R. Davis, J. M. White, J. Assay, and A. Kupperman, Private communication to J. C. Polanyi, Ref. 36.
109. W. C. Price, *Proc. Roy. Soc., Ser. A*, **167**, 216 (1938).
110. P. Fink and C. F. Goodeve, *Proc. Roy. Soc., Ser. A*, **163**, 592 (1937).

111. D. Porret and C. F. Goodeve, *Proc. Roy. Soc., Ser. A*, **165**, 31 (1938).
112. A. D. Walsh, *J. Chem. Soc.*, **1953**, 2260–2331, 2321.
113. G. Herzberg, *Molecular Spectra and Molecular Structure. III, Electronic Spectra and Electronic Structure of Polyatomic Molecules*, Van Nostrand, New York, 1966.
114. R. J. Donovan, F. G. M. Hathorn, and D. Husain, *J. Chem. Phys.*, **49**, 953 (1968).
115. R. J. Donovan, F. G. M. Hathorn, and D. Husain, *Trans. Faraday Soc.*, **64**, 3192 (1968).
116. S. J. Arnold, N. Finlayson, and E. A. Ogryzlo, *J. Chem. Phys.*, **44**, 2529 (1966).
117. M. A. A. Clyne and D. H. Stedman, *Trans. Faraday Soc.*, **62**, 2164 (1966).
118. G. C. Pimentel, *Sci. Amer.*, **214**, 32 (April, 1966).
119. E. G. Niemann and M. Klenert, *Appl. Opt.*, **7**, 295 (1968).
120. C. Zener, *Phys. Rev.*, **37**, 556 (1931).
121. J. M. Jackson and N. F. Mott, *Proc. Roy. Soc., Ser. A*, **137**, 703 (1932).
122. K. F. Herzfeld and T. A. Litovitz, *Absorption and Dispersion of Ultrasonic Waves*, Academic Press, New York, 1959.
123. A. B. Callear and R. J. Oldman, *Trans. Faraday Soc.*, **63**, 2888 (1967).
124. A. B. Callear, in *Photochemistry and Reaction Kinetics*, P. G. Ashmore, F. S. Dainton, and T. M. Sugden, Eds., Cambridge Univ. Press, London, 1967, p. 133.
125. G. Herzberg, *Spectra of Diatomic Molecules*, Van Nostrand, New York, 1965.
126. V. K. Bykhovskii and E. E. Nikitin, *Opt. Spectrosc.*, **16**, 11 (1964).
127. E. E. Nikitin, *Combust. Flame*, **10**, 381 (1966).
128. A. C. G. Mitchell and M. W. Zemansky, *Resonance Radiation and Excited Atoms*, Cambridge Univ. Press, London, 1934, p. 247.
129. J. O. Hirschfelder, C. F. Curtiss, and R. B. Bird, *Molecular Theory of Gases and Liquids*, Wiley, New York, 1954.
130. P. G. Dickens, J. W. Linnett, and O. Sovers, *Disc. Faraday Soc.*, **33**, 52 (1962).
131. R. J. Donovan and D. Husain, *Trans. Faraday Soc.*, **62**, 2987 (1966).
132. E. Wigner, *Z. Phys. Chem.*, **B23**, 28 (1933).
133. R. J. Cvetanovic and S. Sato, *J. Amer. Chem. Soc.*, **81**, 3223 (1959).
134. M. A. A. Clyne and D. H. Stedman, *Chem. Phys. Lett.*, **1**, 36 (1967).
135. T. C. Clark, M. A. A. Clyne, and D. H. Stedman, *Trans. Faraday Soc.*, **62**, 3354 (1966).
136. M. A. A. Clyne and D. H. Stedman, *Trans. Faraday Soc.*, **64**, 2698 (1968).
137. V. Kondratiev and A. Leipunsky, *Z. Phys.*, **50**, 366 (1925).
138. Y. Uchida, *Sci. Pap. Inst. Phys. Chem. Res.* (Tokyo), **30**, 71 (1936).
139. N. Davidson, R. Marshall, A. E. Larsh, and T. Carrington, *J. Chem. Phys.*, **19**, 1311 (1951); R. Marshall and N. Davidson, *J. Chem. Phys.*, **21**, 659 (1953); M. I. Christie, R. G. W. Norrish, and G. Porter, *Proc. Roy. Soc., Ser. A*, **216**, 153 (1953); K. E. Russell and J. Simons, *Proc. Roy. Soc., Ser. A*, **217**, 271 (1953); D. L. Bunker and N. Davidson, *J. Amer. Chem. Soc.*, **80**, 5085, 5091 (1958); G. Porter and J. A. Smith, *Nature*, **184**, 449 (1959); G. Porter and J. A. Smith, *Proc. Roy. Soc., Ser. A*, **261**, 28 (1961); J. Engleman and N. Davidson, *J. Amer. Chem. Soc.*, **82**, 4770 (1960); G. Porter, L. G. Szabo, and M. G. Townsend, *Proc. Roy. Soc., Ser. A*, **270**, 493 (1962); G. Porter, *Discussions Faraday Soc.*, **33**, 198 (1962).
140. M. I. Christie, A. J. Harrison, R. G. W. Norrish, and G. Porter, *Proc. Roy. Soc., Ser. A*, **231**, 446 (1955).
141. R. C. Strong, J. C. W. Chien, P. E. Graf, and J. E. Willard, *J. Chem. Phys.*, **26**, 1287 (1957).
142. W. G. Givens and J. E. Willard, *J. Amer. Chem. Soc.*, **81**, 4773 (1959). M. R. Basila and R. L. Strong, *J. Phys. Chem.*, **67**, 521 (1963).

143. H. Hiroaka and R. Hardwick, *J. Chem. Phys.*, **36**, 1715 (1963).
144. R. W. Diesen and W. J. Felmlee, *J. Chem. Phys.*, **39**, 2115 (1963).
145. V. I. Vedeneyev, L. V. Gurvich, V. N. Kondratiev, V. A. Medvedev, and Ye. L. Frankevich, *Bond Energies, Ionisation Potentials and Electron Affinities*, Arnold, London, 1962.
146. S. W. Benson, *J. Chem. Phys.*, **38**, 1945 (1963); M. C. Flowers and S. W. Benson, *J. Chem. Phys.*, **38**, 882 (1963); D. B. Hartley and S. W. Benson, *J. Chem. Phys.*, **39**, 132 (1963); H. Teranishi and S. W. Benson, *J. Amer. Chem. Soc.*, **85**, 2887 (1963).
147. G. K. Rollefson and M. Burton, *Photochemistry and the Mechanism of Chemical Reactions*, Prentice-Hall, New York, 1946.
148. E. W. R. Steacie, *Atomic and Free Radical Reactions*, Volumes I and II, Reinhold, New York, 1954.
149. J. R. Majer and J. P. Simons, "Photochemical Processes in Halogenated Compounds," in Advances in Photochemistry, Volume 2, W. A. Noyes, G. S. Hammond, and J. N. Pitts, Jr., Eds., Wiley-Interscience, New York, 1954, p. 137.
150. S. Aditya and J. E. Willard, *J. Chem. Phys.*, **44**, 418 (1966).
151. M. L. Christie, R. S. Roy, and B. A. Thrush, *Trans. Faraday Soc.*, **55**, 1149 (1959).
152. R. A. Durie, F. Legay, and D. A. Ramsay, *Can. J. Phys.*, **38**, 444 (1960).
153. R. N. Hazeldine, *J. Chem. Soc.*, **1953**, 1764.
154. C. C. Kiess and C. H. Corliss, *J. Res. Nat. Bur. Stand.*, **A63**, 1 (1959).
155. J. L. Tech, *J. Res. Nat Bur. Stand.*, **A67**, 505 (1963).
156. L. J. Radziemski and V. Kaufman, *J. Opt. Soc. Amer.*, **59**, 424 (1969).
157. D. Husain and J. R. Wiesenfeld, *Rev. Sci. Instr.*, **41**, 1146 (1970).

The Photochemistry of α-Dicarbonyl Compounds[†][‡]

BRUCE M. MONROE,§ Gates and Crellin Laboratories of Chemistry, California Institute of Technology, Pasadena, California 91109

CONTENTS

I. Introduction	77
II. Excited States of α-Dicarbonyl Compounds	78
III. Photochemical Reactions of α-Diketones	80
A. Glyoxal	80
B. Biacetyl	80
C. Hexafluorobiacetyl	85
D. Benzil	85
E. Long Chain Aliphatic α-Diketones	87
F. *cis*-α-Diketones	90
IV. Photochemical Reactions of α-Keto Acids	95
A. Pyruvic Acid	95
B. Phenylglyoxylic Acid	97
C. Long Chain α-Keto Acids	98
V. Photochemical Reactions of α-Keto Esters	99
A. Pyruvates and Phenylglyoxylates	99
B. Diethyl Oxalate	102
C. Methyl-*o*-benzyloxyphenylglyoxylate	102
VI. Conclusion	103
VII. Addendum	103
References	105

I. INTRODUCTION

The photochemical reactions of α-dicarbonyl containing compounds have recently been the subject of a number of investigations. This review will consider reactions that take place in oxygen-free solution and will contrast them with reactions that have been observed in the vapor state.

[†] Supported by the Directorate of Chemical Sciences, Air Force Office of Scientific Research, Contract No. AF49(638)-1479.

[‡] Contribution No. 3934 from the Gates and Crellin Laboratories of Chemistry, California Institute of Technology, Pasadena, California, 91109.

[§] Current address: Explosives Department, Experimental Station, E. I. du Pont de Nemours and Company, Wilmington, Delaware, 19898.

The photochemistry of an α-diketone was first studied by Klinger who in 1886 reported the formation of a precipitate containing the elements of benzoin and benzil when an ether solution of benzil was exposed to sunlight.[1] This reaction was also studied by Ciamician and Silber,[2] by Benrath,[3] and by Cohen,[4] who showed that the product was a thermally unstable pinacol which decomposed to benzoin and benzil.

Other scattered reports of α-dicarbonyl photoreactions appear in the early literature. Porter, Ramsperger, and Steel[5] and Bowen and Horton[6] examined the photochemistry of biacetyl. Norrish and Griffiths[7] and Kirkbridge and Norrish[8] examined the vapor phase decomposition of glyoxal. Dirscherl[9,10] and Lieben, Löwe, and Bäuminger[11] studied the photoreaction of pyruvic acid in aqueous solution. The photoaddition of benzil to acenaphthene was reported by Oliveri-Mandala, Caronna, Giacalone, and Deleo.[12,13]

Recently, more quantitative and systematic studies, beginning with those of the vapor phase reactions of biacetyl,[14] have been carried out on a variety of α-diketones, α-ketoacids, and α-ketoesters both in the vapor phase and in solution. This work has shown that, although formation of acyl radicals by cleavage of the carbon–carbon bond between the two carbonyl groups is often the major reaction in the vapor phase, these reactions often take a different course in solution. Irradiation of an α-dicarbonyl compound in deoxygenated solution in the presence of a hydrogen atom donor leads almost invariably to products that can be rationalized as arising from an initial hydrogen abstraction by the photoexcited α-dicarbonyl system followed by further reactions of the resulting radicals. When intramolecular abstraction is possible, the process is generally favored over intermolecular abstraction. In the absence of abstractable hydrogen atoms, no reaction at all is often observed. Only those α-diketones that are reasonably strained or can decarbonylate to yield reasonably stable free radicals produce any significant amount of cleavage products on irradiation in solution. In the presence of a suitable substrate, cycloaddition may also take place.

II. EXCITED STATES OF α-DICARBONYL COMPOUNDS

In order to understand the photoreactions of α-dicarbonyl compounds it is necessary to have some knowledge of their excited states. We will not exhaustively review the large amount of work that has been done on these compounds, particularly biacetyl, but merely touch upon those points of interest to photochemists. More detailed discussions are available in various review articles.[14–19]

The absorption spectrum of an aliphatic α-diketone consists of a weak $\pi^* \leftarrow n$ band ($\epsilon \sim 20$–50) in the visible and a stronger band in the ultraviolet. A number of studies have examined the effect of various structural modifications of the molecule on the position of the visible maximum.[20-22] This maximum has been particularly useful for the determination of quantum yields of α-diketone photoreactions.[23,24] Since most reaction products and most commonly used sensitizers and solvents do not absorb at these long wavelengths, the disappearance of the starting compound is easily followed by monitoring the change in optical density at the visible maximum. Because the products normally do not absorb visible light, it is often possible to reduce or eliminate secondary reactions by the use of visible rather than ultraviolet exciting light for preparative scale reactions.

The α-diketones show both phosphorescence and fluorescence emission, not only in a glass at 77°K but also in fluid solutions at room temperature, a property which has made these compounds useful in energy transfer studies.[25-29] Fluorescence, however, is quite weak with intersystem crossing and decay through the triplet state being the principle mode of decay. The absolute fluorescence yields of biacetyl and benzil in solution are reported to be 0.22% and 0.27%, respectively, while the measured phosphorescence yield

TABLE I
Triplet Energies of α-Dicarbonyl Compounds

Compounds	Energy, kcal/mole	Glass	Reference
Biacetyl	55.6	EPA[a]	40
	57.2	EPA	35
	54.9	MCIP[b]	35
2,3-Pentanedione	55.6	EPA	40
	57.2	EPA	35
	54.7	MCIP	35
Pivalil	51.6	EPA	40
Camphorquinone	51.6	EPA	40
Benzil	54.3		28
	53.7	MCIP	35
	57.3	EPA	35
	53.0	EPA	40
Anisil	55.2		28
Phenylglyoxal	62.5	MCIP	35
1-Phenyl-1,2-propanedione	54.0	EPA	40
Ethyl phenylgloxylate	61.9	MCIP	39
	63.0	EPA	39
α-Ketodecanoic acid	64.3		34

[a] Ether-isopentane-ethanol (5:5:2, by volume).
[b] Methylcyclohexane-isopentane (5:1, by volume).

of benzil is 3.4%.[30] Reported intersystem crossing yields include: biacetyl, 0.97; 2,3-pentanedione, 0.98; 2,3-heptanedione, 0.98; 2,3-octanedione, 0.95[31]; and benzil, 0.92.[32] Although phosphorescence has been reported for pyruvic acid,[33] α-ketodecanoic acid,[34] and ethyl phenylgloxylate,[35] no thorough studies of the emission of α-keto acids or α-keto esters have been carried out.

As might be expected, the excited states of α-diketones are somewhat lower in energy than those of the analogous monoketones, a property that has led to their use as quenchers and low energy sensitizers in photosensitization studies.[36,37] However, it has been shown that their use in these studies can lead to spurious results since α-diketones may participate in various hydrogen abstraction[38] and hydrogen transfer[39] reactions as well as transfer energy. From phosphorescence spectra taken in glasses at low temperature the triplet energies of a number of α-dicarbonyl compounds have been estimated. Some of these values are given in Table I.

III. PHOTOCHEMICAL REACTIONS OF α-DIKETONES

A. Glyoxal

The vapor phase photoreactions of glyoxal have been thoroughly studied.[18,41] The products include formaldehyde, hydrogen, carbon monoxide, and polymeric material. This work has been reviewed.[18,42]

B. Biacetyl

The photochemistry of biacetyl has been extensively studied, both in the vapor phase and in solution. In the vapor phase the products include carbon monoxide, ethane, methane, acetone, ketene, and 2,3-pentanedione. It has been shown that the primary process is cleavage of the carbon–carbon bond between the two carbonyl groups to yield acyl radicals, which on further reaction give the observed products.[14,43]

However, irradiation of biacetyl in solution gives products derived from an initial hydrogen abstraction by a photoexcited biacetyl unless the reaction is carried out in a solvent containing no abstractable hydrogen atoms.[24,44] Irradiation of biacetyl in perfluorinated solvents, conditions under which no hydrogen abstraction should take place, yields small amounts of carbon monoxide and ethane, but the quantum yield for their production is much lower than is observed at low pressure in the vapor state.[24] The quantum yield for the disappearance of biacetyl in perfluoro-n-octane at 100° with 366 nm exciting light is reported to be 2.4×10^{-3}; this compares with a value of 0.21 for biacetyl disappearance in the vapor state at $1.8 \times 10^{-3} M$

at 100° with 313 nm exciting light.[14] In addition the quantum yields for biacetyl disappearance in solution show wavelength and temperature dependencies although these differences are not as great as those observed in the vapor phase reaction.

These observations are readily explained by consideration of the energies involved. From their studies of the wavelength dependence of the vapor phase dissociation of biacetyl, Noyes, Mulac, and Matheson concluded that 65 kcal/mole was barely enough energy to break the carbon–carbon bond.[43] However, on the basis of phosphorescence studies in low temperature glasses, the lowest triplet of biacetyl lies somewhere in the 55–57 kcal/mole range.[35,40] Since biacetyl intersystem crosses with high efficiency, it is apparent that cleavage of the photoexcited biacetyl will be observed only if it is competitive with collisional deactivation to the lowest triplet state. At low pressure in the gas phase, collisional deactivation is slower than dissociation. The low quantum yields of dissociation observed from the irradiation of biacetyl in perfluorinated solvents indicate that in solution deactivation is much faster than bond cleavage. In addition, in the gas phase the lowest triplet state, once formed, can do little but decay to the ground state or undergo reaction with another excited molecule to produce a higher excited state, but in solution it has other pathways of reaction open to it. It can abstract a hydrogen atom from the solvent or any hydrogen donor present in solution. The same differences in reaction between the vapor and condensed phases are also observed with monoketones.[45]

Presumably the same situation is true for the other α-diketones. Although their bond energies are not known, the energy necessary to break the carbon–carbon bond must be about that necessary to break the carbon–carbon bond of biacetyl. Phosphorescence measurements put most of the α-diketone triplets in the 50–60 kcal/mole energy range (Table I). Only those α-diketones that appear to be highly strained or can produce reasonably stable free radicals on decarbonylation give significant amounts of products derived from carbon–carbon bond ruptures. In these cases it might be expected that the energy necessary to break the carbon–carbon bond is about the same as or less than the triplet energy.

$$CH_3-\underset{\underset{O}{\|}}{C}-\underset{\underset{O}{\|}}{C}-CH_3 + CH_3-\underset{\underset{OH}{|}}{CH}-CH_3 \xrightarrow{h\nu}$$

1

$$CH_3-\underset{\underset{O}{\|}}{C}-\underset{\underset{OH}{|}}{\overset{\overset{CH_3}{|}}{C}}-\underset{\underset{OH}{|}}{\overset{\overset{CH_3}{|}}{C}}-\underset{\underset{O}{\|}}{C}-CH_3 + CH_3-\underset{\underset{O}{\|}}{C}-CH_3$$

2
98%

Irradiation of biacetyl (1) in isopropyl alcohol gives acetone plus a nearly quantitative yield of a mixture of diastereomeric pinacols in approximately equal amounts (2).[44,46] The mechanism of this reaction is shown in the following scheme in which B is biacetyl, AH_2 is isopropyl alcohol, and A is acetone. Superscripts indicate the multiplicity of the excited states.

$$B \xrightarrow{h\nu} B^1 \quad \text{excitation}$$
$$B^1 \longrightarrow B \quad \text{nonradiative decay}$$
$$B^1 \longrightarrow B + h\nu' \quad \text{fluorescence}$$
$$B^1 \longrightarrow B^3 \quad \text{intersystem crossing}$$
$$B^3 \longrightarrow B \quad \text{nonradiative decay}$$
$$B^3 \longrightarrow B + h\nu'' \quad \text{phosphorescence}$$
$$AH_2 + B^3 \longrightarrow AH\cdot + BH\cdot \quad \text{hydrogen abstraction}$$
$$AH\cdot + B \longrightarrow A + BH\cdot \quad \text{hydrogen transfer}$$
$$BH\cdot + BH\cdot \longrightarrow HBBH \quad \text{coupling}$$

This reaction scheme is identical with that written for the photoreduction of benzophenone in isopropyl alcohol[47,48] except that it is necessary to include fluorescence and nonradiative decay of the excited singlet here since biacetyl, unlike benzophenone,[32] does not undergo intersystem crossing with unit efficiency.[31] Although no photosensitization experiments have been carried out to demonstrate that the triplet state of biacetyl is the abstracting state, the intermediacy of ketone triplets in hydrogen abstraction is well documented.[48] Isopropyl alcohol and other hydrogen donors quench the phosphorescence of biacetyl,[49] presumably because of chemical reaction, but have no effect on fluorescence.[50] Electron paramagnetic resonance studies have shown that the biacetyl ketyl radical (3) is formed in irradiated solutions of biacetyl.[51,52] The transfer of a hydrogen atom from an acetone ketyl radical to a ketone is well-known from studies of ketone photoreductions.[47,48] A value for the rate constant for the coupling of 3 has been determined.[52] This value, $3.9 \times 10^8 M^{-1} \sec^{-1}$, is of the same order of magnitude as that observed for the coupling of benzophenone ketyl radical, $1.1 \times 10^8 M^{-1} \sec^{-1}$.[53]

The photoreactions of biacetyl have also been studied in a number of other solvents which contain abstractable hydrogen atoms[4,6,44,54] as well as in aqueous solutions containing hydrogen atom donors.[55] In the most thorough study, the products from irradiation of biacetyl in five different solvents were determined.[44] These products are shown in Table II. In all cases the products can be explained as arising from an initial hydrogen abstraction by a photoexcited biacetyl followed by further reaction of the radicals thus formed. It is not necessary to invoke formation of acyl radicals by direct cleavage of biacetyl. The following scheme indicates the reactions.

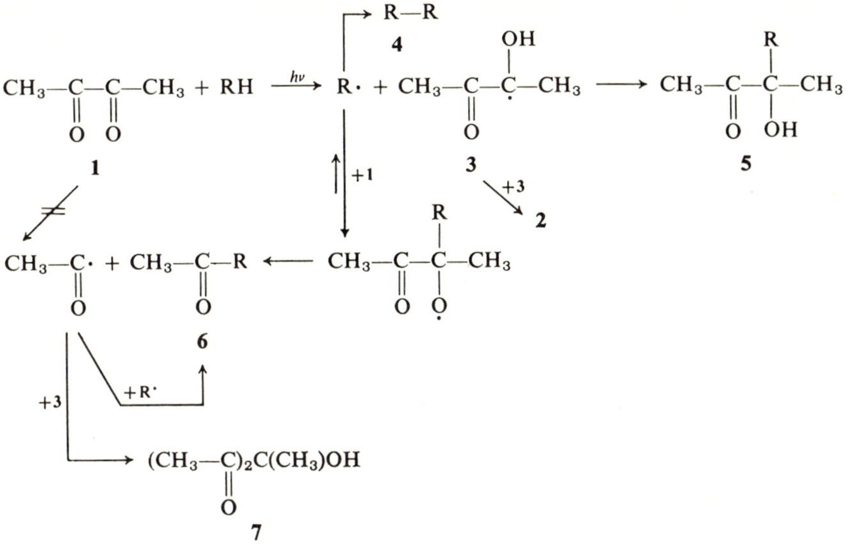

THE PHOTOCHEMISTRY OF α-DICARBONYL COMPOUNDS

$$CH_3-\underset{\underset{O}{\|}}{C}-\underset{\underset{O}{\|}}{C}-CH_3 + RH \xrightarrow{h\nu} R\cdot + CH_3-\underset{\underset{O}{\|}}{C}-\underset{\underset{\cdot}{|}}{\underset{CH_3}{C}}-CH_3 \longrightarrow CH_3-\underset{\underset{O}{\|}}{C}-\underset{\underset{OH}{|}}{\underset{R}{C}}-CH_3$$

1   R—R (4)   3   5

$$CH_3-\underset{\underset{O}{\|}}{C}\cdot + CH_3-\underset{\underset{O}{\|}}{C}-R \longleftarrow CH_3-\underset{\underset{O}{\|}}{C}-\underset{\underset{\cdot O}{|}}{\underset{R}{C}}-CH_3$$

6   2

$$\longrightarrow (CH_3-\underset{\underset{O}{\|}}{C})_2C(CH_3)OH$$

7

Electron paramagnetic resonance studies have shown that the biacetyl ketyl radical (3) and the radical derived from dioxane are formed when biacetyl is irradiated in dioxane solution.[51] Nonphotochemically generated radicals derived from dioxane and cyclohexane add to biacetyl to yield acylated product.[56] If cleavage were occurring in addition to or instead of hydrogen abstraction, one would expect to find at least moderate yields of acylated products in all solvents instead of just those which produce the least stable solvent radical.

A similar mechanism can be written for the formation of phenyl acetone in an irradiated aqueous solution of biacetyl and phenylacetic acid.[55] Abstraction of a benzilic hydrogen from phenylacetic acid by a photoexcited biacetyl would produce 3 and a benzylic radical which can add to a second biacetyl molecule. Cleavage of this intermediate would yield a β-keto acid which, on

TABLE II
Products from Irradiation of Biacetyl in Various Solvents

Solvent	2	4	5	6	7
Cyclohexane	36%		10%	24%	7%
Diethyl ether	36	4%	10	16	5
Dioxane	18		4	39	13
Cyclohexene	37	50	59		
Ethylbenzene	50	41	50		

loss of CO_2, would give the observed phenyl acetone. Although the radical derived from ethyl benzene was not observed to add to biacetyl,[44] addition of the benzilic radical to biacetyl would be favored in this reaction since there was a two-hundredfold excess of biacetyl.

In studies of the vapor phase photolysis of biacetyl it was observed that a new product, which quenched both the phosphorescence and primary dissociation of biacetyl, was formed. A strong absorption at 275 nm, which was associated with the quenching activity, was also observed. Since no new product which displayed significant quenching activity could be isolated, it was concluded that the quenching was due to the enol of biacetyl (8).[57] This assignment was supported by the disappearance of the absorption at 275 nm when ICl was added to an irradiated aqueous solution of biacetyl.[58]

$$CH_3-\underset{\underset{O}{\|}}{C}-\underset{\underset{}{|}}{\overset{OH}{C}}=CH_2$$

8

This reaction can be viewed as an internal hydrogen abstraction which takes place through a highly strained four-membered transition state instead of the usual six-membered one. The formation of the enol is wavelength dependent and is retarded by triplet quenchers.[58-60] In water, excitation into the second excited singlet of biacetyl formed the enol with a quantum yield of 0.10, while excitation into the first excited singlet formed the enol with a quantum yield of 0.01, possibly via a higher excited state formed by a triplet-triplet reaction. Therefore, the authors conclude that the second excited triplet is the state that isomerizes to the enol.

The observation of this isomerization raises several interesting questions. The high quantum yields reported for the formation of the enol in water plus the observation of its characteristic absorption on irradiation of biacetyl in methanol or hexane suggests that the isomerization is fast enough to compete effectively with collisional deactivation and that enol formation can take place in solvents containing abstractable hydrogen atoms. This would lead to the prediction that the quantum yields for the reactions of biacetyl in such solvents as isopropyl alcohol, methanol, or toluene should show an unusual wavelength effect. Irradiation into the second excited singlet of biacetyl should produce the enol, which is an effective triplet quencher, so that most of the triplet reaction would be quenched. Irradiation into the first excited singlet, however, would produce little or no enol so a "normal" reaction should be observed.

One further reaction of biacetyl should be noted. It has been found that phenols and aromatic amines, including triphenyl amine which contains no abstractable hydrogen atoms, quench the fluorescence of biacetyl at rates approaching diffusion control, but that the biacetyl remains unchanged.[50] A reversible electron abstraction reaction has been proposed to explain this observation. This reaction would be similar to the amine quenching of hydrocarbon fluorescence.[61]

C. Hexafluorobiacetyl

Irradiation of hexafluorobiacetyl in the vapor phase produces a 2:1 mole ratio of carbon monoxide and hexafluoroethane, products consistent with an initial carbon–carbon bond cleavage.[62] However, vapor phase irradiation of hexafluorobiacetyl in the presence of a large excess of 2,3-dimethylbutane vapor or in 2,3-dimethylbutane solution gave less than 1% carbon monoxide and trifluoromethane. No trifluoroacetaldehyde or hexafluoroacetone was produced in the latter reaction. Instead a complex mixture of products, which was not separated and identified but whose infrared showed the presence of a hydroxyl band and a diminished carbonyl band, was obtained. This observation is consistent with product formation via hydrogen abstraction.

D. Benzil

Although benzil was the first α-dicarbonyl compound to be investigated,[1-4] its photoreactions have not been as thoroughly studied as those of biacetyl. By 1916 it had been established that a precipitate was formed when a solution of benzil in ether,[1,3] ethanol,[2] aldehydes,[3] or alkyl substituted benzenes[3] was exposed to sunlight. The thermal and photochemical instability of the precipitate caused some confusion about its structure until Cohen demonstrated that it was pinacol **10** which decomposed to benzil and benzoin on heating.[4]

$$\phi-\underset{\underset{O}{\|}}{C}-\underset{\underset{O}{\|}}{C}-\phi \xrightarrow[\text{sunlight}]{RH} \phi-\underset{\underset{O}{\|}}{C}-\underset{\underset{OH}{|}}{\overset{\overset{\phi}{|}}{C}}-\underset{\underset{OH}{|}}{\overset{\overset{\phi}{|}}{C}}-\underset{\underset{O}{\|}}{C}-\phi \xrightarrow{\Delta}$$

     9   10

$$9 + \phi-\underset{\underset{O}{\|}}{C}-\underset{\underset{OH}{|}}{CH}-\phi$$

Recently a complete product study has been carried out for benzil irradiated in cyclohexane.[63] A complex mixture containing benzaldehyde, phenylcyclohexyl ketone, benzoin benzoate, benzoin, benzoic acid, and smaller amounts of other compounds containing from three to six benzyl groups was obtained.

$$\phi-\underset{\underset{O}{\|}}{C}-\underset{\underset{O}{\|}}{C}-\phi \xrightarrow{\underset{h\nu}{C_6H_{12}}}$$

9

$$\phi-\underset{\underset{O}{\|}}{C}-H + \phi-\underset{\underset{O}{\|}}{C}-\bigcirc + \phi CO_2H + \phi-\underset{\underset{H}{|}}{\underset{\underset{O}{\|}}{C}}-\underset{\underset{H}{|}}{\underset{\underset{O-\underset{\underset{O}{\|}}{C}-\phi}{|}}{C}}-\phi + \phi-\underset{\underset{O}{\|}}{C}-\underset{\underset{OH}{|}}{\underset{\underset{H}{|}}{C}}-\phi$$

6.5%† 5.7%   19.5%   15.6%   9.3%

The authors suggest that the formation of benzaldehyde, phenylcyclohexyl ketone, and benzoin benzoate requires the photochemical cleavage of benzil to benzoyl radicals. However, the formation of these products and the formation of benzoin can be explained by a scheme in which hydrogen abstraction from cyclohexane by a photoexcited benzil is the initial step and the additional steps are analogous to those observed when biacetyl is irradiated in cyclohexane.[44] Addition of cyclohexyl radical to benzil followed by cleavage would yield phenyl cyclohexyl ketone and the benzoyl radical which could abstract from the solvent to form benzaldehyde. Disproportionation of the benzil ketyl radical (**11**) would give benzoin. Coupling of the benzoyl radical and **II** would give benzoin benzoate.

$$9 \xrightarrow{\underset{h\nu}{C_6H_{12}}} \phi-\underset{\underset{O}{\|}}{C}-\underset{\underset{OH}{|}}{\overset{\cdot}{C}}-\phi + \bigcirc \xrightarrow{+9} \phi-\underset{\underset{O}{\|}}{C}-\underset{\underset{O}{\overset{\cdot}{|}}}{C}-\phi$$

with branch to $\phi-\underset{\underset{O}{\|}}{C}-\underset{\underset{OH}{|}}{\underset{\underset{H}{|}}{C}}-\phi$ and downward to:

$$\overset{\cdot}{\bigcirc} + \phi-\underset{\underset{O}{\|}}{C}-H \longleftarrow \phi-\underset{\underset{O}{\|}}{C}\cdot + \phi-\underset{\underset{O}{\|}}{C}-\bigcirc$$

11

This rationalization is supported by the following evidence. Benzil is unchanged on irradiation in oxygen-free benzene, indicating that cleavage does not take place in solution.[5] Photoexcited benzil is known to abstract hydrogen atoms. The benzil ketyl radical has been observed from flash photolysis of benzil in ethanol[64] and by electron paramagnetic resonance studies of irradiated solutions of benzil in isopropyl alcohol.[52] Since cyclohexane is a

† Per cent yields based on number of benzoyl groups consumed.

moderately good hydrogen atom donor,[65] this abstraction should take place. In fact the presence of benzoin in the product mixture practically requires the intermediacy of **11**. The decay constant for **11**, measured in isopropyl alcohol, is $3.3 \times 10^8 M^{-1} \text{sec}^{-1}$, a value consistent with a coupling rather than a disproportionation reaction.[52,53]

A detailed product study of the irradiation of benzil in cumene and in isopropyl alcohol has shown that a complex product mixture, similar to that observed from irradiation of benzil in cyclohexane, is obtained.[63] However, from the data in these solvents, as in the case of cyclohexane, it is difficult to draw any firm mechanistic conclusions since it is not known if all the observed products are primary photoproducts or if many of them are formed via secondary reactions of the primary photoproduct or products. The complexity of the product mixture may be due to photodecomposition of **10**.

One puzzling observation is the formation of large quantities of benzoic acid in all three solvents. The authors suggest that since only a trace of oxygen can be present during irradiation, it must be formed from some unstable product which decomposes during work up.

Benzil forms an adduct with acenaphthene when a mixture of the two is irradiated in benzene solution, presumably by cross coupling of the radicals produced by abstraction of a hydrogen atom from acenaphthene by benzil.[12,13,66] The other products of this reaction were not reported.

Benzil undergoes a cycloaddition reaction with stilbene and 1,1-diphenylethylene to form adducts containing the 2,3-dihydro-[1,4]-dioxin ring system and with visnagin to form an oxetane. These reactions and the cycloaddition reactions of the analogous *o*-quinones are reviewed by Schönberg.[67]

E. Long Chain Aliphatic α-Diketones

The photochemistry of a number of long chain aliphatic α-diketones has been reported.[46,68] Included in the study were 2,3-pentanedione, 3,4-hexanedione, 4,5-octanedione, 5,6-decanedione, 2,7-dimethyl-4,5-octanedione, and 1,2-cyclodecanedione. Without exception, irradiation of these compounds in cyclohexane or benzene yielded cyclobutanols, presumably by an internal abstraction-cyclization mechanism. The reaction is quite selective, producing only one of the two possible cyclobutanols. The hydrogen atom is always abstracted by the "distant" rather than by the "near" carbonyl group.

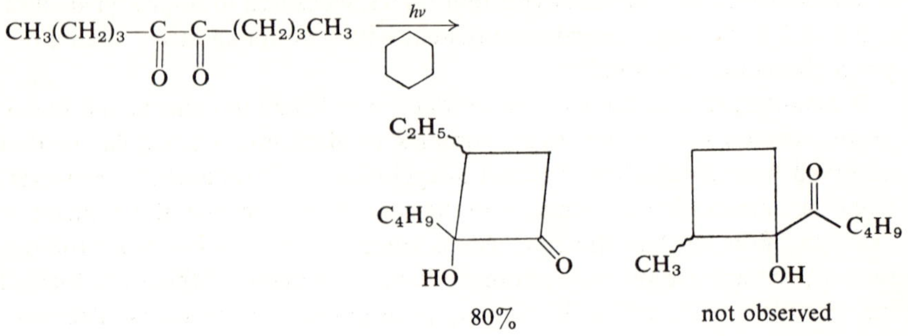

$$CH_3(CH_2)_3-\underset{\underset{O}{\|}}{C}-\underset{\underset{O}{\|}}{C}-(CH_2)_3CH_3 \xrightarrow{h\nu}$$

80%   not observed

A similar reaction is given by 1,2-cyclodecanedione, a compound which might have been expected to give some transannular abstraction.[69]

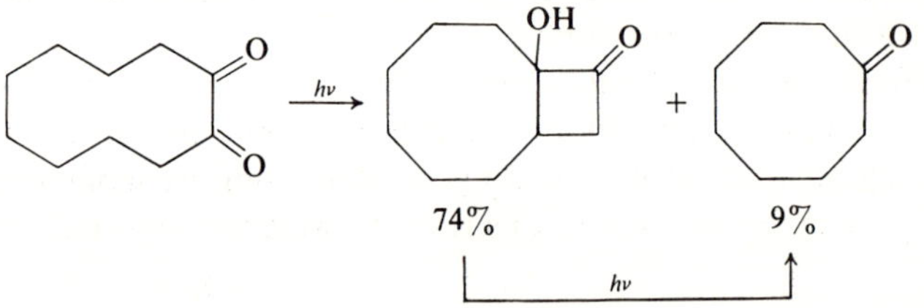

74%   9%

Cyclization remains the preferred mode of reaction when the irradiation is carried out in a solvent containing abstractable hydrogen atoms. Irradiation of 4,5-octanedione in butanal gave 92% of cyclized product plus smaller amounts of abstraction products. Since the products account for more than 100% of the diketone consumed, it must be assumed that some of the

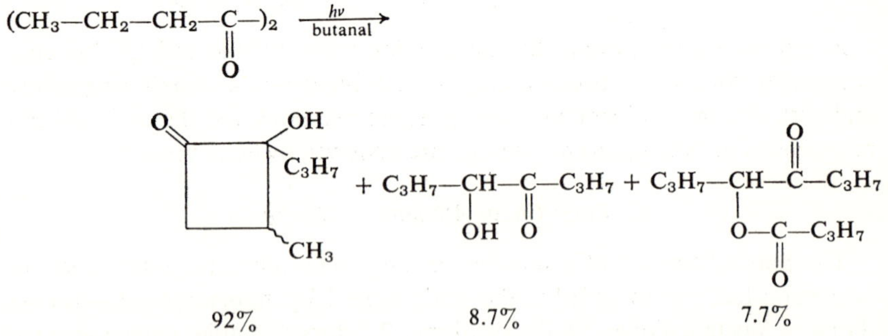

$$(CH_3-CH_2-CH_2-\underset{\underset{O}{\|}}{C}-)_2 \xrightarrow[\text{butanal}]{h\nu}$$

$$+ C_3H_7-\underset{\underset{OH}{|}}{CH}-\underset{\underset{O}{\|}}{C}-C_3H_7 + C_3H_7-\underset{\underset{O-\underset{\underset{O}{\|}}{C}-C_3H_7}{|}}{CH}-\underset{\underset{O}{\|}}{C}-C_3H_7$$

92%   8.7%   7.7%

butanoyl radicals generated by hydrogen abstraction combine to form more diketone. When 3,4-hexanedione was irradiated in propanal, the yield of cyclization product was reduced to 23%, indicating that the methyl hydrogen

THE PHOTOCHEMISTRY OF α-DICARBONYL COMPOUNDS

atoms are harder to abstract than those of the methylene groups of 4,5-octanedione. In each case the acyloin rather than the pinacol was isolated.

$$(CH_3-CH_2-\underset{O}{\overset{\|}{C}}-)_2 \xrightarrow[\text{propanal}]{h\nu}$$

$$\underset{23\%}{C_2H_5-\underset{O}{\overset{\|}{C}}-\underset{OH}{\overset{|}{CH}}-C_2H_5} \quad + \quad \underset{23\%}{\square\text{(cyclobutane with }=O, C_2H_5, OH)} \quad + \quad \underset{54\%}{C_2H_5-\underset{O}{\overset{\|}{C}}-\underset{\underset{O}{\overset{\|}{O-C-C_2H_5}}}{\overset{|}{CH}}-C_2H_5}$$

23%   23%   54%

This would indicate either that disproportionation is the preferred route of termination for the ketyl radicals or that the pinacol is unstable and disproportionates to acyloin and ketone under the reaction conditions.

Some work has been done on the mechanism of this reaction. Since oxygen, naphthalene, and anthracene exhibit a quenching effect and benzophenone sensitizes the reaction, the triplet state was shown to be the abstracting state. Although it was not determined if the singlet state were also involved as in the Norrish Type II cleavage of aliphatic ketones,[70] these compounds undoubtedly have high intersystem crossing yields, so that only a very small singlet contribution is possible. Several unsensitized quantum yields have been measured, and a sensitized quantum yield has been determined relative to an unsensitized one. The authors suggest that the high values that were observed imply that a chain mechanism is operative in this system.

$$C_4H_9-\underset{O}{\overset{\|}{C}}-\underset{O}{\overset{\|}{C}}-C_4H_9 \xrightarrow[\text{ethanol}]{h\nu} \square\text{(cyclobutane with }=O, C_2H_5, C_4H_9, OH)$$

$$\Phi = 1.01 \pm 0.06$$

$$C_3H_7-\underset{O}{\overset{\|}{C}}-\underset{O}{\overset{\|}{C}}-C_3H_7 \xrightarrow{\phi H, h\nu} \square\text{(cyclobutane with }=O, CH_3, C_3H_7, OH)$$

$$\Phi_{\phi_2 CO \text{ sensitized}} = 1.6 \times \Phi_{\text{unsensitized}}$$
($\Phi_{\text{unsensitized}}$ not determined)

A chain mechanism seems most unreasonable for an intramolecular reaction of this type. Although the experimental procedures were not reported in detail, these values appear to be open to question. Pinacol formation, a process that removes two diketone molecules per quantum of light, is possible in ethanol so that a high value might be obtained in this solvent.[47] Since no product study was reported for this reaction, it is not possible to determine if this reaction took place. However, the products observed from irradiation in aldehydes strongly suggest that some intermolecular abstraction would be expected. The ratio of sensitized to unsensitized quantum yields was determined by irradiation of two tubes, one with and one without sensitizer, in parallel. This method is valid only when the same amount of light is absorbed by the material in each tube. Since depletion of the diketone in the tube without sensitizer would reduce the amount of light absorbed in this tube, these values should be redetermined before the chain hypothesis is accepted.†

Studies on the luminescence of long chain aliphatic α-diketones have shown that decreased phosphorescence is observed as the length of the aliphatic side chain is increased.[31] The series which was studied included biacetyl, 2,3-pentanedione, 2,3-heptanedione, and 2,3-octanedione. Since it was assumed that no photoreactions occur, the decreased phosphorescence was attributed to an increased probability of vibrational deactivation of the excited state as the length of the side chain is increased. However, photochemical reaction appears to be an inviting alternative explanation for the decreased phosphorescence. 2,3-Pentanedione is known to undergo an intramolecular cyclization reaction[68] and 2,3-heptanedione and 2,3-octanedione would be expected to undergo the same reaction and with a somewhat higher quantum yield since methylene hydrogen atoms rather than methyl hydrogen atoms would be abstracted.‡

F. cis-α-Diketones

The photoreactions of a number of α-diketones in which the carbonyl groups are held in a somewhat rigid *cis*-configuration have been reported. These compounds undergo a variety of reactions depending upon the strain of the system and the solvent in which the reaction is carried out.

Camphorquinone (**12**), which has been studied in several solvents, gives typical hydrogen atom abstraction reactions. It undergoes photoreduction in isopropyl alcohol[39,71,72] and forms mixtures of solvent addition and reduction products when irradiated in methanol,[39,71] aldehydes,[73] and xylene.[23] No transformation is observed in benzene or carbon tetrachloride.[74]

† See Addendum.
‡ See Addendum.

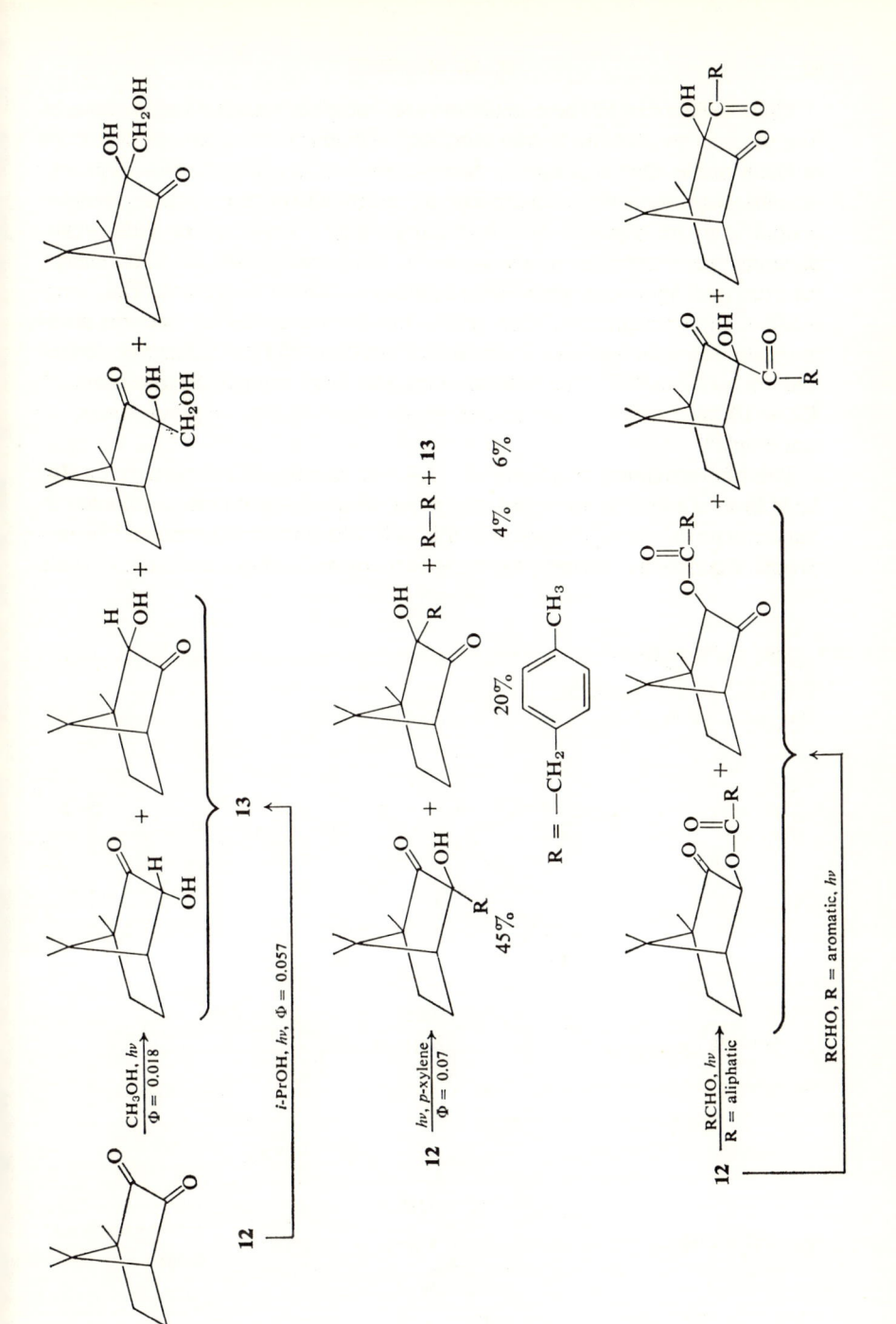

The mechanisms of these reactions are completely analogous to those of biacetyl, except that the termination step is disproportionation of two ketyl radicals rather than a coupling. Sensitization of the reduction in isopropyl alcohol with *m*-methoxyacetophenone, a nonabstracting ketone, demonstrated that the triplet is the abstracting state.[39] Sensitization with benzophenone gave complex kinetics due to its participation in both energy transfer and hydrogen abstraction-hydrogen transfer reactions. The intermediate camphorquinone ketyl radical has been detected by electron paramagnetic resonance studies of irradiated solutions of **12** in isopropyl alcohol and in *p*-xylene.[39,52] The rate constant for ketyl radical disappearance,[52] $4.2 \times 10^7 M^{-1} \sec^{-1}$, is typical of those observed for disproportionation reactions.[76]

The photochemical reactions of 1,1,4,4-tetramethyl-2,3-dioxotetralin (**14**) have been studied in methanol, isopropyl alcohol, cyclohexane, toluene, *n*-butyraldehyde, methyl formate, acetic acid, dioxane, and benzene.[75] In isopropyl alcohol the reduced ketone (**15**) is formed almost exclusively, while the reduced ketone and solvent addition products make up the major fraction of the product mixture in the other solvents. The ketyl radical has been detected by examination of irradiated solutions of 14 in isopropyl alcohol and p-xylene.[52] Its decay constant, $4.2 \times 10^7 M^{-1} \text{sec}^{-1}$, is typical of a disproportionation reaction.[76]

However, the origin of some products is obscure. The formation of 18 from the irradiation in methanol is particularly hard to rationalize except by complex schemes involving radical attack, cleavage, and internal abstraction. Since the yellow color of 14 disappears when it is dissolved in methanol, this product may arise from irradiation of the hemiketal rather than from the diketone. Varying amounts of the decarbonylation product (16) were obtained in different solvents, the maximum being the 15% observed in methanol and in benzene. This product can be envisioned as arising via carbon–carbon bond rupture, decarbonylation, and recombination. Such a process would be favored because this α-diketone is reasonably strained, and decarbonylation can produce a benzyl radical. Varying quantities of anhydride (17) were obtained in several of these solvents. Although the authors believe this product is due to the diffusion of oxygen into their reaction vessels, they appear to have taken every precaution to exclude it. The formation of the acid 19, in particular, from the irradiation of the diketone in cyclohexane is reminiscent of the benzoic acid formed from the irradiation of benzil in cyclohexane. Both benzil and 14 possess aromatic systems that might be attacked by intermediate radicals to form unstable compounds which, on work up, could react with oxygen to give the observed products.

A partial product study has shown that 3,3-dimethyl-1,2-indandione (20) undergoes photoreduction when exposed to sunlight in isopropyl alcohol, ethanol, or methanol solution.[77]

Product studies have been carried out on the irradiations of several highly strained cis-α-diketones. Although there have been no mechanistic studies, the observed products can be rationalized by a mechanistic scheme involving initial carbon–carbon bond cleavage and decarbonylation.

The highly strained [4.4.2]propella-3,8-diene-11,12-dione (21) undergoes decarbonylation when illuminated by Pyrex-filtered sunlight in methanol, carbon tetrachloride, pentane, or benzene solution.[78]

A similar decarbonylation is observed when **22** is irradiated with visible light either in benzene solution or in the crystalline state.[79] Four other similar bicyclo[2,2,2]-octadiene-2,3-diones have been reported to undergo loss of two molecules of CO on irradiation in benzene solution.[80]

On irradiation, 3,4-di-*t*-butylcyclobutanedione loses only one molecule of CO to yield di-*t*-butylcyclopropanone.[81]

3,3,-Diphenyl-1,2-indandione (**23**) yields 9-phenyl-10-anthrone when irradiated in benzene solution.[82] One possible rationalization of the observed product is shown.

THE PHOTOCHEMISTRY OF α-DICARBONYL COMPOUNDS 95

The highly strained benzocyclobutene-1,2-dione (**24**) forms dimers when irradiated in 2:1 pentane-methylene chloride[83] or cyclohexane[84] and forms adducts when irradiated in ethanol or in the presence of olefins. These products have been interpreted in terms of a carbene intermediate (**25**).

IV. PHOTOCHEMICAL REACTIONS OF α-KETO ACIDS

A. Pyruvic Acid

Irradiation of pyruvic acid (**26**) in the vapor phase produces acetaldehyde and carbon dioxide.[85] The most reasonable mechanism for the reaction appears to be a concerted decarboxylation via a cyclic four- or five-membered transition state. Although the four-membered transition state has an analogy

$$CH_3-\underset{O}{\underset{\|}{C}}-\underset{O}{\underset{\|}{C}}-OH \xrightarrow[5\,mm,\,80-85°]{h\nu,\,366\,nm} CH_3-\underset{O}{\underset{\|}{C}}-H + CO_2 + CO + CH_4$$

26   65%   100%   1-2% 1-2%

in the photoenolization of biacetyl,[58] the five-membered transition state,

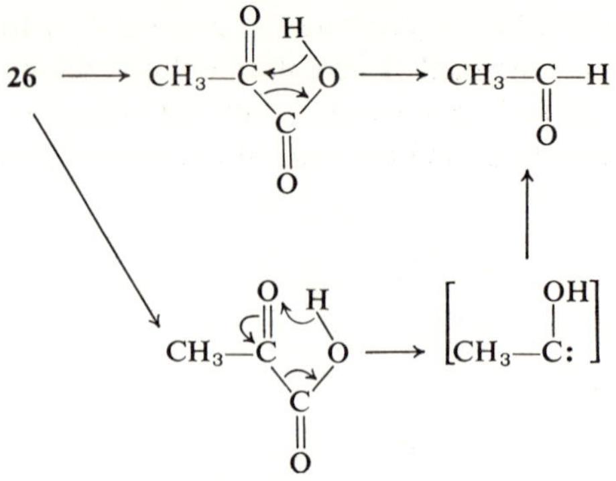

26 ⟶

$$CH_3-\overset{O}{\underset{\underset{O}{\overset{\|}{C}}}{\overset{\|}{C}}}\overset{H}{\underset{}{\diagdown}}O \longrightarrow CH_3-\overset{O}{\underset{\|}{C}}-H$$

$$CH_3-\overset{O}{\underset{\underset{O}{\overset{\|}{C}}}{\overset{\|}{C}}}\overset{H}{\underset{}{\diagdown}}O \longrightarrow [CH_3-\overset{OH}{\underset{}{C:}}]$$

followed by an intermediate hydroxy carbene, is favored because it is less strained and because initial bonding to oxygen appears to be more likely than bonding to carbon. Since up to 50 mm of nitrogen, oxygen, or ethylene had no effect on the reaction, it must be very rapid and, hence, may also involve an upper excited state. The concerted mechanism was supported by the appearance of a C_2H_4O peak in the mass spectrum of pyruvic acid.[86]

Like those of biacetyl, the photoreactions of pyruvic acid in solution are entirely different from the reaction in the vapor state. Pyruvic acid is an excellent hydrogen abstractor which is photoreduced to dimethyltartaric acid (**27**) in isopropyl alcohol, methanol, *t*-butyl alcohol, chloroform, or diethyl ether.[87,88] No reaction was observed in benzene, but photoreduction was

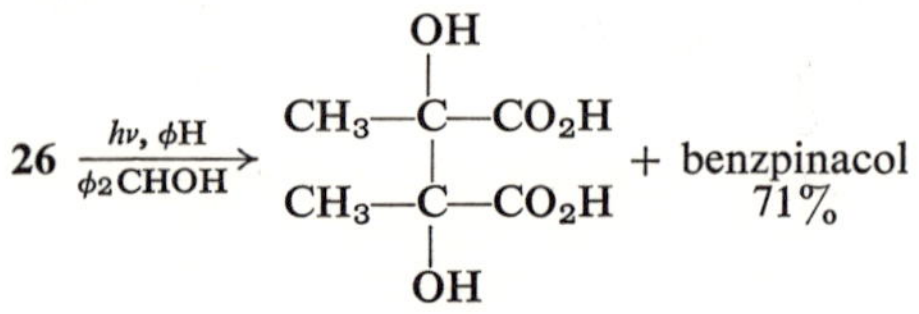

$$26 \xrightarrow[\phi_2 CHOH]{h\nu,\, \phi H} \begin{array}{c} OH \\ | \\ CH_3-C-CO_2H \\ | \\ CH_3-C-CO_2H \\ | \\ OH \\ 65\% \\ \text{(mixture of isomers)} \\ \mathbf{27} \end{array} + \text{benzpinacol} \\ 71\%$$

$$26 \xrightarrow[h\nu]{CH_3OH} \mathbf{27} + CH_3-\underset{\underset{CH_2OH}{|}}{\overset{\overset{OH}{|}}{C}}-CO_2H + \text{acetoin} + CO_2 \\ 37\% \qquad\qquad 4.1\% \qquad \text{trace} \quad 5.5\%$$

$\Phi_{-26} = 0.96 \pm 0.08$

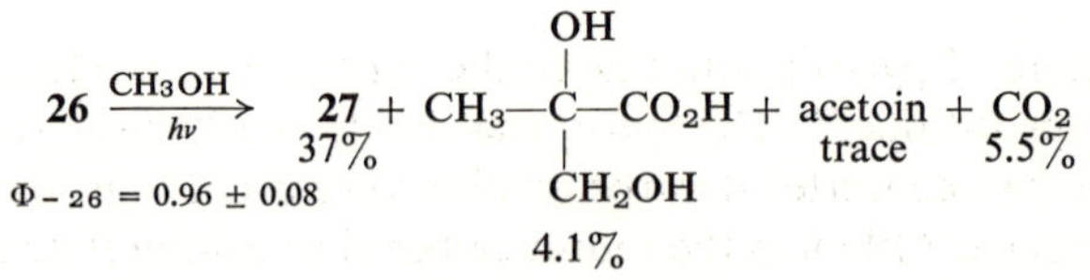

observed when benzhydrol was added. In methanol some of the solvent addition product plus a trace of acetoin were obtained in addition to **27**. The other products from irradiation in the other solvents have not been determined.

The reaction scheme must be entirely analogous to that written for bi-

THE PHOTOCHEMISTRY OF α-DICARBONYL COMPOUNDS 97

acetyl. Quenching of the reaction by 1,3-pentadiene, a poor singlet quencher,[89] indicates that the triplet state is involved.[88] The intermediate ketyl radical (28) has been observed by electron paramagnetic resonance examination of an irradiated solution of pyruvic acid in isopropyl alcohol.[90]

$$CH_3-\underset{\cdot}{\underset{|}{C}}-\underset{}{\overset{O}{\overset{\|}{C}}}-OH$$
$$\underset{OH}{}$$

28

When irradiated in water solution, pyruvic acid undergoes a remarkable transformation which converts it to acetoin and carbon dioxide.[9-11,87,91]

$$26 \xrightarrow[\Phi = 0.79 \pm 0.05]{h\nu,\ H_2O} CO_2 + CH_3-\underset{OH}{\underset{|}{CH}}-\underset{O}{\overset{\|}{C}}-CH_3 + CH_3-\underset{O}{\overset{\|}{C}}-H$$

                                  85%   not accurately   trace
                                                 determined

$$\xrightarrow{} {}^3\!\left[CH_3-\overset{O\cdots H}{\underset{\underset{O}{\overset{\|}{C}}}{\overset{|}{C}}}\underset{}{\overset{}{O}}\right] \longrightarrow {}^3\!\left[CH_3-\underset{OH}{\overset{\cdot\cdot}{C}}\right]$$

The mechanism of this reaction is obscure. One suggested mechanism, analogous to the vapor phase reaction, involves concerted decarboxylation of the pyruvic acid to yield a triplet hydroxy carbene which can either dimerize or attack another molecule of pyruvic acid to yield the observed product.[91] Dimerization seems to be the less likely process since the carbene can rearrange to acetaldehyde or react with water. Further, this mechanism predicts that acetoin will be formed when pyruvic acid is irradiated in any solvent that does not possess readily abstractable hydrogen atoms, such as benzene, a solvent in which no reaction is observed. One possible explanation of this discrepancy is that the solvation of the pyruvic acid is extremely different in benzene and in water. However, the specific role that the water plays in the reaction has not been determined.

B. Phenylglyoxylic Acid

The photoreactions of phenylglyoxylic acid (28) are similar to those of pyruvic acid. In isopropyl alcohol it is photoreduced to diphenyl tartaric acid.[92] The intermediate ketyl radical has been observed by electron paramagnetic resonance spectroscopy.[90] Irradiation of **28** in water gives benzaldehyde as the major product rather than benzoin.[87]

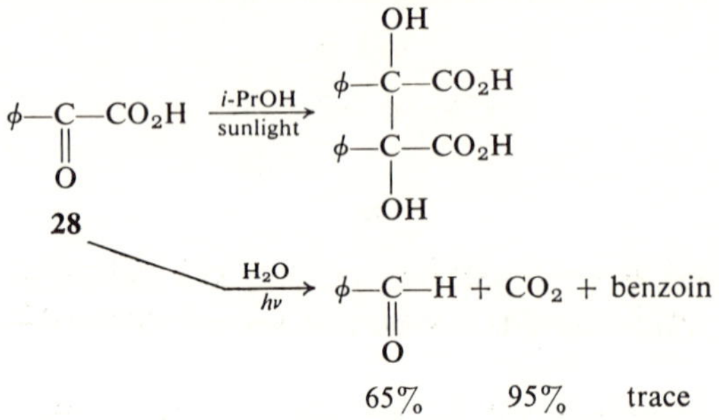

$$\phi\text{—C(=O)—CO}_2\text{H} \xrightarrow{i\text{-PrOH, sunlight}} \phi\text{—C(OH)(CO}_2\text{H)—C(OH)(CO}_2\text{H)—}\phi$$

28

$$\xrightarrow[h\nu]{\text{H}_2\text{O}} \phi\text{—CHO} + \text{CO}_2 + \text{benzoin}$$

65%   95%   trace

C. Long Chain α-Keto Acids

Unlike the long chain α-diketones, α-keto-decanoic acid does not form a cyclobutanol when irradiated in benzene solution.[34] The products are 1-

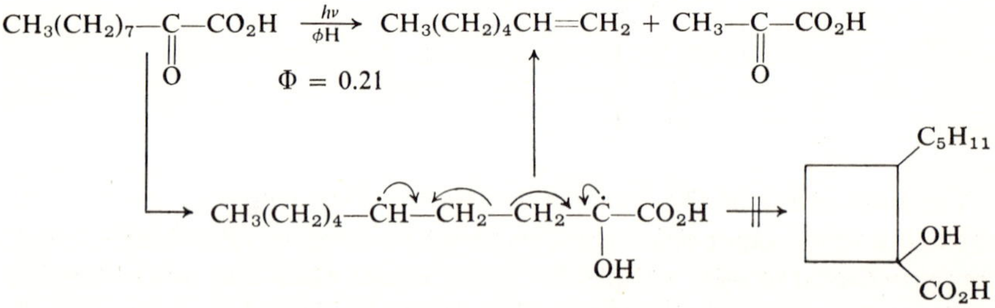

$$\text{CH}_3(\text{CH}_2)_7\text{—C(=O)—CO}_2\text{H} \xrightarrow[\phi\text{H}]{h\nu} \text{CH}_3(\text{CH}_2)_4\text{CH=CH}_2 + \text{CH}_3\text{—C(=O)—CO}_2\text{H}$$

$$\Phi = 0.21$$

$$\longrightarrow \text{CH}_3(\text{CH}_2)_4\text{—CH—CH}_2\text{—CH}_2\text{—C(OH)—CO}_2\text{H} \xrightarrow{\parallel\!\!\!\!/}$$

(cyclobutane with C$_5$H$_{11}$, OH, CO$_2$H substituents)

heptene and pyruvic acid, the products of a Norrish Type II elimination. Quenching with 1,3-cyclohexadiene gave a linear Stern-Volmer plot in which the quantum yield of elimination did not level out at high concentrations. Thus the reaction proceeds exclusively by the triplet state with no singlet contribution as is observed in the Norrish Type II photoelimination of aliphatic mono-ketones.[70]

The irradiation of α-keto-pentanoic acid in water is reported to produce butyroin,[10] a reaction similar to that of pyruvic acid in water. The low yield may have been due to competing Type II photoelimination.

$$\text{CH}_3\text{—CH}_2\text{—CH}_2\text{—C(=O)—CO}_2\text{H} \xrightarrow[\text{H}_2\text{O}]{h\nu} \text{CH}_3\text{—(CH}_2)_2\text{—CH(OH)—C(=O)—(CH}_2)_2\text{—CH}_3$$

30–40%

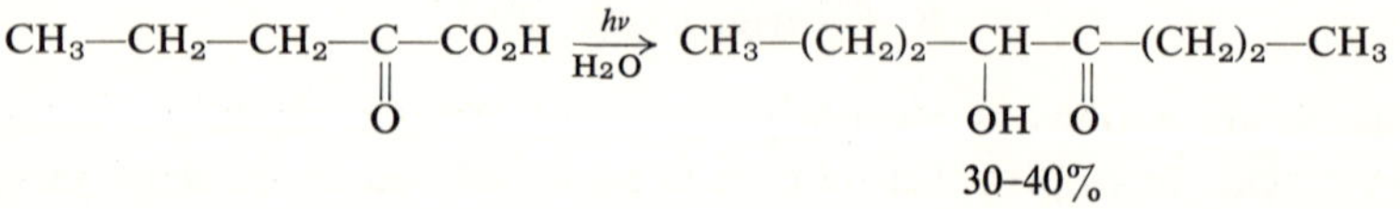

V. PHOTOCHEMICAL REACTIONS OF α-KETO ESTERS

A. Pyruvates and Phenylglyoxylates

Irradiation of ethyl or isopropyl pyruvate in the vapor phase produces a complex mixture of products which does not fit any simple photochemical mechanism.[93] The initial step of the reaction appears to be formation of radicals by a carbon–carbon bond cleavage.

$$CH_3-\underset{O}{\underset{\|}{C}}-\underset{O}{\underset{\|}{C}}-OEt \xrightarrow{h\nu} CH_3-\underset{O}{\underset{\|}{C}}\cdot + CO + \cdot OEt$$
$$\mathbf{29}$$

In solution these esters undergo a variety of transformations which are dependent on the reaction conditions. In benzene, decomposition to carbon monoxide and carbonyl compounds is observed either upon direct irradiation[94] or with benzophenone sensitization.[33] In cyclohexane a complex product mixture is obtained.[95] Addition of solvent to the carbonyl group is observed when the reaction is carried out in cyclohexene.[54] At room temperature photoreduction takes place when the reaction is carried out in a secondary alcohol.[96,97] However, in the case of the phenylglyoxylates quite a different reaction is observed when the reaction is carried out at elevated temperatures. The ester is reduced to the mandelate ester of the solvent alcohol, and the alcohol moiety of the ester is oxidized to the corresponding carbonyl compound. The pyruvates have not been studied at an elevated temperature.

$$\mathbf{29} \xrightarrow[\Phi\text{-ester} = 0.17 \pm 0.02]{h\nu,\ \phi H} CH_3-\underset{O}{\underset{\|}{C}}-H + CO$$
$$\qquad\qquad\qquad\qquad\qquad 60\%\quad 80\%$$

$$\phi-\underset{O}{\underset{\|}{C}}-\underset{O}{\underset{\|}{C}}-OEt \xrightarrow[\Phi\text{-ester} = 0.056 \pm 0.005]{h\nu,\ \phi H} \phi-\underset{O}{\underset{\|}{C}}-H + CH_3-\underset{O}{\underset{\|}{C}}-H + CO$$
$$\mathbf{30} \qquad\qquad\qquad\qquad\qquad\qquad 30\%\quad\ 25\%\quad 30\%$$

$$\text{(1-naphthyl)}-\underset{O}{\underset{\|}{C}}-\underset{O}{\underset{\|}{C}}-OCH_2CH_3 \xrightarrow[\Phi\text{-ester} < 0.01]{h\nu,\ \phi H} \text{No reaction}$$
$$\mathbf{31}$$

$$29 + \text{[cyclohexene]} \xrightarrow{h\nu} \text{[cyclohexenyl]}-\underset{\underset{CO_2Et}{|}}{\overset{\overset{CH_3}{|}}{C}}-OH$$

$$30 \xrightarrow{\frac{C_6H_{12}}{h\nu}} \phi-\underset{O}{\overset{\|}{C}}-OEt + \phi-\underset{O}{\overset{\|}{C}}-H + \phi-\underset{OH}{\overset{H}{\underset{|}{C}}}-\underset{O}{\overset{\|}{C}}-OEt + \phi_2 + \phi_2CO +$$

$$\text{[cyclohexyl]}-\underset{O}{\overset{\|}{C}}-OEt + (\text{[cyclohexyl]}-)_2 + CH_3-\underset{O}{\overset{\|}{C}}-H + \text{[cyclohexyl]}-O-CH_3 + CO_2$$

$$\underbrace{\hspace{8cm}}_{\text{small amounts}}$$

$$29 \cdot \xrightarrow[h\nu,\ 15-20°]{i\text{-PrOH}} \underset{\underset{OH}{|}}{\overset{\overset{OH}{|}}{\underset{CH_3-C-CO_2Et}{CH_3-C-CO_2Et}}} + CH_3-\underset{OH}{\overset{H}{\underset{|}{C}}}-CO_2Et + CH_3-\underset{O}{\overset{\|}{C}}-CH_3$$

$$30 \xrightarrow[\sim 30°]{2\text{-butanol}} \underset{\underset{OH}{|}}{\overset{\overset{OH}{|}}{\underset{\phi-C-CO_2Et}{\phi-C-CO_2Et}}} + CH_3-\underset{O}{\overset{\|}{C}}-C_2H_5$$

$$30 \xrightarrow[78°]{2\text{-butanol}} \phi-\underset{OH}{\overset{H}{\underset{|}{C}}}-\underset{O}{\overset{\|}{C}}-O-\underset{C_2H_5}{\overset{CH_3}{\underset{|}{C}}}-H + CH_3-\underset{O}{\overset{\|}{C}}-H$$

$$\phi-\underset{O}{\overset{\|}{C}}-\underset{O}{\overset{\|}{C}}-O-\text{[cyclohexyl]} \xrightarrow[40°]{EtOH} \phi-\underset{OH}{\overset{H}{\underset{|}{C}}}-\underset{O}{\overset{\|}{C}}-OEt + \text{[cyclohexanone]} + \underset{\underset{OH}{|}}{\overset{\overset{OH}{|}}{\underset{\phi-C-CO_2-\text{[cyclohexyl]}}{\phi-C-CO_2-\text{[cyclohexyl]}}}}$$

Not all of these reactions can be fitted to a single simple reaction scheme. It is apparent that in cyclohexene and in isopropyl alcohol at room temperature the first step is hydrogen abstraction from the solvent followed by radical coupling. Most of the major products from the reaction in cyclohexane can be accounted for by a reaction scheme similar to that invoked to account for the products observed from the irradiation of biacetyl and benzil in cyclo-

THE PHOTOCHEMISTRY OF α-DICARBONYL COMPOUNDS

hexane. Some of these products may be secondary reaction products since no effort was made to prevent further reaction.

A direct cleavage of the ester to radicals similar to that observed in the vapor state[94] and an internal hydrogen abstraction followed by radical formation[98] have been proposed to explain the reaction in benzene. The internal hydrogen abstraction mechanism appears to be the more reasonable. Direct cleavage would yield ethoxy radicals, but no products derived from them are observed. The low quantum yield of ester disappearance observed for ethyl

$$29 \not\to CH_3-\overset{\cdot}{\underset{O}{C}} + CO + \cdot OEt$$

$$\to CH_3-\underset{OH}{\overset{\cdot}{C}} \overset{\cdot}{\underset{O}{C}} \overset{\cdot}{O} \overset{\cdot}{C}H-CH_3 \to CH_3-\underset{OH}{\overset{}{C}}\colon + CO + H-\underset{O}{\overset{}{C}}CH_3$$

α-naphthyl glyoxylate (31) can be explained on the basis of the low reactivity of naphthyl ketones in hydrogen abstraction reactions[99] although the alternative explanation, that the triplet state is so low lying that the ester does not have enough energy for carbon–carbon bond rupture, works equally well.

The change of reaction in isopropyl alcohol with temperature is striking and certainly deserving of further study. It has been explained as a change

$$30 \xrightarrow[30°]{\text{external}\atop \text{abstraction}} \phi-\underset{OH}{\overset{\cdot}{C}}-CO_2Et \longrightarrow \begin{array}{c} \phi-\underset{}{\overset{OH}{C}}-CO_2Et \\ \phi-\underset{OH}{\overset{}{C}}-CO_2Et \end{array}$$

$$\xrightarrow[78°]{\text{internal}\atop \text{abstraction}} \phi-\underset{OH}{\overset{\cdot}{C}}\overset{\cdot}{\underset{O}{C}}\overset{\cdot}{O}\overset{\cdot}{C}H-CH_3 \longrightarrow \phi-\underset{OH}{\overset{}{C}}=C=O + H-\underset{O}{\overset{}{C}}-CH_3$$

$$\phi-\underset{OH}{\overset{H}{\underset{}{C}}}-\underset{O}{\overset{}{C}}-OR \xleftarrow{ROH} \rightleftarrows \phi-\underset{O}{\overset{}{C}}-\underset{O}{\overset{}{C}}-H$$

from external to internal hydrogen abstraction.[96] Type II cleavage of the resulting diradical would produce the carbonyl derivative of the ester alcohol and a hydroxy ketene, the enol of phenylglyoxal. Addition of the solvent alcohol yields the observed product.

One disturbing aspect of this mechanism is the complete absence from the product mixture of phenylglyoxal, which might have been expected from ketonization of the intermediate hydroxy ketene. In the absence of an alcohol, in benzene for example, this mechanism would predict that phenylglyoxal should be the product since the hydroxy ketone would ketonize. This prediction can be easily checked.

B. Diethyl Oxalate

Irradiation of diethyl oxalate in cyclohexane produces a variety of products which must be formed by complex radical processes.[100,101] Cleavage of the ester to form radicals is a reasonable first step since hydrogen abstraction by

$$\begin{array}{c} CO_2Et \\ | \\ CO_2Et \end{array} \xrightarrow{C_6H_{12}} C_6H_{11}-\underset{\underset{O}{\|}}{C}-CO_2Et + C_6H_{11}-OCO_2Et + EtOH + acetone$$

55%

+ diethyl ether + $C_6H_{11}-C_6H_{11}$ + CO, CO_2, C_2H_4, CH_4, C_2H_6, H_2, and propane

an ester carbonyl is not a likely process. Similar products were obtained from irradiations carried out in dioxane, benzene, and toluene.[101]

Diethyl oxalate is also reported to form an oxetane with 1,1-diphenylethene and 2-phenyl propene.[102]

$$\begin{array}{c} CO_2Et \\ | \\ CO_2Et \end{array} + \begin{array}{c} \phi \diagdown \diagup R \\ \| \\ CH_2 \end{array} \xrightarrow{3660 \text{ Å}}$$

15.7% (R = ϕ)
36% (R = CH_3)

C. Methyl-o-benzyloxyphenylglyoxylate

Methyl-o-benzyloxyphenylglyoxylate (**32**) cyclizes on irradiation to give an isomeric mixture of 2-phenyl-3-carboxymethyl-3-hydroxyl-2,3-dihydrobenzofurans (**33**), apparently by an internal hydrogen abstraction-cyclization mechanism.[103] This abstraction must take place through a seven-membered

32 $\xrightarrow{h\nu}$ **33**   >90%

transition state rather than by the usual six-membered one, even though abstraction of a hydrogen from the methyl group of the ester could take place through a six-membered transition state. A similar cyclization is shown by benzyl-*o*-benzyloxyphenylglyoxylate (**34**).[104]

$$\underset{\mathbf{34}}{\underset{}{\text{[structure: benzene ring with }-C(=O)-CO_2CH_2\phi\text{ and }-OCH_2\phi\text{ substituents]}}}$$

VI. CONCLUSION

The photoreactions of α-dicarbonyl compounds are quite different in the vapor and condensed phases. In the vapor phase, carbon–carbon bond cleavage is the preferred mode of reaction; but in the condensed phase, many of the observed reactions can be rationalized by a mechanism involving hydrogen abstraction. Internal hydrogen abstraction, when possible, is generally preferred over abstraction from the solvent. With the exception of diethyl oxalate, which undergoes photoreactions typical of an ester, only those compounds that are reasonably strained or can yield reasonably stable free radicals give decarbonylation products. In the presence of suitable substrates, cycloaddition reactions have also been observed.

VII. ADDENDUM

Almgren has studied the emission spectra of *o*-anisil, benzil, and biacetyl in low temperature glasses containing varying amounts of isopropyl alcohol.[105] Tsai and Charney have assigned the $\pi^* \leftarrow n$ transitions of camphorquinone on the basis of its solid state phosphorescence excitation spectrum.[106] Rate constants for quenching biacetyl triplets in benzene by a variety of quenchers have been determined,[107] and further studies on the quenching of biacetyl emission by amines, phenols, and ethers have been reported.[108,109] Singh, Scott, and Sopchyshyn have carried out flash photolysis of camphorquinone and biacetyl, both in isopropyl alcohol and in solvents that do not contain readily abstractable hydrogen atoms, and have recorded both their triplet–triplet absorption spectra and the absorption spectra of the ketyl radicals derived from these ketones.[110]

Turro and Lee have investigated the intramolecular cyclization of several long chain aliphatic α-diketones.[111] The quantum yields of cyclobutanol formation (2,3-pentanedione, $\Phi = 0.06$; 2,3-heptanedione, $\Phi = 0.5$; and 5-methyl-2,3-hexanedione, $\Phi = 0.6$) are less than one, indicating that cyclization does not take place by a chain process, but are large enough to account for the decrease in phosphorescence with increasing chain length (see Section III.E).

Burkoth and Ullman have reported that irradiation of **35** yields a mixture of cyclobutanol **36** and cyclopentanol **37**.[112] While the authors interpret the

product ratio as a measure of the ratio of the rates at which the hydrogen atoms at C-1 and C-7 are abstracted by the carbonyl at C-4, a hydrogen migration in the intermediate biradical would be required to obtain the cyclopentanol **37**. An alternative explanation is that **37** is formed by abstraction of the hydrogen at C-8 instead of the hydrogen at C-1 by the C-4 carbonyl, so that the product ratio is related to the relative ease with which abstraction may take place via a six- or a seven-membered transition state. In either case, however, the product ratio is the ratio of the quantum yields for formation of these products but not necessarily the ratio of the abstraction rates. Only if the two different biradical intermediates cyclize with the same efficiency will the product ratio represent the abstraction rate ratio.[113]

Bishop and Hamer found that acyclic α,β-unsaturated 1,2-diketones form cyclopentanol derivatives in high yield, while β,γ-unsaturated derivatives form oxetanes by internal cycloaddition.[114] Unexpectedly, the γ,δ-unsaturated derivatives also gave oxetanes after an initial migration of the double bond to the β,γ position. The formation of oxetanes such as **38** was observed in the camphorquinone sensitized dimerization of butadiene.[115] Photocycloadditions of α-diketones to various olefins have been studied by several groups.[116]

Rubin has investigated the benzophenone sensitized addition of xylene to camphorquinone.[117] Enhanced quantum yields of camphorquinone disappearance and changes in product ratios are observed with decreasing camphorquinone concentration, indicating that the reaction is "chemically sensitized." The photoreduction and photocyclization of the α-keto esters derived from dehydroginkgolide A have been reported by Nakadaira, Hirota, and Nakanishi.[118] In the presence of certain nucleophiles such as water and phenols, coumarandiones undergo decarbonylation to give salicylic acid derivatives.[119] Rubin has reviewed the photochemistry of the α-diketones and that of the analogous o-quinones.[120]

REFERENCES

1. H. Klinger, *Chem. Ber.*, **19**, 1862 (1886).
2. G. Ciamician and P. Silber, *Chem. Ber.*, **34**, 1530 (1901).
3. A. Benrath, *J. Prakt. Chem.*, **87**, 416 (1913); *Chem. Abstr.*, **7**, 2545 (1913).
4. W. D. Cohen, *Chem. Weekblad*, **13**, 590 (1916); *Chem. Abstr.*, **10**, 2092 (1916).
5. C. W. Porter, H. C. Ramsperger, and C. Steel, *J. Amer. Chem. Soc.*, **45**, 1827 (1923).
6. E. J. Bowen and A. T. Horton, *J. Chem. Soc.*, **1934**, 1505.
7. R. G. W. Norrish and J. G. A. Griffiths, *J. Chem. Soc.*, **1928**, 2829.
8. F. W. Kirkbridge and R. G. W. Norrish, *Trans. Faraday Soc.*, **27**, 405 (1931).
9. W. Dirscherl, *J. Physiol. Chem.*, **188**, 225 (1930); *Chem. Abstr.*, **24**, 3212 (1930).
10. W. Dirscherl, *J. Physiol. Chem.*, **219**, 177 (1933); *Chem. Abstr.*, **27**, 4773 (1933).
11. F. Lieben, L. Löwe, and B. Bäuminger, *Biochem. Z.*, **271**, 209 (1934); *Chem. Abstr.*, **28**, 5758[8] (1934).
12. E. Oliveri-Mandala, G. Caronna, and E. Deleo, *Gazz. Chim. Ital.*, **68**, 327 (1938); *Chem. Abstr.*, **33**, 570[3] (1939).
13. E. Oliveri-Mandala, A. Giacalone, and E. Deleo, *Gazz. Chim. Ital.*, **69**, 104 (1939); *Chem. Abstr.*, **33**, 5386[3] (1939).
14. W. A. Noyes, Jr., G. B. Porter, and J. E. Jolly, *Chem. Rev.*, **56**, 49 (1956).
15. J. W. Sidman, *Chem. Rev.*, **58**, 689 (1958).
16. R. M. Hochstrasser and G. B. Porter, *Quart. Rev.*, **14**, 146 (1960).
17. W. A. Noyes, Jr. and I. Unger, *Pure Appl. Chem.*, **9**, 461 (1964).
18. J. N. Pitts, Jr. and J. K. S. Wan, in *The Chemistry of the Carbonyl Group*, S. Patai, Ed., Wiley-Interscience, New York, 1966, pp. 871–873.
19. P. J. Wagner and G. S. Hammond, in *Advances in Photochemistry*, Vol. 5, W. A. Noyes, Jr., G. S. Hammond, and J. N. Pitts, Jr., Eds., Wiley-Interscience, New York, 1968, pp. 109–111.
20. N. J. Leonard, A. J. Kresge, and M. Oki, *J. Amer. Chem. Soc.*, **77**, 5078 (1955) and earlier papers in this series.
21. G. Sandris and G. Ourisson, *Bull. Chim. Soc. France*, **25**, 350 (1958).
22. K. Alder, H. K. Schäfer, H. Esser, H. Krieger, and R. Reubke, *Ann. Chem.*, **593**, 23 (1955).
23. M. B. Rubin and R. G. La Barge, *J. Org. Chem.*, **31**, 3283 (1966).
24. S. A. Greenberg and L. S. Forster, *J. Amer. Chem. Soc.*, **83**, 4339 (1961).

25. F. Wilkinson, in *Advances in Photochemistry*, Vol. 3, W. A. Noyes, Jr., G. S. Hammond, and J. N. Pitts, Jr., Eds., Wiley-Interscience, New York, 1964, pp. 253–256.
26. J. T. Dubois and B. Stevens, in *Luminescence of Organic and Inorganic Materials*, H. P. Kallmann and G. M. Spruck, Eds., Wiley, New York, 1962, pp. 115–131.
27. W. A. Noyes, Jr., and I. Unger, in *Advances in Photochemistry*, Vol. 4, W. A. Noyes, Jr., G. S. Hammond, and J. N. Pitts, Jr., Eds., Wiley-Interscience, New York, 1966, pp. 62–64.
28. H. H. Richtol and A. Belorit, *J. Chem. Phys.*, **45**, 35 (1966).
29. J. S. E. McIntosh and G. B. Porter, *J. Chem. Phys.*, **48**, 5475 (1968).
30. H. L. J. Bäckström and K. Sandros, *Acta Chem. Scand.*, **14**, 48 (1960); C. A. Parker and T. A. Joyce, *Chem. Commun.*, **1968**, 1421.
31. H. H. Richtol and F. H. Klappmeier, *J. Chem. Phys.*, **44**, 1519 (1966).
32. A. A. Lamola and G. S. Hammond, *J. Chem. Phys.*, **43**, 2129 (1965).
33. G. S. Hammond, P. A. Leermakers, and N. J. Turro, *J. Amer. Chem. Soc.*, **83**, 2395 (1961).
34. T. R. Evans and P. A. Leermakers, *J. Amer. Chem. Soc.*, **90**, 1840 (1968).
35. W. G. Herkstroeter, A. A. Lamola, and G. S. Hammond, *J. Amer. Chem. Soc.*, **86**, 4537 (1964).
36. G. S. Hammond and P. A. Leermakers, *J. Phys. Chem.*, **66**, 1148 (1962).
37. G. S. Hammond, J. Saltiel, A. A. Lamola, N. J. Turro, J. S. Bradshaw, D. O. Cowan, R. C. Counsell, V. Vogt, and C. Dalton, *J. Amer. Chem. Soc.*, **86**, 3197 (1964).
38. L. M. Coyne, Ph.D. Thesis, California Institute of Technology, 1967, pp. 96–110; *Dissertation Abstr.*, **28**, 515-B (1967).
39. B. M. Monroe and S. A. Weiner, *J. Amer. Chem. Soc.*, **91**, 450 (1969); B. M. Monroe, *Intra-Science Chem. Reports*, **3**, 283 (1969).
40. T. R. Evans and P. A. Leermakers, *J. Amer. Chem. Soc.*, **89**, 4380 (1967).
41. C. S. Parmenter, *J. Chem. Phys.*, **41**, 658 (1964).
42. J. G. Calvert and J. N. Pitts, Jr., *Photochemistry*, Wiley, New York, 1966, p. 376.
43. W. A. Noyes, Jr., W. A. Mulac, and M. S. Matheson, *J. Chem. Phys.*, **36**, 880 (1962).
44. W. G. Bentrude and K. R. Darnall, *Chem. Commun.*, **1968**, 810.
45. R. Srinivasan, in *Advances in Photochemistry*, Vol. 1, W. A. Noyes, Jr., G. S. Hammond, and J. N. Pitts, Jr., Eds., Wiley-Interscience, New York, 1963, pp. 83–113.
46. W. H. Urry and D. J. Trecker, *J. Amer. Chem. Soc.*, **84**, 118 (1962).
47. J. N. Pitts, Jr., R. L. Letsinger, R. P. Taylor, J. M. Patterson, G. Recktenwald, and R. B. Martin, *J. Amer. Chem. Soc.*, **81**, 1068 (1959).
48. N. J. Turro, *Molecular Photochemistry*, Benjamin, New York, 1965, pp. 137–158.
49. H. L. J. Bäckström and K. Sandros, *Acta Chem. Scand.*, **12**, 823 (1958).
50. N. J. Turro and R. Engel, *Mol. Photochem.*, **1**, 143 (1969).
51. H. Zeldes and R. Livingston, *J. Chem. Phys.*, **47**, 1465 (1967).
52. S. A. Weiner, E. J. Hamilton, Jr., and B. M. Monroe, *J. Amer. Chem. Soc.*, **91**, 6350 (1969).
53. S. A. Weiner and G. S. Hammond, Unpublished results.
54. P. W. Jolly and P. de Mayo, *Can. J. Chem.*, **42**, 170 (1964).
55. E. J. Baum and R. O. C. Norman, *J. Chem. Soc.*, B, **1968**, 227.
56. W. G. Bentrude and K. R. Darnall, *J. Amer. Chem. Soc.*, **90**, 3588 (1968).
57. D. S. Weir, *J. Chem. Phys.*, **36**, 1113 (1962).
58. J. Lemaire, *J. Chem. Phys.*, **71**, 2653 (1967).

59. D. Phillips, J. Lemaire, C. S. Burton, and W. A. Noyes, Jr., in *Advances in Photochemistry*, Vol. 5, W. A. Noyes, Jr., G. S. Hammond, and J. N. Pitts, Jr., Eds., Wiley-Interscience, New York, 1968, pp. 355–356.
60. J. Lemaire, M. Niclause, X. Deglise, J. André, G. Persson, and M. Bouchy, *C. R., Acad. Sci., Paris, Ser. C*, **267**, 33 (1968).
61. A. Weller, *Pure Appl. Chem.*, **16**, 115 (1968).
62. I. M. Whittemore and M. Szwarc, *J. Phys. Chem.*, **67**, 2492 (1963).
63. D. L. Bunbury and C. T. Wang, *Can. J. Chem.*, **46**, 1473 (1968); D. L. Bunbury and T. T. Chuang, *Ibid.*, **47**, 2045 (1969).
64. A. Beckett, A. D. Osborne, and G. Porter, *Trans. Faraday Soc.*, **60**, 873 (1964).
65. C. Walling and M. J. Gibian, *J. Amer. Chem. Soc.*, **87**, 3361 (1965).
66. P. de Mayo and A. Stoessl, *Can. J. Chem.*, **40**, 57 (1962).
67. A. Schönberg, *Preparative Organic Photochemistry*, Springer-Verlag, New York, 1968, pp. 118–125, 419–420.
68. W. H. Urry, D. J. Trecker, and D. A. Winey, *Tetrahedron Lett.*, **1962**, 609.
69. M. Barnard and N. C. Yang, *Proc. Chem. Soc.*, **1958**, 302.
70. P. J. Wagner and G. S. Hammond, *J. Amer. Chem. Soc.*, **88**, 1245 (1966).
71. B. M. Monroe, S. A. Weiner, and G. S. Hammond, *J. Amer. Chem. Soc.*, **90**, 1913 (1968).
72. H. Berg, *Z. Chem.*, **2**, 237 (1962); *Chem. Abstr.*, **59**, 2508c (1963).
73. M. B. Rubin, R. G. LaBarge, and J. M. Ben-Bassat, *Israel J. Chem.*, **5**, 39p (1967); M. B. Rubin and J. M. Ben-Bassat, *Tetrahedron*, **26**, 3579 (1970).
74. J. Meinwald and H. O. Klingele, *J. Amer. Chem. Soc.*, **88**, 2071 (1966).
75. G. E. Gream, J. C. Paice, and C. C. R. Ramsey, *Aust. J. Chem.*, **20**, 1671 (1967).
76. S. A. Weiner, E. J. Hamilton, Jr., and G. S. Hammond, Unpublished results; G. S. Hammond and S. A. Weiner, *Intra-Science Chem. Reports*, **3**, 241 (1969).
77. C. F. Koelsch and C. D. Le Claire, *J. Org. Chem.*, **6**, 516 (1941).
78. J. J. Bloomfield, J. R. Smiley Irelan, and A. P. Marchand, *Tetrahedron Lett.*, **1968** 5647.
79. D. Bryce-Smith and A. Gilbert, *Chem. Commun.*, **1968**, 1702.
80. J. Strating, B. Zwanenburg, A. Wagenaar, and A. C. Udding, *Tetrahedron Lett.*, **1969**, 125.
81. Ae. de Groot, D. Oudman, and H. Wynberg, *Tetrahedron Lett.*, **1969**, 1529.
82. J. Riguady and N. Paillous, *Tetrahedron Lett.*, **1966**, 4825.
83. H. A. Staab and J. Ipaktschi, *Tetrahedron Lett.*, **1966**, 583; *Ibid.*, *Chem. Ber.*, **101**, 1457 (1968).
84. R. F. C. Brown and R. K. Solly, *Tetrahedron Lett.*, **1966**, 1691.
85. G. F. Vesley and P. A. Leermakers, *J. Phys. Chem.*, **68**, 2364 (1964).
86. N. J. Turro, D. S. Weiss, W. F. Haddon, and F. W. McLafferty, *J. Amer. Chem. Soc.*, **89**, 3370 (1967).
87. P. A. Leermakers and G. F. Vesley, *J. Amer. Chem. Soc.*, **85**, 3776 (1963).
88. D. S. Kendall and P. A. Leermakers, *J. Amer. Chem. Soc.*, **88**, 2766 (1966).
89. L. M. Stephenson and G. S. Hammond, *Pure Appl. Chem.*, **16**, 125 (1968).
90. T. Fujisawa, B. M. Monroe, and G. S. Hammond, *J. Amer. Chem. Soc.*, **92**, 542 (1970).
91. P. A. Leermakers and G. F. Vesley, *J. Org. Chem.*, **28**, 1160 (1963).
92. A. Schönberg, N. Latif, R. Moubasher, and A. Sina, *J. Chem. Soc.*, **1951**, 1364.
93. P. A. Leermakers, M. E. Ross, G. F. Vesley, and P. C. Warren, *J. Org. Chem.*, **30**, 914 (1965).

94. P. A. Leermakers, P. C. Warren, and G. F. Vesley, *J. Amer. Chem. Soc.*, **86**, 1768 (1969).
95. T. Tominga, Y. Odaira, and S. Tsutsumi, *Bull. Chem. Soc. Japan*, **39**, 1824 (1966).
96. E. S. Huyser and D. C. Neckers, *J. Org. Chem.*, **29**, 277 (1964).
97. N. C. Yang and A. Morduchowitz, *J. Org. Chem.*, **29**, 1654 (1964).
98. D. C. Neckers, *Mechanistic Organic Photochemistry*, Reinhold, New York, 1967, pp. 179–182.
99. N. C. Yang, D. S. McClure, S. L. Murov, J. J. Houser, and R. Dusenberry, *J. Amer. Chem. Soc.*, **89**, 5466 (1967).
100. Y. Odaira, T. Tominaga, T. Sugihara, and S. Tsutsumi, *Tetrahedron Lett.*, **1964**, 2527.
101. T. Tominaga, Y. Odaira, and S. Tsutsumi, *Kogyo Kagaku Zasshi*, **69**, 2290 (1966); *Chem. Abstr.*, **66**, 94522j (1967).
102. T. Tominaga, Y. Odaira, and S. Tsutsumi, *Bull. Chem. Soc. Japan*, **40**, 2451 (1967).
103. S. P. Pappas, B. C. Pappas, and J. E. Blackwell, Jr., *J. Org. Chem.*, **32**, 3066 (1967).
104. S. P. Pappas, J. E. Alexander, and R. D. Zehr, Jr., *J. Amer. Chem. Soc.*, **92**, 6927 (1970).
105. M. Almgren, *Photochem. Photobiol.*, **9**, 1 (1969).
106. L. Tsai and E. Charney, *J. Phys. Chem.*, **73**, 2462 (1969).
107. R. B. Cundall, G. B. Evans, and E. J. Land, *J. Phys. Chem.*, **73**, 3982 (1969).
108. N. J. Turro and R. Engel, *J. Amer. Chem. Soc.*, **91**, 7113 (1969).
109. N. J. Turro and R. Engel, *Mol. Photochem.*, **1**, 235 (1969).
110. A. Singh, A. R. Scott, and F. Sopchyshyn, *J. Phys. Chem.*, **73**, 2633 (1969).
111. N. J. Turro and T. J. Lee, *J. Amer. Chem. Soc.*, **91**, 5651 (1969).
112. T. L. Burkoth and E. F. Ullman, *Tetrahedron Lett.*, **1970**, 145.
113. P. J. Wagner, *J. Amer. Chem. Soc.*, **90**, 5896 (1968).
114. R. Bishop and N. K. Hamer, *Chem. Commun.*, **1969**, 804.
115. W. L. Dilling, R. D. Kroening, and J. C. Little, *J. Amer. Chem. Soc.*, **92**, 928 (1970).
116. H. S. Ryang, K. Shima, and H. Sakurai, *Tetrahedron Lett.*, **1970**, 1091; G. E. Gream, M. Mular, and J. C. Paice, *Tetrahedron Lett.*, **1970**, 3479; Y. L. Chow, T. C. Joseph, H. H. Quon, and J. N. S. Tam, *Can. J. Chem.*, **48**, 3045 (1970).
117. M. B. Rubin, *Tetrahedron Lett.*, **1969**, 395; M. B. Rubin and Z. Hershtik, *Chem. Commun.*, **1970**, 1267.
118. Y. Nakadaira, Y. Hirota, and K. Nakanishi, *Chem. Commun.*, **1969**, 1469.
119. W. M. Horspool and G. D. Khandelwal, *Chem. Commun.*, **1970**, 257.
120. M. B. Rubin, *Fortschr. Chem. Forsch.*, **13**, 251 (1969).

Photo-Fries Rearrangement and Related Photochemical [1,j]-Shifts (j = 3, 5, 7) of Carbonyl and Sulfonyl Groups

DANIEL BELLUŠ, *Polymer Institute of Slovak Academy of Sciences, Bratislava 9, Czechoslovakia*[†]

CONTENTS

I. Introduction.	109
II. Photorearrangement of Aryl Esters of Carboxylic Acids	114
A. Excitation and Primary Photophysical Processes	114
B. Secondary Chemical Processes	119
1. Dissociative Path A.	119
2. Concerted Path B	126
C. Photodecarboxylation: A Competitive Side Reaction	134
D. Evaluation of Quantum Yields	136
E. Photostability of Photorearrangement Products	139
F. Conclusion	139
III. Photorearrangement of Aryl Esters of Sulfonic Acids	140
IV. Photorearrangement of N-Aryl Amides of Carboxylic Acids	141
V. Photorearrangement of N-Aryl Lactams.	145
VI. Photorearrangement of N-Aryl Amides of Sulfonic Acids	146
VII. Photorearrangement of Enol Esters of Carboxylic Acids	146
VIII. Reversible Photorearrangement of Enol Lactones and Nonenolizable β-Diketones.	149
IX. Photorearrangement of Enamides of Carboxylic Acids.	152
References	155

I. INTRODUCTION

Only eight years ago Anderson and Reese[1] observed that catechol monoacetate (**9**) in ethanolic solution rearranges upon irradiation with ultra violet light to two isomeric dihydroxyacetophenones **10** and **11**. Since that time a series of similar photorearrangements has been discovered as witnessed by a great number of original papers, dissertations, and reviews.[2] Sufficient insight gained by this work now allows a number of correlations to be drawn that were not evident previously. It is the aim of this article to critically review the

† Present address: Central Research Laboratories, J. R. Geigy AG., 4000 Basel, Switzerland.

present literature of the field and to correlate a number of previously unrelated observations.

From the chemical point of view the unifying feature of all discussed rearrangements is 1,3 migration (or 1,5 and 1,7 migrations) of a group X, as illustrated by the simplified equations **1 → 2** and **3 → 4**.

The double bond between carbon C_2 and C_3 can be either isolated, conjugated, or aromatic. The latter case, shown by **1 → 2**, represents the "genuine" photo-Fries rearrangement (named according to the acid-catalyzed Fries rearrangement, with which it has in common both starting materials and final products).[3] The photochemical step is followed by fast non-photochemical aromatization of the intermediate cyclohexadienone **5 → 6** and iminocyclohexadiene **7 → 8**. None of the given rearranged products were found when oxygen in compound **1** was replaced by sulfur.[4]

As will be discussed later, from the physical point of view the common feature of the photorearrangements reviewed is that they are not influenced by the presence of triplet quenchers, including oxygen, and sensitizers.

The scope of this article is limited by the following types of photorearrangements (the typical reaction for each type is given without specifying the by-products):

Aryl esters of carboxylic acids (X in general formula **1** = —COR, —COOR or —CONHR; Section II).

Aryl esters of sulphonic acids (X = —SO$_2$R; Section III).

N-*Aryl amides of carboxylic acids* (X = —COR or —COOR; Y = H, alkyl or —COR; Section IV).

N-Aryl lactams [X and Y in the same ring = —CO(CH$_2$)—; Section V].

N-Aryl amides of sulphonic acids (X = —SO$_2$R, Y = H; Section VI).

Enol esters of carboxylic acids (X = —COR; Section VII).

Enol lactones of carboxylic acids [X = —COR— (to C_2) with exocyclic and X = —COR— (to C_3) with endocyclic lactones; Section VIII].

Enamides of carboxylic acids (X = —COR or —COOR; Y = alkyl or —COR; Section IX).

According to the definition (Eqs. 1 → 2 and 3 → 4), it would also be proper to discuss at this point the photorearrangements of phenyl and vinyl ethers (X = alkyl, usually substituted by unsaturated group, or X = phenyl)[21,22]; of ketene dialkyl acetals (X = alkyl, compound 1 substituted at C_2 with an alkoxy group)[23,24]; of arylcyanates (X = —CN)[25]; of N-chloroacetanilides (X = —Cl, Y = COCH$_3$)[26]; of N-nitro-N-alkylarylamines (X = NO$_2$, Y = alkyl)[27]; and of N-aryl-N-chlorophosphoramidates (X = —Cl, Y = —P(O)(OR)$_2$).[28] The photorearrangement of these compounds, however, is not included, though 1,3 shifts of group X are also operative. It is impossible to define the limitations of the reaction as applied to other similar systems at this time; however, this will undoubtedly be an active and fruitful area of photochemistry in the future.

References are included, with some exceptions, till the end of 1968.

II. PHOTOREARRANGEMENT OF ARYL ESTERS OF CARBOXYLIC ACIDS

A. Excitation and Primary Photophysical Processes

Before we start to discuss the primary photophysical processes in aryl ester molecules, it is important to note that the multiplicity of the excited state of the reacting molecule has not been elucidated explicitly. Based upon results from the sensitizing and quenching experiments, some possibilities have been excluded and others have been shown possible. Nevertheless, no unequivocal identification of either a singlet or triplet manifold of a rearranging aryl ester molecule has been achieved. Unfortunately, spectroscopic data are of little assistance to solve this problem. In the absorption spectra the lowest singlet electronic levels, π,π^* and n,π^*, are overlapping and, usually, show a single broad absorption band at 300–310 nm. Some authors report[29,30] that aryl ester solutions exhibited neither measurable fluorescence nor phosphorescence.

1-Naphthyl acetate (51), however, does exhibit phosphorescence emission. The energy of the lowest triplet level,[8] $E_T = 60.1$ kcal mole^{-1}, and the quantum yield for intersystem crossing, $\phi_{ST} = 0.29$,[31] are known. In the presence of ethyl bromide, the phosphorescence of 51 becomes more intensive and the lifetime of the triplet state decreases; the addition of ethyl bromide, however, does not influence the rate of photorearrangement of 51.[8] Attempts to sensitize the rearrangement of 51 with benzophenone ($E_T = 68.5$ kcal mole^{-1} [32a]) or triphenylene ($E_T = 66.6$ kcal mole^{-1} [32a]) on irradiation with light of wavelengths greater than 300 nm were not successful. With irradiation in the absorption region of 51, anthracene ($E_T = 42$ kcal mole^{-1} [32b]) acted only as an optical filter without decreasing the rate of photorearrangement of 51.

The quenching experiments by photorearrangements of 4-methylphenyl acetate (55)[33,34] and 4-methylphenyl-N-methyl carbamate (17)[7] demonstrated that the overall process was unaffected by naphthalene ($E_T = 60.9$ kcal mole^{-1} [32a]) and ferric acetylacetonate,[35] which are triplet responsive additives. The photorearrangement of 55 and 4-methylphenyl benzoate (60) in dioxane is insensitive to the presence of biacetyl[4] ($E_T = 54.9$ kcal mole^{-1}, $E_S = 62$ kcal mole^{-1} [36]), which, in some cases, is active as a quencher of photorearrangements proceeding via the singlet state.[36]

It is typical that the photorearrangement 57 → 58 proceeds readily.[37] This seems to rule out the triplet state of the rearranging molecule 57, reasoning that the ferrocene part of the molecule would very probably act as a triplet quencher,[35] as is the case with other ferrocene derivatives structurally able to undergo triplet photoreactions.[38] On the other hand, the reaction 55 → 56 cannot be sensitized with acetophenone[33] ($E_T = 73.6$ kcal mole^{-1} [32a]).

| 59 R = H | 61 R = H | 63 | 23 R = H |
| 60 R = CH$_3$ | 62 R = CH$_3$ | | 19 R = CH$_3$ |

Alternatively, there is evidence pointing to the existence of an effective energy transfer from excited benzene to an aryl ester 59. The results of irradiation of 59 in some transparent solvents and in benzene are presented in Table I. The calculation of the spectral data of the reaction systems (accounting for the emission characteristics of mercury lamp, the absorption spectrum of 59 and of benzene) shows that ca. 97–98% of light absorbed by 59 is of wavelength 253.7 nm. From the light quanta of this wavelength absorbed by the benzene solution of 59, only 1–4% are absorbed directly by the molecules of 59. If benzene were acting as an optical filter only and not as a sensitizer as well, the conversion of ester and the product yields should be lower than those observed in solvents transparent at 253.7 nm.

TABLE I
Comparison of the Yields of **61**, **63** and **23** on Uv-Irradiation and γ-Radiolysis of Phenyl Benzoate (**59**)

	Conc. **59**, M	Solvent	Temp., °C	Conversion of **59**, %	**61**	**63**	**23**	Ref.
Uv-irradiation	0.13	Benzene	52	4	13	20	a	37
	0.22	Cyclohexane	55	4	11	29	a	37
	0.10	Ether	a	92	18	18	44	39
	0.04	Ethanol	30	a	20	28	14	40
γ-Radiolysis	0.05	Benzene	40	17	8	Traces	14	41
	0.05	Toluene	40	18	14	Traces	63	41

[a] Not given.

The discussed energy transfer from the solvent was studied in more detail by γ-radiolysis of 4-methylphenyl acetate (**55**).[41] From the kinetic analysis of the results, the following conclusions can be drawn:

(1) The dependence of $1/G$ of the formation of products **56** and **19**, plotted against the reciprocal concentration of **55**, are differently curved, i.e., the ratio of the yields of **56** and **19** changes with the concentration of **55**. This points out that at least two different excited states of the donor (benzene) are transferred, differing in k_{t55}/k_q values (k_{t55} = rate constant of energy transfer from an excited benzene molecule to **55**, k_q = over-all constant of deactivation of excited benzene molecule by other pathways) which influence the reactions of the acceptor **55** in different ways.

(2) The addition of p-terphenyl quenches the rearrangement of **55** in benzene. However, this quenching effect is considerably smaller (by ca. 15 times) than that expected for deactivation of the first excited singlet state of benzene. One may therefore conclude that energy transfer from benzene to **55** must occur from energetically higher excited states of benzene rather than from the lowest singlet level. In this connection it is of interest that Braun, Kato, and Lipsky[42] found the internal conversion efficiencies in excited benzene (the efficiency with which the absorbing molecule converts internally from the upper electronic states to the emitting S_1 state) to be significantly less than unity.

The sensitized photo-Fries rearrangement of **55** in benzene, toluene, and in concentrated polystyrene solution in dioxane is effectively quenched with biacetyl.[4] This phenomenon must again be attributed to quenching of the aromatic energy donor, because in pure dioxane the photorearrangement of **55** is not influenced by biacetyl (*vide supra*).

From the facts given so far it seems most plausible (though not unequivocally proven) that the aryl ester molecule in some of its excited singlet states (after direct or sensitized excitation) reorganizes the electrons at a much higher rate than in other experimentally observable photophysical processes (fluorescence, collisional deactivation with quencher molecule). For example, the fluorescence rate constant k_f is 10^7–10^{10} sec^{-1},[43a] and the rate constant for quenching limited by collisional diffusion is in the range of 10^{10} liter mole^{-1} sec^{-1}, but the rate constant k_{dis} of dissociation of polyatomic molecules from the excited singlet is about 10^{12} sec^{-1},[44] or even 10^{13}–10^{14} sec^{-1}.[45]

From the benzene sensitized rearrangement of **55** it can be concluded that this *very fast electron reorganization in the excited molecules of aryl esters proceeds from two or more excited singlet states.*[41] Thus, from the energetically higher excited state dissociative electron reorganization can take place with preference or exclusively (leading, e.g., to the products of homolytic cleavage of the ester —O—CO— linkage; Path A, see Section II.B.1), whereas from the energetically lower excited state nondissociative electron reorganization such as concerted [1, 3]-shifts prevails (leading, e.g., to the products of *ortho*-rearrangement; Path B, see Section II.B.2).

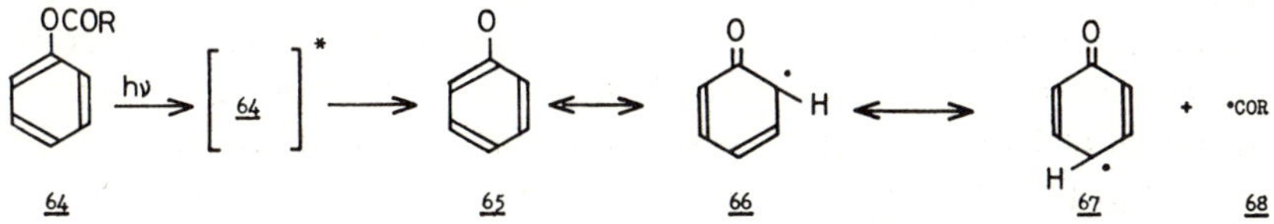

Path A

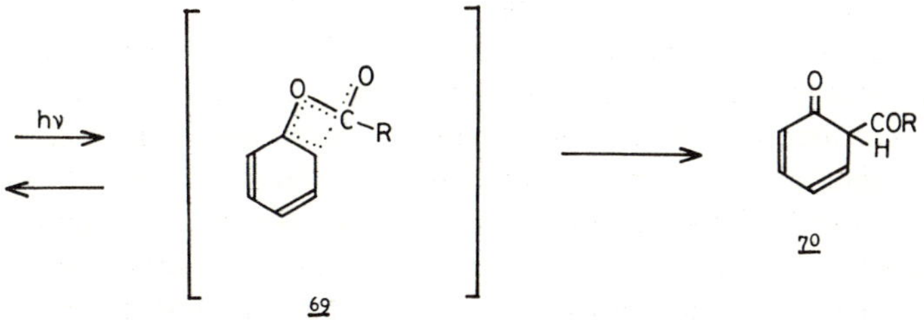

Path B

Attributing the higher and the lower singlet state to Path A and Path B, respectively, is very delicate, but nevertheless could probably be rationalized on the following basis.

TABLE II
The Influence of the Polarity of Solvent on the Distribution of Photorearrangement Products. The Photolysis of Phenyl Benzoate (59) in Ether–Methanol Mixture[39]

Methanol, vol %	Conversion of 59, %	Yields, mole % 61	63	23	Ratio 61/63
0	92	18	18	44	0.50
30	93	26	22	32	0.55
50	96	38	20	23	0.65
70	98	46	21	17	0.71
100	99	59	20	13	0.75

Plank[39] found a pronounced influence of solvent polarity on the photorearrangement of phenyl benzoate (59) (Table II). Polar solvents such as methanol favor the rearrangement, and nonpolar solvents favor phenol formation. It was also found that both benzene and ether (nonpolar solvents) gave essentially the same results, but were vastly different in their hydrogen-donating ability. It is readily apparent that this effect is not merely a solvent dependent partitioning between an intramolecular rearrangement-recombination process and an intermolecular hydrogen abstraction process in which only radicals arising by Path A would be involved. The higher contribution of Path B in the polar solvents could be tentatively attributed to a lowering of the excited singlet state energy of the aryl ester in polar solvents. However, changes in energy are minimal, e.g., the "red shift" in **59** accounts for only 1.3 kcal mole^{-1} ($\lambda_{max}^{ether} = 229.4$ nm, $\lambda_{max}^{MeOH} = 231.8$ nm). At the same time it would be thereby indicated that both Path A and Path B proceed from π,π^* excited singlet state. This would make Plank's results consistent with both solvent polarity and hydrogen-abstracting ability.

In the γ-radiolysis of benzene solutions of aryl esters, "preexcitation" of the donor (benzene) molecules to higher energetic singlet levels and a proportional increase of both the probability of sensitized excitation of aryl ester to a higher energetic singlet level and of reactions according to Path A are highly probable. Experiments (Table I) check with this expectation. The ratio of phenol versus *ortho*-rearrangement product on sensitized γ-radiolysis,[41] as well as on electron irradiation of diphenyl carbonate in the solid state,[46] is much higher than with comparable photorearrangements.

While the possibility of a very rapid electron reorganization from a low energy triplet state is ruled out by the observed insensitivity of the photo-Fries rearrangement to paramagnetic quenchers (*vide supra*) and the usual rate of intersystem crossing ($k_{ST} = 10^6$–10^{11} sec^{-1} [43a]), the remaining alternative to singlet processes for both Path A and Path B, i.e., reactions from

TABLE III
Spin Densities for Aryloxy Radicals

Radical from	p_{oxygen}	p_{ortho}	p_{meta}	p_{para}	Ref.
Phenol	a	0.28	−0.075	0.45	47
	0.170	0.154	−0.022	0.338	48
4-Methylphenol	a	0.25	−0.06	0.44	47
2,6-Dimethylphenol	a	0.24	−0.07	0.40	47
Aniline	0.435	0.192	−0.061	0.281	48

[a] 20–25% of the unpaired electron spin is associated with both the oxygen center and the carbon atom to which it is directly attached.

"hot" vibrational ground states, has been neither substantiated nor disproved experimentally.

Still open also is the problem of *para*-products formation. Practically all comparable data are given in Table II. They exhibit a surprising independence of yields of *para*-product (63) from the polarity of the solvent. The not-too-convincing explanation that both Path A and Path B are equally participating in *para*-product formation is at hand. It is clear that more quantitative data are necessary to elucidate this problem. Experimentally found spin densities of photoexcited phenoxy radicals (Table III) would allow a preferential recombination of radicals in the *para*-position (cf. Eqs. 67 + 68 → 73).

B. Secondary Chemical Processes

The processes leading to product formation from radicals 65–68 (Path A) and from 70 (Path B) were added arbitrarily[32b] to the primary photochemical processes.

1. Dissociative Path A. Path A then proceeds by further chemical reactions. Aryloxy radicals are mesomeric systems (65 ↔ 67) in which it is evident from the magnitudes of esr coupling constants, as well as from knowledge of the chemical reactivity, that the unpaired electron must be largely associated with the π-electron system of the aromatic ring (Table III). In the absence of any rapid reactions following —O—CO— bond fission, recombination with or without rearrangement would be the expected fate of radical species formed in the primary process. This is the well characterized "cage recombination" and it is manifested chemically by a characteristic pattern of the products formed.

As suggested by Kobsa,[29] the radicals enclosed in a solvent cage recombine upon collision either to the original ester (**65** + **68** → **64**) or to 2-acyl-3,5-cyclohexadienone (**66** + **68** → **70**), followed by rapid isomerization to 2-hydroxyaryl ketone **71**. Analogously, the formation of 4-hydroxyaryl ketone **73** can be anticipated.

If radicals diffuse from the solvent cage, fragmentation products are formed. Abstraction of hydrogen from the solvent by a phenoxy radical results in phenol, which can almost always be observed among the photoproducts of aryl esters in solution. Chemical evidence for the reaction of phenoxy radical with solvent is the formation of nearly stoichiometric amounts of 4-methylphenol and acetone from the irradiation of 4-methylphenyl benzoate (**60**) in isopropyl alcohol.[34]

The fate of the free acyl radical **68** and radical **74** is not known. Most probably it is a constituent of polymer deposits on the wall of the irradiation vessel which hitherto have not been identified more definitely.[29] Moreover, the identification of methane and carbon monoxide among the gaseous products of the photolysis of 4-methylphenyl acetate (**55**) provides evidence for the existence of the acetyl fragment. This intermediate is expected to decarbonylate to give carbon monoxide and a methyl radical, which in turn abstracts hydrogen from the solvent.[34]

Direct experimental evidence relating to the existence of free radicals **65–67** and **68** among the photoproducts of aryl esters has not yet been obtained. On the photolysis of diaryl carbonates, no features attributable to trapped radical parts were seen.[49] In the esr spectra obtained at 77°K only a single weak line at $g = 2.005$ due to phenoxy radical was detected.[49] A much more intensive spectrum of the phenoxy radical was observed by γ-irradiation of diaryl carbonates. The observed difference in intensity of the spectra was rationalized by the contribution of phenoxy radicals arising by secondary dissociation[49] of aryloxy-carbonyl radicals ·COOC$_6$H$_5$. This radical from γ-irradiation is doubtlessly energetically more "hot" than the same radical produced by uv-irradiation, and consequently more apt to decarbonylation. Similarly, no optical or esr evidence for metastable transients has been found in the photolysis of solid polyaryl carbonate and its dilute glassy solution[50] at temperatures of −150 to −170°.

Let us now consider means of estimating the relative yields of recombination reactions **66 + 68** and **67 + 68** along with hydrogen transfer reaction from the solvent in the photoproduct formation. Experiments of Sandner and Trecker[33,34] concerning the influence of solvent viscosity on the rearrangement products of 4-methylphenyl acetate **(55)** yield indirect information on the participation of these recombination reactions (Table IV). All solvents used are assumed to be of related polarity, and consequently one can expect the same contribution of photophysical processes (*vide supra*) of both Path A and Path B. The decrease of the quantum yield of 4-methylphenol **(19)** caused by decreasing diffusion-ability from the solvent cage with an increasing viscosity is easy to understand. It was reported that in solid solvents, e.g., polyethylene, phenol was not formed.[30]

Which, then, are the reactions of a pair of radicals remaining in the solvent cage? The almost independent quantum yield of the formation of **56** from the solvent viscosity favors the possibility that this *ortho*-rearrangement product is formed mostly by other than a radical recombination mechanism, i.e., probably via the concerted Path B. However, one might assume that in the difference to the diffusion from the solvent cage (eventually in the difference to the *para*-product formation, i.e., **67 + 68**), the increasing viscosity is

TABLE IV
The Influence of the Viscosity of Solvents on the Quantum Yields of Photorearrangement Products. The Photolysis of 4-Methylphenyl Acetate (55)[33]

Solvent	Viscosity, cP, 30°C	Quantum yields of 2-hydroxy-5-methyl acetophenone (56)	Formation of 4-methyl-phenol (19)
Ethanol	1.00	0.17	0.45
iso-Propanol	1.73	0.17	0.09
tert-Butanol	3.00	0.18	0.07
tert-Butanol, O$_2$	3.00	0.17	0.04
Dioxane + H$_2$O	—	0.17	0.06
Carbowax 400	76	0.17	0.03
Carbowax 600	109	0.16	0.02

unable to influence the *ortho*-rearrangement. The recombination of the radicals in question, i.e., **66** + **68**, requires only a minimum amount of molecular movement in the solvent cage. Therefore, the observed independence of *ortho*-rearrangement product formation (Table IV) cannot be assumed as the only unambiguous proof for Path B.

Schutte and Havinga[51] compared the formation of 2-hydroxy-5-methoxy acetophenone (**75**) by photo-Fries rearrangement of 4-methoxyphenyl acetate (**76**) and of 4-methoxyphenyl acetate-carboxyl-^{14}C (**77**) in ethanol in order to evaluate the ^{14}C-isotope effect. The authors claimed that "from the C—O stretching frequency (1205 cm^{-1}) the maximum isotope effect for bond rupture in the ground state may be computed to be about 1.13. From calculations of the bond orders of the ground state and the lowest singlet excited state a considerable difference in C—O bond strength is not to be expected, and for the photodissociation a maximum isotope effect of 1.13 seems a reasonable guess." It was found, however, that within the experimental error of 2%, the ratio of the reaction rates of the photorearrangement **76** (k_{12}) and **77** (k_{14}) is unity (average $k_{12}/k_{14} = 1.007 \pm 0.018$). Except for the fact that dissociation of the C—O linkage probably does not take place from the lowest energetic level of the excited singlet state (see Section II.A), this ^{14}C-experiment afforded an independent proof that in the formation of the *ortho*-rearranged product, Path A and recombination **66** + **68** are either minimal or not operative. In the terminology of the transition state theory, the absence of an isotope effect might be interpreted as indicating that the dissociating bond is still largely intact in the transition state. Unfortunately, it has not been reported whether an isotope effect is operating in 4-methoxyphenol formation.

In both viscosity[33,34] and ^{14}C-isotopic[51] experiments, it has not been possible to monitor the formation of *para*-rearranged product because of structural reasons (substituted *para*-position). Such studies would be desired to support the assumption that in *para*-rearrangement both Path A and Path B participate equally (*vide supra*).

78 → hv → **79**

In the heretofore reported experiments, as in almost all photo-Fries rearrangements, no *meta*-rearranged products have been found. Traces of 3-methoxyacetophenone (yield <0.3%) were found upon the irradiation of phenylacetate and after methylation of the reaction mixture.[40] Finnegan and

Mattice[37] observed the photorearrangement of 3,4-benzocumarin (**78**) to 4-hydroxyfluorenone (**79**) with an exceptionally low yield of 0.08% of the initial **78**. Here, structural features determine the migration of the aroyl group to the *meta*-position with regard to the phenolic oxygen atom. For the eventual recombination, the *meta*-position of a phenoxy radical is extremely deactivated (Table III).

Lactones of ω-(2-hydroxyphenyl)-alkane carboxylic acids (**86**) react on uv-irradiation in a quite different way. A transient state of type **69** is not possible on steric grounds and, indeed, for compounds of this type solvolysis was shown to be characteristic[40,52-54] instead of photorearrangement. The solvolysis can be rationalized by the primary homolytic cleavage of the linkage —O—CO—. Plank[52] explained the incorporation of two deuterium atoms into the aliphatic chain of ω-(2-hydroxyphenyl)-propionic acid (isolated as lactone **85**) by uv-irradiation of 3,4-dihydrocumarine (**80**) in a dioxane–D_2O mixture assuming homolytic dissociation **80** → **81**, followed by ketene formation (**82**), deuterolysis to **83**, and relactonization to **84**. In the parallel dark experiment no deuterated **85** was formed.

On irradiation of **86** ($n = 2$ and 3) in alcoholic solutions, methyl- (86%), ethyl- (74%), and isopropyl- (52%) esters of the corresponding ω-(2-hydroxyphenyl)-alkane carboxylic acids are formed.[40,54] *tert*-Butanol, however, fails to yield any alcoholysis product whatsoever. A considerable influence of ring substituents on the quantum yield of ethanolysis of the lactones **86** and the failure in finding any carbon-bound deuterium in the ester when the photolysis was carried out in CH_3OD (contradicting Plank's results,[52] *vide supra*) led to the postulation of specific reaction pathways for the photoinduced alcoholysis.[54] It was suggested that a charge-separated structure of the excited singlet **87** is formed, which then undergoes rapid rebonding and internal conversion to (5.3)- and (5.4)-spirodiketone **88**, respectively. The subsequent rapid alcoholysis would yield the corresponding esters. The assumption is that **88** is formed (attempts to isolate **88** have been unsuccessful) either via **87 → 88** or by direct recombination of acyl radical with an *ortho*-alkyl substituted *ortho*-phenoxy radical of type **66**. According to the spin densities (Table III), radicals of these types are equally suited to undergo radical recombination as the unsubstituted **66**. Nevertheless, 2-alkyl-2-acylcyclohexa-2,4-dienones of the type **70** (2-H replaced by alkyl) have never been isolated from a photo-Fries reaction mixture. It might be due to the already mentioned general improbability of a radical *ortho*-recombination **66 + 68**, to steric hindrance in Path B or to the photochemical reactivity of this type of compound.[55]

Migration or elimination of alkyl groups from the phenolic moiety, so characteristic for the ionic Fries rearrangement,[3,56] has not been encountered in the photochemical transformations. However, it has been established that in contrast to the ionic rearrangement, the photo-Fries rearrangement proceeds in good yield also to an *ortho*-position occupied by chlorine (cf. **89 → 90**).[29] According to Kobsa's view, radical addition of a benzoyl radical in the *ortho*-position takes place in this reaction, affording the intermediate **91** in which chlorine is exchanged for a proton from the solvent (e.g., from traces of water). Other authors are, however, of the opinion that chlorine elimination from the *ortho*-position of the diene intermediate **91** might be preceded by photoelimination of halogen from the phenol nucleus, as is often described with chlorophenols[57] and other aromatic halogenides.[22,58] The formation of 4-hydroxybenzophenone (**63**) (with a quantum yield of 0.007), in addition to other products on photolysis of 4-chlorophenyl benzoate (**92**) in ethanol, is explained by this reaction pathway.[59]

The methoxy group can also be eliminated from both *ortho*- and *para*-positions of a phenol nucleus in photo-Fries rearrangement.[60] Although in this case the concerted process giving rise to an intermediate of type **91** was suggested, the actual way of elimination and the fate of the departing methoxy group are not known. The possibility was considered that the methoxy group is a source of formaldehyde which contributes to the polymers always observed in these reactions.

It has been mentioned that phenol is formed via Path A, by diffusion of radicals from the solvent cage and hydrogen abstraction from the solvent. This process is undoubtedly favored (and the yield of phenol is increased) when the phenoxy radical **65** already loses its counterpart in the solvent cage, i.e., when it loses the acyl radical **68** as a consequence of its decarbonylation. From the hitherto reported results it can be assumed that decarbonylation is significant and proceeds very readily under two conditions. It occurs (1) if the acyl radical formed possesses excess energy ("hot" radical) due to excitation of high energy, e.g., by γ-radiolysis,[41,46] and (2) if the alkyl or aryl radicals formed by the decarbonylation of the acyl radical are exceptionally stable.[61]

An unambiguous proof that decarbonylation of the acyl radical takes place at least in part in the solvent cage has been obtained by the photolysis of the 3,5-di-*t*-butylphenyl ester of (*S*)-(+)-2-methylbutanoic acid (**125**) in dioxane.[62] 3,5-Di-*t*-butylphenyl-*sec*-butyl ether (**127**), isolated in 3% yield, was shown to be completely racemic. This means that the *sec*-butyl radical had to exist as an independent, free-rotating radical intermediate in the solvent cage. The relatively high yield of ether, the assumed lifetime of radicals **65–68**, and radical concentration at the usual solution concentrations (10^{-3} to 10^{-1}

mole liter^{-1}) along with the irradiation intensity (10^{14}–10^{18} quanta liter^{-1} sec^{-1} at 253.7 nm) firmly exclude diffuse recombination outside the solvent cage. Consistent with this assumption is the previously conducted cross-experiment pointing out the intramolecular character of the photo-Fries rearrangement.[37]

The aryl esters of formic and oxalic acid (exclusively) undergo very efficient decarbonylation to form phenol.[5] It is noteworthy that aryl esters of formic acid, when irradiated in the presence of olefins, are added to the latter in the same way as phenols.[63]

2. Concerted Path B. The observation that the quantum yields of the *ortho*-rearrangement are independent of the solvent viscosity (Table IV),[33,34] the completely intramolecular nature of the rearrangement,[37] and the dependence of the *ortho*-product yields on the substituents at the phenol nucleus or the acyl moiety[29,30,59] (Fig. 1, Table V) led several authors to conclude that the photo-Fries rearrangement to the *ortho*-position is a concerted [1,3]-shift of an acyl group. This assumption was substantiated by finding that *ortho*-product and phenol formation is sensitized differently by excited benzene[41] and depends on the polarity of solvent[39,64] (Table II) but not on its hydrogen abstracting ability.[39]

Following Path B, the crucial step is a transient state **69**, representing a very tightly bound intermediate which proceeds to a product unaffected by its reaction environment.[33] Anderson and Reese[40] assumed these transient states to be bridged biradicals, **93** and **94**, with a neutral charge (**93** for *ortho*- and **94** for *para*-rearrangement, respectively). The biradical intermediate could then collapse to dienones **70** and **72** which would enolize to the corresponding aromatic hydroxyketones **71** and **73**. Coppinger and Bell[30] suggested a photoactivated charge-transfer complex for this intermediate. Lacking the means to distinguish between these possibilities, Sandner and Trecker[33] represented the 4-ring intermediate or transition state as **69**. In an analogous manner Bradshaw et al.[60] described the 6-ring transient state for the *para*-rearrangement as **95**. Inspection of the molecular models shows that, provided aromaticity of the nucleus is destroyed, 6-ring transient states of type **94** or **95** are possible. However, it is rather difficult to imagine formation of **98** (arising in 48% yield[8] on photolysis of **96** and proven to arise on γ-radiolysis of **96**[41]) via an intermediate bridged structure between the *amphi*-positions of naphthalene. An 8-ring transient state cannot be constructed from the molecular models.

It seems attractive (and this author feels this is a correct approach) to attempt to explain concerted rearrangements via Path B as photochemical [1,*j*]-sigmatropic shifts on the basis of Woodward-Hoffman rules.[65,66] Even the first try to interpret photo-Fries rearrangement in such a manner brings some stimulating ideas.[34] According to the general Woodward-Hoffman rules it can be deduced that photochemical rearrangement of this type can be predicted on the basis of the symmetry of the unoccupied molecular orbitals of a phenoxy radical.

Diagrams AS, 1SA, and 2SA (S = MO symmetric with respect to the plane, A = MO asymmetric with respect to the C_2 axis) represent the symmetries of three lowest unoccupied molecular orbitals of a phenoxy radical. The order of increasing energy AS < 1SA < 2SA is chosen according to an extended Hückel calculation. Consideration of the symmetry properties of these MO's led Sandner, Hedaya, and Trecker[34] to the following conclusions:

"Sigmatropic, suprafacial migration is allowed to both *ortho* and *meta* positions in the first excited state, since the AS molecular orbital has a node along the C_2 axis. However, we can exclude rearrangement to the *meta* position since it would result in an unstable biradical product. Furthermore, antarafacial rearrangement obviously is sterically excluded. The symmetry of 1SA, which corresponds to the HOMO of the second excited state, also allows suprafacial migration to *ortho* and *meta* positions, and the latter is again excluded. Rearrangement to both *ortho* and *para* positions is allowed for the third excited state involving the 2SA molecular orbital. One clear value of this formulation is that it forms the basis for further predictions. First, the *ortho/para* ratio should be wavelength dependent. Second, the rearrangement should occur generally for phenols and simple phenyl ethers.... *We propose on this basis that the photo-Fries rearrangement occurs by a concerted, symmetry-allowed process.*"

This theoretical conclusion that *ortho-rearrangements* might occur from more than one transient state had already been predicted from the experimental results of benzene sensitized Fries rearrangement.[41] It should be noted, however, that both the second and the third excited states (1SA and 2SA) do not fulfill the steric requirements for either antarafacial or suprafacial [1,5]-shifts (cf. Refs. 66 and 67). An alternative possibility for *para*-rearrangement could be two subsequent [1,3]-suprafacial shifts in the third excited singlet state (2SA). The alternative of two photochemical shifts per one absorbed photon, though on a different model, has been ruled out.[68] Similarly, formal [1,7]-sigmatropic shift **96** → **98** would require sterically impossible transient state. It is thus evident that the interpretation of photo-Fries rearrangement as a concerted symmetry allowed process calls for more theoretical studies based on appropriately designed experiments.

$$X-\text{C}_6\text{H}_4-O-\overset{O}{\underset{\|}{C}}-\text{C}_6\text{H}_4-Y$$

100

The results of the viscosity experiments (Table IV) indicate that the phenolic and the acyl moiety of the excited aryl ester molecule remain associated at all stages of the rearrangement, which is not due to a cage effect but rather to attractive forces between them. There are some data at hand which are based on the photorearrangement of aryl esters of the type **100** with different electron-donating and electron-withdrawing groups X and Y (Table V). Satisfactory correlation of the data can be made if one considers the general substituent effect of the electron distribution of **100** in the ground state. From the transient state representation **69** it can be assumed that its formation will be positively influenced by factors increasing the electron density in the phenol nucleus and pulling electrons away from the carboxyl carbon. This leads to the concept of "electrophilic photorearrangements," similar to the acid-catalyzed Fries reaction, where the requirements for the rearrangement are that the carboxy carbon has to be electron deficient and the *ortho*-position of the phenol ring has to be electron rich. Undoubtedly, the electron density of the photoexcited state should be known for a full understanding of the photorearrangement. This information is not yet available.

The results in Table V support this ground state electron distribution effect. The data in italics show that the relative quantum yields are CN < Cl < CH$_3$ with respect to the electron-donating substituent X, and CN > Cl > CH$_3$ with respect to electron-withdrawing substituent Y. Unfortunately, the different quantities in which the quantum yields are expressed and the different reaction media do not allow a direct comparison of the results. Kobsa's data[29] on *ortho*-hydroxyphenyl ketone formation in the first column can be correlated with Hammett's σ constants (Eq. 1, the fraction is proportional to the quantum yield).

$$\log \frac{\alpha}{1-\alpha} = 0.105 + 0.391\sigma \qquad (1)$$

The values presented in the second column show the quantum yields of the aryl ester conversion and can be correlated with other values in Table V, only because in ethanol at 20°, 70–80% of the ester is converted to *ortho*-rearranged product.[59] It was shown that the given data for *para*-substituted esters could be also correlated with Hammett's σ constants (Eq. 2).

$$\log \phi_{-\text{ester}} = -0.657 - 0.347\sigma^- \qquad (2)$$

The best correlations were obtained with the values of σ^- constants generally applied to correlating chemical reactions which involve the phenolic oxygen.[69] It is typical of the adverse influence of substituents that Eqs. (1) and (2) are significantly different in ρ values. The two nitro derivatives in both do not lie on the correlation lines. It was found that nitro-substituted aryl esters preferentially undergo alcoholysis and reduction.[70]

According to the reaction medium the results in the third column of the Table V should be approached in a different way. A rigid polyethylene film suppresses phenol formation almost entirely by keeping the radicals in the solvent cage. Furthermore, all substituents decrease the quantum yield relative to hydrogen. This effect cannot be explained solely by a Hammett-type treatment of electronic effects. The effect of the substituent size is particularly pronounced in the case of the *p*-isooctyl substituted ester whose quantum yield is reduced drastically when compared to that of the *p*-methyl substituted

TABLE V
Quantum Yields of the Photorearrangement of Type 100 Aryl Esters

Substituent	—[a]	mole einstein^{-1} [b]	mole cm^{-2} sec^{-1} × 10^8 [c]
p-H	1.18; 158.8[d]	0.239	213.3
p-CH$_3$	*1.55*	0.245; 0.261[e]	*150.8*
p-C(CH$_3$)$_3$	46.6[d]	—	—
p-OH	—	0.292	—
p-NH$_2$	0.57	—	—
p-OCH$_3$	—	0.223	108.6
p-Cl	2.08; 61.1[d]	0.199	*109.6*
p-i-C$_8$H$_{17}$	—	—	63.7
p-C$_6$H$_5$	—	—	75.0
p-CN	2.40; 135.6[d]	—	67.4
p-COCH$_3$	—	0.108	—
p-CHO	—	—	32.0
p-CO$_2$C$_2$H$_5$	—	—	22.0
p-NO$_2$	0.44	0.004	15.1
m-CH$_3$	—	0.319	125.0
m-OH	—	0.223	—
m-OCH$_3$	—	0.242	52.7
m-Cl	—	0.280	81.7
m-CF$_3$	—	—	42.7
m-F	—	—	114.2
m-CN	—	—	43.7
m-J	—	—	52.9
m-NO$_2$	—	0.040	9.82
m,m-(NO$_2$)$_2$	0.1	—	—

[a] Quantity not given, but from the context the given values are proportional to the quantum yield of *ortho*-rearrangement[29]; 0.05 M solutions, benzene, 300 W mercury arc, 245–330 nm, 30°.

[b] Values given for aryl ester decrease,[59] 0.002 M, ethanol, 20 W low-pressure mercury arc, 253.7 nm, 20°.

[c] Quantity proportional to the quantum yield of *ortho*-product formation,[30] molded polyethylene films, quartz, medium-pressure mercury arc.

[d] Ref. 30.   [e] Ref. 71.

ester. In order to rationalize the decrease of quantum yields by alkyl substituents as well as by electron-withdrawing substituents, Coppinger and Bell[30] corroborated some very complicated approaches. These are based on the combination of the Hammett-type treatment of electronic effects with the "quasiequilibrium" model, considering the number of internal degrees of freedom in the polyatomic molecule. The correlation of the best fitting equation with the experimental results is shown in Figure 1.

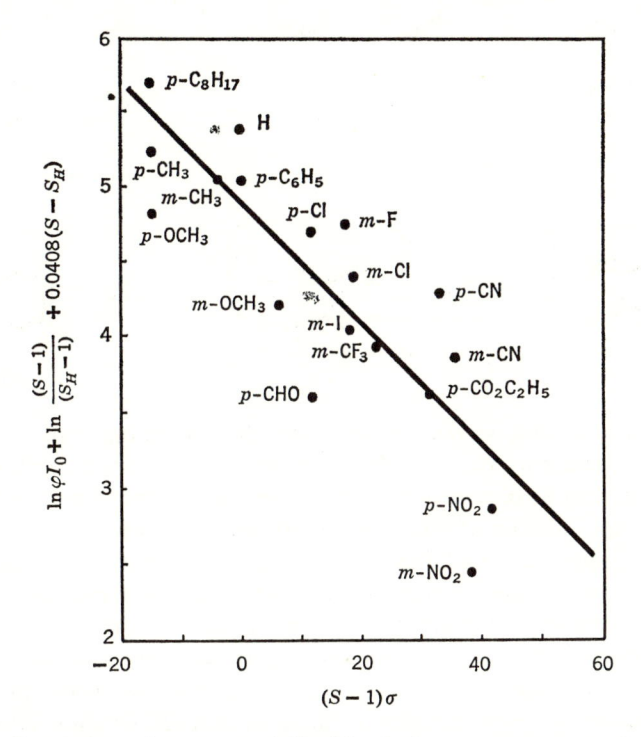

Fig. 1. Correlation of quantum yields (φ) of photorearrangement of substituted phenyl esters of 4-hydroxy-3,5-di-*t*-butylbenzoates with both electronic effect (σ) and internal degree of freedom (S) in aryl ester molecule.

The observation that the yield of *ortho*-rearranged product depends on the bulk of *para*-substituent is unique. It would be interesting to find out whether this also holds true with photorearrangement in a low viscosity solvent, e.g., in an experiment where only the substituent size would be changed preserving the electronic effect of the substituents. One can argue that the presented photorearrangement in polyethylene film is almost a photoreaction in the solid state. Such reactions occur with the minimum amount of atomic or molecular movement.[72] The formation of transient states of type **69** undoubtedly requires certain molecular movement. The bulkier the *para*-substituent is, the more hindered the formation of the transient state **69** in the polyethylene matrix could be on account of deactivation by collisional energy transfer to the matrix.

<u>101</u> R = OC_6H_5
<u>104</u> R = NHC_6H_5

<u>102</u> R = OH
<u>105</u> R = NH_2

<u>103</u> R = OH
<u>106</u> R = NH_2

<u>107</u>   <u>108</u>   <u>109</u>

A special case of a decrease in aryl ester photorearrangement as a consequence of decreasing electron deficiency on the carboxylic carbon in the excited state can be observed in the photorearrangement of phenyl salicylate (**101**) and salicyl anilide (**104**).[73] When **101** was irradiated in methanol, the expected products **102** and **103** were formed in 28 and 32% yields, respectively. The percentage conversion was 68% based on recovered starting material. However, when the irradiation was carried out in *n*-hexane, the yields of **102** and **103** dropped to 4 and 1%, respectively, and the percentage conversion decreased to 9%. This solvent effect is due to the presence of intramolecular hydrogen bonding in these substances. The low efficiency of the photorearrangement in a nonpolar solvent is explained by rapid decay of the excited state in the form of a tautomeric shift via hydrogen bonding involving a six-membered ring (cf. **107** → **109**), as is the case with 2-hydroxybenzophenones.[43b] On the other hand, in a polar solvent which itself is capable of hydrogen bonding, the intramolecular hydrogen bond is broken to a great extent,[74] thereby preventing the tautomerization and favoring rearrangement.

Similarly, it was found that the rate of photorearrangement **14** → **16** in dichloroethane is approximately 2.5 times lower as that of **13** → **14** under the same conditions.[6] This decrease could also be attributed to a tautomerism according to equation **107** → **109** participating in the primary photochemical processes with **14**. In this connection it is noteworthy that in contrast to all other aryl esters, fluorescence of an ethanolic solution was only found for phenyl salicylate. As mentioned above, neither fluorescence nor phosphorescence for other aryl esters has been reported.

The cyclization of 2-biphenylyl carbonate **110** can probably serve as an analogy to the [1,5]-shift of the acyl group.[76] The formation of 6-ring transient state **111** is preferred for steric reasons. Following Path B, an analogous but much more strained 4- or 6-ring transient state collapses to the enone, with a simultaneous cleavage of the bond between phenolic oxygen and carboxyl carbon. In **111** this step is apparently substituted by bond fission between the carboxylic carbon and the alkyl bearing oxygen. This step is very efficient, and dibenz-α-pyrone (**112**) is formed in 85% yield. Proceeding via strained 4- and 6-ring transient states, ethyl phenyl carbonate (**113**) affords[77] "normal" rearrangement products **114**, **115**, and phenol (**23**).

Photo-Fries rearrangement of aryl esters or its participation in addition to other photochemical reactions was reported in numerous papers concerning low-molecular[78] or macromolecular[79,83] compounds.

In a single case acyl migration on the heteroaromatic nucleus is described,[30] cf. **116** → **117**. 3-Aroyloxy-substituted pyridine rearranges with approximately half the quantum yield of **116**; 2-substituted pyridines do not rearrange at all.

C. Photodecarboxylation: A Competitive Side Reaction

Ortho- and para-rearrangement and phenol formation on uv-irradiation of aryl esters are accompanied in several cases by decarboxylation,[37,60,62,64,80,81] represented for 3,5-di-t-butylphenyl benzoate by the equation 118 → 119–122. It was shown that this reaction cannot be sensitized,[64] but the dramatic differences in product distribution could be observed by changing of the solvent.[60,64] The results in Table VI indicate that in polar solvents the decarboxylation process is minimized while the formation of the photo-Fries rearrangement 119 is enhanced. The reverse appears to be true when nonpolar ethereal solvents are used. A considerable amount of biaryls are formed, and hence this reaction may prove useful for the preparation of biaryls and alkylaryls.

Finnegan and Knutson[62,80] suggest that photoexpulsion of carbon dioxide proceeds in a concerted fashion via a transition state schematically illustrated by 123. Such a mechanism requires that the configuration of the α-carbon in the R group be retained during the transformation. This is indeed the case when 3,5-di-t-butylphenyl ester of (S)-(+)-2-methylbutanoic acid (125) is irradiated in dioxane. From the retained configuration of (S)-(+)-2-(3,5-di-t-butylphenyl)-butane (126) it follows that the photochemical loss of carbon dioxide from 125 occurs with a predominant retention of configuration, and therefore the proposal of a concerted mechanism for this reaction is strongly supported.

The reversed dependence of this concerted decarboxylation reaction, as well as the concerted ortho-rearrangement on the polarity of solvents (cf. Tables II and VI), and the competition of decarboxylation to phenol formation in nonpolar solvents, are striking. Lack of information on the influence of substituents, viscosity, sensitization, or possible quenching in direct comparison with the formation of rearranged products and phenol do not yet

TABLE VI
Irradiation of 118:[a] Solvent Effect on the Product Distribution[64]

Solvent	118	119	120	121 + 122 (ratio 121/122)
Ethanol	22	30	48	—
2-Propanol	29	34	37	—
Dimethylformamide	35	51	9	—
n-Hexane	10	55	8	15 (0.5)
Tetrahydrofuran	6	53	12	28 (0.65)
Ethyleneglycol dimethyl ether	2	41	5	51 (0.70)
Dioxane	9	42	2	48 (0.60)
Diethyl ether	—	56	—	44 (0.52)
Diethyl ether–ethanol (9:1)	8	43	34	15 (0.67)

[a] 0.021M Solutions of 118, 450 W medium-pressure mercury arc, quartz reactor, 3hr of irradiation.

allow us reasonable speculations on the nature of the excited state responsible for this interesting photoreaction of aryl esters.

Compound **122** represents the product of a transposition of the substituents on the benzene ring. Such products are quite characteristic for this photodecarboxylation reaction. Their origin could be explained *a priori* by the well-known photoisomerization of substituted benzenes[32c] of type **121** and **126**, which are formed primarily. Alternatively, it has been suggested[60] that the phototransposition reaction occurs during the course of the decarboxylation. This supposition is based on two facts. First, no photo-Fries rearranged products were observed in which the ring substituents had "migrated." This indicates that the transposition did not take place in the starting esters before they reacted. Second, the ratio of decarboxylated products 2,5-dimethylanisole (**129**) to 2,6-dimethylanisole (**130**) is constant during irradiation of 2-methoxy-4-methyl acetate (**128**). This shows that the phototransposition **129** → **130** did not take place after the formation of **129**. Therefore it is assumed that both **129** and **130** are originating from a transient state with the benzvalene structure **124** which exists along with the "normal" transient state **123**.

The comparison of the published structures of the irradiated aryl esters with the yields of photodecarboxylation products points out that increasing size and probably also the increasing number of alkyl substituents on a phenolic nucleus positively influence the yields.

D. Evaluation of Quantum Yields

From the point of view of quantum yields calculation, photo-Fries rearrangement, including phenol and products formation, represents a photoreaction (expressed by Eq. 3) in which all products absorb intensively in the absorption region of the starting phenyl ester A.

$$A \xrightarrow{h\nu} B_1 + B_2 + B_3 + \cdots + B_n \tag{3}$$

The exact calculation of quantum yields is possible only (1) if products B_1 to B_n are photochemically very stable when compared with A, and (2) if the products do not exhibit any influence upon the excited state of compound A. Both these conditions are satisfactorily fulfilled in the photo-Fries rearrangement, particularly for small conversions of the ester (up to 20% conversion of A).

Outlined below is the exact method of calculation of quantum yields of the disappearance of compound A and the formation of the compounds B_1 to B_n, even if only their analytical concentrations at a few time intervals are known and their molar extinction coefficients are not known (for more details see Ref. 71).

The rate of disappearance of compound A in a photochemical reaction (3) is given by Eq. (4)

$$-dc_A/dt = \varphi_A I_A \tag{4}$$

where c_A is the concentration of compound A (mole liter^{-1}) at the time t, I_A is the number of quanta of monochromatic light absorbed by compound A per unit time interval and per unit of volume (einstein liter^{-1} sec^{-1}), and φ_A is the quantum yield of disappearance of compound A (mole einstein^{-1}), defined as φ_A = number of reacted molecules of A per number of quanta absorbed by A.

The total amount of radiation absorbed in a system with $n + 1$ absorbing components (cf. Eq. 3) being I_0, only fraction I_A is absorbed by compound A (cf. Eq. 4). For the rate of disappearance of the compound A we obtain therefore

$$-\frac{dc_A}{dt} = \frac{\varphi_A I_0 \epsilon_A c_A}{\epsilon_A c_A + \sum_{i=1}^{n} \epsilon_{B_i} c_{B_i}} [1 - 10^{-(\epsilon_A c_A + \sum_{i=1}^{n} \epsilon_{B_i} c_{B_i})d}] \quad (5)$$

where $\epsilon_A, \epsilon_{B_1}, \epsilon_{B_2}, \ldots, \epsilon_{B_n}$ are the corresponding molar extinction coefficients (liter mole^{-1} cm^{-1}) and d is the pathlength of the irradiation cell (cm).

The differential Eq. (5) can be solved provided that (1) the absorbing products $B_1, B_2, \ldots, B_n$ are photostable, and hence Eq. (6) holds at any moment of irradiation

$$c_{A_0} - c_A = c_{B_1} + c_{B_2} + \cdots + c_{B_n} \quad (6)$$

and (2) that the ratios

$$c_{B_i}/(c_{A_0} - c_A) = k_i \quad (7)$$

are constant for all i's in a competitive photoreaction according to Eq. (3). Furthermore, a nonabsorbing solvent† and complete absorption of the incident irradiation by the reaction mixture‡ have to be assumed if Eq. (5) is to be solved. If the exponential member on the right side of the differential Eq. (5) were not neglected (i.e., for the case of partial absorption only), an analytically unsolvable integral would be obtained (computer calculation can be employed here; see Appendix in Ref. 82 and considerations in Ref. 30).

After integration of Eq. (5) and some adjustments, *the equation for the calculation of quantum yield of disappearance of compound A possesses the form*

$$\varphi_A = \frac{2.3 c_{A_0} \alpha}{2.3 c_{A_0} - \beta} \quad (8)$$

Constant α is the intersection point with the axis of ordinates and β is the slope of a straight-line equation:

$$\frac{c_{A_0} - c_A}{I_0 t} = \alpha - [\beta \ln (c_{A_0}/c_A)]/I_0 t \quad (9)$$

---

† If this condition is not fulfilled, then a constant must be added in the denominator of Eq. 10, which does not alter the consecutive steps.
‡ I.e., $10^{-(\epsilon_A c_A + \sum_{i=1}^{n} \epsilon_{B_i} c_{B_i})d} \ll 1$.

From Eqs. (8) and (9) it is evident that for the calculation of φ_A it is necessary to know only I_0 and the immediate concentrations of compound A at some intervals t of the irradiation, which themselves are necessary for the calculation of α and β.

The quantum yield of formation of the ith compound B is

$$\varphi_{B_i} = c_{B_i} \Big/ \int_0^t I_A \, dt \qquad (10)$$

By comparing Eqs. (4) (in its integral form), (7), and (10) we obtain a general equation for the calculation of the quantum yield of formation of the ith product B in a competitive photochemical reaction

$$\varphi_{B_i} = k_i \varphi_A \qquad (11)$$

It is necessary to know the analytical concentrations of compounds B_i and A in order to determine k_i and, consequently, also φ_{B_i}.

This method of quantum yields calculation was verified in more detail in studying the photorearrangement of 4-methylphenyl benzoate (**60**) (in ethanol with 253.7 nm).[71] The quantum yield of ester decrease, $\varphi_A = 0.231$, and 2-hydroxy-5-methylbenzophenone (**62**), $\varphi_{B, ortho} = 0.19$, and 4-methylphenol (**19**) formation, $\varphi_{B, phenol} = 0.07$, were determined. Additional examples are calculations of the photorearrangements of poly-4-benzoyloxystyrene (in dioxane with 253.7 nm) ($\varphi_A = 0.082$, $\varphi_{B, ortho} = 0.055$) and poly-4-methylphenyl acrylate ($\varphi_A = 0.208$, $\varphi_{B, ortho} = 0.102$).[83]

The expression on the left side of Eq. (9) represents the so-called apparent quantum yield φ_A' of the disappearance of compound A. The quantum yield of disappearance of A can be approximately determined also by graphic extrapolation of the dependence φ_A' vs. $I_0 t$, because for $I_0 t = 0$, φ_A' equals φ_A. However, the higher the ratio $\sum_{i=1}^{n} k_i \epsilon_{B_i} / \epsilon_A$, the steeper the beginning of the curve and, consequently, the less accurate is the extrapolation for zero radiation dose.

For the initial phase of the photo-Fries rearrangement, I_A can be determined approximately by subtracting the absorption due to the strongest absorbing product of the reaction mixture, provided its concentration and molar extinction coefficient are known. By means of this "initial rate approach" Humphrey[6] determined the quantum yields of the photorearrangement of **13** ($\varphi_{13 \to 14} = 0.14$), of **14** ($\varphi_{14 \to 16} = 0.056$), and of poly-2,2-propanebis(4-phenyl carbonate) ($\varphi_{over-all} = 0.1$–0.14) in dichlorethane at 270.5 nm.†

---

† Private communication from Dr. J. S. Humphrey. To obtain the data of quantum yields, the values of quantum efficiency from Ref. 6 must be multiplied by a factor of 3.3.

E. Photostability of Photorearrangement Products

In general the products of *ortho*-rearrangement, 2-hydroxyphenyl ketones, are very stable under the conditions used for the rearrangement. In a few experiments a slight decrease in the course of prolonged irradiation could be demonstrated.[64,71] It is well known that 2-hydroxyphenyl ketone phosphoresces very weakly and does not undergo photoreduction[84,85] or photoelimination,[32d,86] presumably because of rapid enolization in the excited state.[74]

The photochemical behavior of *para*-rearrangement products depends on the polarity of the solvent. Porter and Suppan[84] reported that the photochemical disappearance of 4-hydroxybenzophenone is very slow ($\varphi = 0.02$) in isopropanol, while it shows high reactivity ($\varphi = 0.9$) in cyclohexane leading to the corresponding pinacol. They suggested that the lowest excited state of 4-hydroxybenzophenone may be a charge-transfer triplet in isopropanol and an n,π^*-triplet in cyclohexane. Therefore, the only experiment where the yield of *para*-rearranged photoproduct in different polar solvents was determined quantitatively (Table II), is not unequivocal and allows several interpretations.

Very little information is available concerning the photochemical behavior of phenols in common organic solvents. On prolonged irradiation during photo-Fries rearrangement in ethanol, their relative amounts are decreased slightly.[71] Though oxygen apparently does not influence the course of the photorearrangement, the simultaneous presence of moisture and oxygen in the medium may cause effective photooxidation of phenols on irradiation[87] at 253.7 nm.

The possible subsequent phototransformations of diaryls or alkylaryls (which are products of aryl ester photodecarboxylations) were discussed in Section II.C.

F. Conclusion

The overwhelming majority of the results obtained studying photo-Fries rearrangements can be rationalized by the formulation of two parallel primary photophysical processes, starting from two energetically different excited singlet states of the aryl ester molecule. Path A, presumably from the energetically higher level, leads to dissociation of the ester —O—CO— bond, giving rise to a pair of radicals, viz., acyl and phenoxy radicals (**64 → 65-68**). Path B, probably from the energetically lower level, seems to operate by a concerted [1,3]- or [1,5]-shift of an acyl group (**64 → 70**).

Ortho-Rearranged Products. They seem to be formed for the major part or exclusively via concerted Path B with subsequent aromatization (**70 → 71**).

Their yields are (1) higher in polar and lower in nonpolar solvents; (2) increased by the presence of electron donating substituents in the phenolic nucleus and electron withdrawing substituents on the acyl moiety; and (3) independent of the solvent viscosity. Their formation is possible even if the *ortho*-position of the original aryl ester is substituted with $-I, +M$ substituents (e.g., chlorine, methoxy). They are extraordinarily stable to further photochemical reactions.

Para-*Rearranged Products.* Very little quantitative data concerning their formation has been published. They might be formed by both the dissociative Path A or the concerted Path B. Yields are apparently independent of the solvent polarity. As with the *ortho*-products, their formation is also possible when the *para*-position of the original aryl ester is substituted with $-I, +M$ substituents (chlorine, methoxy). Photochemically they are very stable in polar solvents, whereas in nonpolar ones they can photopinacolize with a considerable quantum yield.

Phenols. Presumably they arise exclusively via dissociative Path A, subsequent radical diffusion from the solvent cage, and abstraction of a hydrogen from the solvent (**65 → 74**). The yields are (1) increased with decreasing viscosity of the reaction medium; (2) higher in nonpolar and lower in polar solvents; (3) practically independent of the hydrogen-donating ability of the solvent; and (4) increased if a radical counterpart of a phenoxy radical, i.e., an acyl radical, decarbonylates in the solvent cage for structural reasons.

Depending on the structure of the original aryl ester, the following compounds can sometimes form in the photoreaction mixtures: *ethers* (after acyl radical decarbonylation and recombination of the radicals remaining in the solvent cage), *reesterified esters* (in alcohol solutions), *aldehydes* (resulting from the abstraction of a hydrogen from the solvent by acyl radical), and *alkylaryls or diaryls* (after photodecarboxylation of the aryl ester). The latter reaction is a concerted reaction with unknown mechanism and the yields are increased (1) by the substitution of the phenol nucleus of the aryl ester with bulky alkyl substituents and (2) by nonpolar solvents.

III. PHOTOREARRANGEMENT OF ARYL ESTERS OF SULFONIC ACIDS

Aryl sulfonates RSO_3Ar (R = C_6H_5, CH_3) were found to undergo photochemical rearrangement upon irradiation with uv-light.[8,9] In the case of phenyl *p*-toluenesulfonate (**20**), products originating from the photo-Fries reaction were identified as 2-hydroxy-4′-methyldiphenyl sulfone (**21**), 4-hydroxy-4′-methyldiphenyl sulfone (**22**), and phenol (**23**).

A heteroaromatic nucleus can serve also as an acceptor for the migrating sulfonyl group. Thus, irradiation of numerous 4-pyrimidinyl esters of alkyl and aryl sulfonic acids (**24**) in alcohol with low-pressure mercury arcs readily gave the corresponding 5-alkylsulfonyl and 5-arylsulfonyl-4-hydroxypyrimidines **25**, respectively, in yields of up to 60%. A number of minor by-products included the parent hydroxypyrimidines **26** (5–30%). In this class of compounds, all attempts to bring about the Fries rearrangement **24 → 25** by conventional procedures, e.g., by heating the 4-pyrimidinyl esters **24** with aluminum chloride alone or in solvents, caused decomposition only.[10]

The primary excitation responsible for this [1,3]-shift of the sulfonyl group is probably π,π^*-excitation of a phenolic part of the sulfone ester molecule. No mechanistic interpretations of this reaction have been published.

IV. PHOTOREARRANGEMENT OF N-ARYL AMIDES OF CARBOXYLIC ACIDS

Not only can the acyl group be exchanged for sulfonyl species without sacrifice of the basic nature of these rearrangements, as seen in the previous section, but the linking heteroatom, oxygen, can also be replaced. *N*-Aryl amides of carboxylic acids are known to cleave and rearrange by action of uv-light[7,11–14,73,88,89] or γ-rays.[41] Unlike phenyl esters, this "photoanilide" rearrangement has not been well studied so far. One can, nonetheless, assume that by analogy with aryl esters both dissociative Path A (**64 → 68**) and concerted Path B (**64 → 70**) are operative. In general, the reaction mixture is much more complicated after the photolysis of *N*-aryl amides of carboxylic acids. More by-products are formed than with aryl esters. This phenomenon can be demonstrated by the photolysis of benzanilide (**128a**). In addition to the "normal" products of *ortho*- and *para*-rearrangement **129a** and **130a** (formed in the approximate ratio 1:1) and aniline **30**, there were also found benzoic acid,[12,89] benzamide[12] and phenanthridone (**131**).[90] **131** is the major product when the irradiation is conducted in benzene in the presence of iodine.[90] **131** also arises upon irradiation of benzene solutions of 2-iodobenzanilide (**132**) (9%) and *N*-benzoyl-2-iodoaniline (**133**) (48% yield).[90] In these experiments, however, the main products of photolysis were 2-phenylbenzanilide (15%) and *N*-benzoyl-2-phenylaniline (59%), respectively, i.e., compounds formed apparently on interaction of the solvent with **132** and **133** (following the homolysis of an aryl C—I bond). No rearranged products have been reported in this case.

128 a R=C$_6$H$_5$
b R=CH$_3$

129   **130**   **30**   **131**

Formation of **131** in a high yield (85%) on the photolysis of ethyl 2-biphenylyl carbamate (**134**) is not surprising on the basis of the discussion for the conversion **110** → **112**. Other carbamates **27**,[7,11] rearranging via sterically strained 4- and 6-ring transient states, afford "normal" rearranged products **28–30**. Here also, as was the case with the irradiation of aryl esters, higher yields of *ortho* and *para* products were obtained in polar solvents (ethanol, *t*-butanol) than in the nonpolar *n*-hexane.[11]

Shizuka and Tanaka studied the rearrangement of acetanilide (**128b**) in more detail.[91] Their results are closely related to the above-mentioned observations on aryl ester photorearrangements. The quantum yields of 2-aminoacetophenone (**129b**) and 4-aminoacetophenone (**130b**) formation in cyclohexane solution at 20° were 0.07 and 0.06, respectively. The quantum yields did not change with variation in irradiation time (i.e., the photorearrangement products are photostable for the same reasons as in the aryl ester rearrangement; see Section II.E), concentration of **128b**, light intensity, or the amount of oxygen in the solution. The addition of 1,3-pentadiene as triplet quencher (E_T = 53 kcal mole^{-1} [92]) also did not affect the quantum yields.

The Japanese authors showed that upon irradiation in the mixed solvent cyclohexane–benzene, the dependences of energy transfer efficiency (from excited benzene to acetanilide) on both the benzene and acetanilide concentrations are the same for **129b** and **130b** (Fig. 2). This important result suggests that **129b** and **130b** are formed from energetically equivalent acetanilide singlets. It is unfortunate that the formation of aniline was not traced: this could shed light on the energetic requirements of its formation. From the kinetic data it was calculated that the energy transfer rate constant (k_3; cf. Eq. 12) is at least 22 times higher than the rate constant of the diffusion-controlled bimolecular reaction in benzene. Assuming that singlet-singlet nonradiative energy transfer proceeds from the lowest excited singlet state of benzene (Eq. 12), the critical transfer distance R ≈ 26 Å was calculated for a resonance transfer with a weak interaction. The possibility of a triplet-triplet energy transfer is excluded because the photorearrangement was not influenced by oxygen, even in benzene.

$$B^*(^1B_{2u}) + 128b(^1A_1) \xrightarrow{k_3} B(^1A_{1g}) + 128b^*(^1B_2) \qquad (12)$$

When irradiated, *N,N*-diacyl substituted arylamines **31** yield both re-

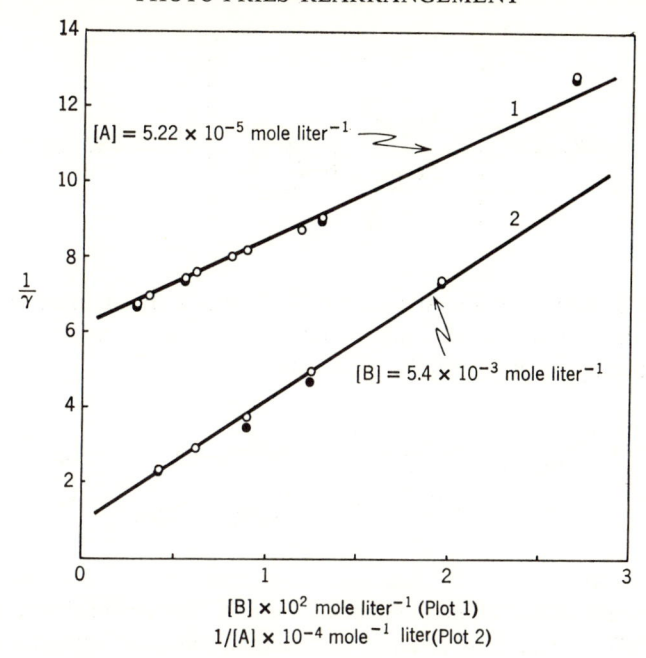

Fig. 2. Plots of reciprocal energy transfer efficiency (γ) against the concentration of benzene (B) and against reciprocal concentration of acetanilide (A) by the photorearrangement of acetanilide in cyclohexane. (○) 2-Aminoacetophenone; (●) 4-aminoacetophenone.

arrangement and cleavage products.[12] It was demonstrated that the product of the primary *ortho*-rearrangement **32** is stable and thus the second acyl group does not migrate to the aromatic nucleus. The reason undoubtedly lies in the deactivation of **32** owing to tautomerism via intramolecular proton transfer (cf. **107** → **109**). For the same reason salicylanilide (**104**) rearranges only in an extremely small yield[73] to give **105** and **106**.

Photolysis of *N*-aryl amides of β-ketocarboxylic acids **135** proceeds in a different way.[93] Thus, phenylisocyanate **136** is formed in 60% yield when in **135** R = phenyl. When R = methyl or propyl, the yield decreases to 18%. Consequently, this reaction belongs only formally to the scope of the discussed rearrangements. It seems that this is a special example of Norrish Type II reaction.[32d] This assumption is supported by the yield dependence of

Ph—NH—COCH$_2$COR $\xrightarrow{h\nu}$ Ph—N=C=O + R—CO—CH$_3$

135   **136**   **137**

136 on the nature of the substituent R. It would be of interest to learn whether this molecular cleavage would be influenced, unlike the photo-Fries rearrangement, by triplet quenchers.

Neither the anilide of cinnamic acid[94] nor the diphenyl substituted acroyl anilide **138**[95] yields any product of rearrangement or cyclodimerization. Upon irradiation of **138** in benzene solution in a Pyrex reactor, only the isomeric β-lactams **139** (2.3%) and **140** (37%), in addition to dihydrocarbostyril **141** (5%), were isolated. The latter is the major product upon irradiation of alkyl substituted acroyl anilides.[96] On the other hand, the closely related phenyl cinnamate rearranges regularly to the *ortho-* and *para-*positions[97] and does not dimerize as the other alkyl esters of cinnamic acid.[98]

No aminothiophenones can be obtained by photorearrangement of *N*-aryl amides of thiocarboxylic acids. The formation of 2-ethoxybenzthiazole (**143**) by oxidative photodehydrogenation of **142** is of some interest.[11] Similarly, 2-phenylbenzthiazole originates by photocyclization of thiobenzanilide in the presence of oxygen.[99]

V. PHOTOREARRANGEMENT OF N-ARYL LACTAMS

The photorearrangement of *N*-aryl lactams **34** to the *ortho*-position offers an interesting possibility for the preparation of cyclic benzo-aza-ketones of type **35**, whereas the rearrangement to the *para*-position should lead to *para*-cyclophane **144** formation. 7-, 8-, and 13-Membered ring *N*-aryl lactams (**34**, $n = 5, 6,$ and 11) were successfully rearranged in ethanol to **35** in chemical yields of 60, 83, and 80%,[13] and in quantum yields of 0.071, 0.11, and 0.082, respectively.[14]

By analogy with the related photorearrangements, one would expect that lactam **34** with proper number of methylene groups should rearrange, in part, to the *para*-azacyclophane **144**. Such a product, however, is not formed from the 13-membered lactam, although eleven methylene groups already allow nonstrained junction to the *para*-position[100] and, according to molecular models, the formation of a transient state appears well possible.

N-Phenyl-β-propiolactam ($n = 2$) and its substituted analogs **145** do not rearrange to **35**. They are rather cleaved[101] to formanil derivatives **146** and ketene (**147**). Also, **34** with $n = 3$ does not afford any rearranged products.[14]

The preparative value of *N*-phenyl lactam rearrangement lies in the possibility of preparing middle and large benz-condensed rings in one step from readily available starting materials.[14]

VI. PHOTOREARRANGEMENT OF N-ARYL AMIDES OF SULFONIC ACIDS

The specific feature of photorearrangement of this class of compounds (cf. **36** → **37** + **30**) is the fact that no *ortho*-rearranged products are formed. The results in a single publication[15] indicate that the yield of *para*-products depends on the substitution of the arylamine nucleus (Table VII).

TABLE VII
Photorearrangement of *N*-Aryl Amides of Sulfonic Acids[15]

	148			Conversion of **148**, %	Yield, %[a]	
	R_1	R_2	R_3	R_4		Diarylsulfone **149**	Arylamine **150**
a	CH_3	H	H	H	28	25	68
b	H	H	H	H	34	12	30
c	H	CH_3	H	H	32	14	25
d	H	H	CH_3	H	27	6	43
e	H	H	H	CH_3	28	—	41

[a] The results are from irradiation in *tert*-butanol with continuous extraction of the products **149** and **150**, using 10% HCl.

Methanesulfoanilide did not rearrange in ethanol and *n*-butanol. In an equimolar mixture of **148a** and **148c**, irradiated in ethanol, crossed reaction products were not formed. This finding favors the intramolecular mechanism for this type of arrangement also.

VII. PHOTOREARRANGEMENT OF ENOL ESTERS OF CARBOXYLIC ACIDS

The [1,3]-shifts of acyl group as described so far concern only cases where the double bond in the general equation **1** → **2** was part of an aromatic nucleus. In enol esters the double bond between C_2 and C_3 is isolated and products of the photorearrangement are β-dicarbonyl compounds. Their yield reaches a maximum if β-dicarbonyl products formed exist predominantly in the enolic form which does not undergo observable further change. Because of this product photostability it is advantageous to irradiate in nonpolar solvents. On the other hand, the enol form acts as a strong filter in the irradiated solution and thus explains the relatively low product yields when compared to the photo-Fries rearrangement of aryl esters. Extremely low yields are obtained if one carbonyl group in the product is a terminal formyl group. Upon irradiation of vinyl benzoate (38) in benzene, benzoylacetaldehyde (39), the primary product of the rearrangement, was isolated in only 6% yield. In transparent solvents 39 decarbonylates completely to acetophenone.[16]

Another subsequent photoreaction of β-diketones, which are enolized only to a small extent and possess a γ-hydrogen, can be Norrish Type II cleavage.[32d] Thus, from cyclohexen-1-yl benzoate (151) the aliphatic diketone 153 was obtained either exclusively[102] or mostly.[103] McIntosh[103] demonstrated in a separate irradiation of the "normal" photorearranged product 152 that 153 was formed by very efficient subsequent Norrish Type II cleavage. An equivocal steric condition for the photoreaction 152 → 153 is an axial position of the benzoyl substituent in 152. It is remarkable that the *cyclohexanone* derivative 152 has an axial acyl group, although it is well known that *cyclohexane* derivatives possessing an equatorial acyl group are thermodynamically much more stable than their axial epimers.[104] In agreement with this, cyclohexyl phenyl ketone and *cis*-4-*t*-butylcyclohexyl phenyl ketone irradiated under the same conditions as 152 did not afford any Norrish Type II cleavage products.[103] There are two other examples observed where the migrating acyl group of steroidal cyclohexenyl esters is directed exclusively to the axial position. Thus, upon irradiation of enolacetate **154** with a fully substituted double bond, thermodynamically less stable isomer with the acetyl group in an axial position in a rigid chair-form of steroid A-ring was the only rearrangement product observed.[105] Similarly, the rearrangement product **160** of dienol acetate **159**, obtained in 17% yield, has the acetyl group also in an axial configuration.[17] The photorearrangement **159** → **161** (13% yield) represents the only known example of a [1,5]-shift in a nonaromatic system.[17]

A fundamental difference exists between the photoreactivity of cyclohexenone enol esters of types **162** and **163**. Steroid **162**, its lowest excited state being identified as π,π^* triplet,[106] was shown to be very stable on irradiation in the $S_0 \to S_{\pi,\pi^*}$ and $S_0 \to S_{n,\pi^*}$ absorption bands in different solvents.[107] No photoproduct was formed. On the other hand, 3-acetoxycyclohexenone **163**, when irradiated[108] in ether solution, afforded the isomeric trione **164** in an unusually high yield of 60%. This result is also interesting in regard to the nature and multiplicity of the excited state of the reacting molecule **163**. While the majority of the photorearrangements reported above are consistent with photoisomerization from the π,π^* singlet state, the effective photoisomerization **163** → **164** was carried out in $S_0 \to S_{n,\pi^*}$ absorption region of **163**.

162

163 → hν → 164

All authors active in the field of enol ester photorearrangements assume a radical mechanism according to Path A and subsequent recombination of acyl and vinyloxy radicals in the β-C radical form. It is evident that in sterically controlled cyclohexenyl ester rearrangements these radicals have to be close in the solvent cage if the radical mechanism is valid. In view of this difficulty an alternative suggestion was advanced, proposing that the excited enol ester cleaves to give two excited radicals which are tightly joined in the solvent cage.[105]

A radical mechanism is, in part, undoubtedly correct as witnessed by the products of hydrogen abstraction from the solvent **156** and by dimers of type **158**. However, experimental material favoring or excluding a concerted shift via a 4-ring transient state of type **69** is lacking.

The photorearrangement of enol esters is strictly intramolecular. Upon irradiation of two different steroidal enol acetates, one of them possessing a trideuterated acetyl group, the original isotope labels were fully retained in the products.[105]

VIII. REVERSIBLE PHOTOREARRANGEMENT OF ENOL LACTONES AND NONENOLIZABLE β-DIKETONES

The equilibration of enol lactones with the corresponding nonenolizable β-diketones, induced by heat, acid, or base, has long been known. In recent years several examples of their reversible photorearrangement of the type **43** ⇌ **44** have been published. Thus, the photorearrangement of enol lactones[18,19,109,111] is the only example of photorearrangement so far known in the sense of the general equation **1** → **2** in which a photostationary state can be reached.† In fact, reality is often less ideal, and, depending on the structure of the lactone, photolytic mixture can be complicated by products arising from competitive and subsequent photoreactions. In some cases the rearrangement is known to proceed in one direction only.[112]

[chemical scheme: 165 → [166 ↔ 167] ⇌ 168]

[chemical scheme: 169 → 170 + 171 + 172]
170: cyclopropyl-C(=O)-CH₃
171: H₂C=C=O
172: CH₂=CH-C(=O)-CH₃

[chemical scheme: 173 → 174, hν/ROH]

No mechanistic studies on this rearrangement have been published as yet. Taking into account the steric situation, it seems unlikely that a concerted [1,3]-shift of the acyl group via a 4-membered ring can operate with enol lactone. However, reversible rearrangement and by-products formation can be well explained by assuming a primary cleavage of the lactone —O—CO— bond to a resonance-stabilized diradical intermediate **166 ↔ 167**. The appreciable diversification of reaction modes of radical **166 ↔ 167** is demonstrated by the reaction products derived from this intermediate. Intramolecular recombination leads to the "normal" rearranged products, i.e., to nonenolizable β-diketones. The formation of (5.3)-spiro diketone **46** is the only example reported with endocyclic enol lactones.[19] Loss of carbon monoxide by fragmentation of bond *a* from the carbonyl part of diradical **166 ↔ 167** leads to ring formation (examples: **45 → 47**; **169 → 170**).[19] The cleavage of bond *b* is also possible (example: **169 → 171 + 172**).[19] Ketoester **174**, isolated after the

† Essentially, photoreversibility between acyclic enol esters and nonenolizable β-diketones appears feasible. So far, however, only photorearrangements of β-diketones[109] and β-ketosulfones,[110] respectively, leading to enol esters have been reported, but not the possible reversed reactions.

photolysis in alcoholic solvents, indicates an intramolecular hydrogen transfer in the rotameric form **168** and ketene formation (cf. related reaction **80** → **85** and **181** → **184**) and its subsequent esterification.[113]

<u>175</u>   <u>176</u>   <u>177</u>

Up to now no reversible photorearrangement of a nonenolizable β-diketone to an endocyclic enol lactone is known, but several examples of photostationary equilibria between β-diketones and exocyclic enol lactones have been established. After irradiation of either **175** or **176** in benzene a photoequilibrium mixture containing **175** + **176** + **177** is formed with a product ratio 3:95:2.[109] With 2-mono- or disubstituted 1.3-indanediones and *cis*- and *trans*-alkylidenenaphthalides **43** ⇌ **44**,[18] it was shown that *the ratio of enol lactone to β-diketone in the photomixtures can be changed* within broad limits just *by changing the wavelength of the uv-light* (Table VIII).

Compounds of type **43** and **44** tend to undergo radical cleavage readily. Carbonyl radicals formed are relatively stable and do not decarbonylate provided that an aromatic nucleus in the β-position exerts a stabilizing effect. Minimal amount of by-products were observed when at least one of the substituents R_1 or R_2 was phenyl. The photochemical conversion **43** ⇌ **44** is otherwise accompanied by tar formation, dimerization, reduction, etc.

The rearrangement of β-diketones is most simply explained assuming the involvement of an n,π^* excited reactive state. Its triplet configuration was demonstrated by the quenching effect of 1,3-pentadiene on the rearrangement **176** → **175**.[109] The nature of the excited state of enol lactone is not

TABLE VIII
Wavelength-Dependent Composition of the Equilibrium Photoreaction Mixture of 2-Diphenyl-1,3-indanedione (**44**; $R_1 = R_2 = C_6H_5$) in Benzene[18b]

	Filter with zero transmission below λ, nm				
	335	325	295	285	[a]
Phtalide **43** ($R_1 = R_2 = C_6H_5$), %	100	80	70	50	25
Indanedione **44** ($R_1 = R_2 = C_6H_5$), %	0	20	30	50	75

[a] Filter transmitting uv radiation between 305–345 nm.

known. On the basis of similarity with the other photoinitiated shifts of acyl groups, a π,π^* singlet may be assumed. Attempts to influence the equilibrium of photochemical interconversions by selective sensitization or quenching represent an obvious task for future experimentation.

IX. PHOTOREARRANGEMENT OF ENAMIDES OF CARBOXYLIC ACIDS

The photochemical rearrangement of enamides of carboxylic acids [20,113–116] is analogous to the photochemical rearrangement of enol esters (Section VII), except that it proceeds more efficiently and in much higher yield. It seems that the latter can be attributed to the high photochemical stability of the photoproducts. The β-iminocarbonyl compound **187** formed immediately after the

[1,3]-acyl shift undergoes a fast prototropic shift to give the 3-amino-2-en-1-one **188** as the final product. This itself can disperse any absorbed energy either via a tautomerism involving hydrogen transfer to the excited carbonyl or by *cis-trans* isomerization. It is interesting to note that in the photorearrangement of *N*-1-propenyl-*N*-1-propylbenzamide (**178**) in methanol the *trans*-isomer **180**, which lacks any intramolecular hydrogen bond, is the major product.[115] The *trans:cis* ratio of 85:15 then represents the photochemical equilibrium rather than the thermodynamic equilibrium. The total yield of **179** and **180** was 85% at a 85% conversion of **178**!

Historically, the photorearrangement of enamides takes second place after the aryl ester photorearrangement. As far back as in 1962 Eschenmoser et al.[20] used this photorearrangement as an important intermediate in the corrine synthesis. Irradiation of the exocyclic enamide **48** (R = —CH$_2$-(CH$_3$)$_2$COCH$_3$) in 0.1% cyclohexane solution employing a low-pressure mercury arc afforded the desired **49** in 50% yield, besides ca. 1% of isomer **50**. The ratio 50:1 of **49** and **50** offers insight into the steric requirements for an eventual 4-ring transient state involved. Undoubtedly the reaction path to produce **50** competes with the [1,3]-shift of the acyclic acyl group to form **49**, but a possible 4-ring transient state between the lactam carbonyl carbon and the terminal methylene is sterically strongly unfavored, if not altogether impossible. Consequently, **50** should arise from the disadvantageous radical mechanism.

A dissociative first step is anticipated exclusively in photolysis of the endocyclic enamide **181** (compare also **80** → **85** and **173** → **174**). The reaction is formulated through a ketene intermediate **182**. The enamine **183** formed then undergoes hydrolysis to give the oxocarboxylic acid derivative **184** as the product.

Some results of Hoffmann and Eicken[116] also suggest that the photorearrangements of **185** and **186** proceed to a great extent by radical mechanism. The formation of **188** on irradiation of *trans*-**185** and *cis*-**186** occurs at about the same rate and is independent of the *cis*-*trans* isomer equilibrium. Upon irradiation of *trans*-**185** in carbon tetrachloride the [1,3]-shift of the acyl group was suppressed, and the intermediate free benzoyl radical was trapped instead as benzoylchloride in 20% yield. At this point the question can be raised whether photolytic [1,3]-shifts of enamides are inter- or intramolecular. To clear this point a 0.04M benzene solution of a 1:1 mixture of **189** and **190** was irradiated. It was found that only 8 ± 2% of [1,3]-acyl migration took place intermolecularly.[116]

It is important to note that a possible photochemical ring expansion by [1,3]- or [1,5]-acyl shifts in the 5-membered vinyl lactams **191** has not been observed,[4,116] nor have [1,5]-shifts of acyl groups involving an aromatic ring been reported for either enol esters or enamides. However, an example of a similar [1,5]-alkyl shift, **193 → 194**, in related ketene dialkyl acetals is known.[24]

In special cases, other photochemical reactions can totally suppress an acyl group migration.[76,117] The primary photochemical process of **195** is fast equilibration with *trans*-isomer **196**. The *cis*-isomer **195** then cyclizes in 65% yield to dehydroaporphane (**197**) in a similar way as does stilbene and its derivatives.[118] *Trans*-isomer **196** cyclizes in 10–21% yield to dehydroprotoberberine (**198**).[76] The latter reaction is analogous to the already discussed cyclization **110 → 112** and **134 → 131** of aromatic carbonates and carbamates.

REFERENCES

1. J. C. Anderson and C. B. Reese, *Proc. Chem. Soc.*, **1960**, 217.
2a. D. Belluš and P. Hrdlovič, *Chem. Rev.*, **67**, 599 (1967).
2b. V. I. Stenberg, in *Organic Photochemistry*, Vol. 1, O. L. Chapman, Ed., Dekker, New York, 1967, p. 127.
3. A. Gerecs, in *Friedel-Crafts and Related Reactions*, Vol. 3, Part 1, G. A. Olah, Ed., Wiley-Interscience, New York, 1964, p. 499.
4. D. Belluš. Unpublished results.
5. W. M. Horspool and P. L. Pauson, *J. Chem. Soc.*, **1965**, 5162.

6. J. S. Humphrey, *Amer. Chem. Soc. Polym. Preprints*, **9**, 453 (1968).
7. D. J. Trecker, C. S. Foote, and C. L. Osborn, *Chem. Commun.*, **1968**, 1034.
8. J. L. Stratenus, Thesis, Leiden, 1966.
9. J. L. Stratenus and E. Havinga, *Rec. Trav. Chim.*, **85**, 434 (1966).
10. B. K. Snell, *J. Chem. Soc., C.*, **1968**, 2367.
11. D. Belluš and K. Schaffner, *Helv. Chim. Acta*, **51**, 221 (1968).
12. R. O. Kan and R. L. Furey, *Tetrahedron Lett.*, **1966**, 2573.
13. M. Fischer, *Tetrahedron Lett.*, **1968**, 4295.
14. M. Fischer and A. Mattheus, *Chem. Ber.*, **102**, 342 (1969).
15. H. Nozaki, T. Okada, R. Noyori, and M. Kawanisi, *Tetrahedron*, **22**, 2177 (1966).
16. R. A. Finnegan and A. W. Hagen, *Tetrahedron Lett.*, **1963**, 365.
17. M. Gorodetsky and Y. Mazur, *J. Amer. Chem. Soc.*, **86**, 5213 (1964).
18a. J. Rigaudy and P. Derible, *Bull. Chem. Soc. France*, **1965**, 3047.
18b. *Ibid.*, **1965**, 3055.
18c. *Ibid.*, **1965**, 3061.
19. A. Yogev and Y. Mazur, *J. Amer. Chem. Soc.*, **87**, 3520 (1965).
20a. A. Eschenmoser, *Pure Appl. Chem.*, **7**, 297 (1963).
20b. E. Bertele, H. Boos, J. D. Dunitz, F. Elsinger, A. Eschenmoser, I. Felner, H. P. Gribi, H. Gschwend, E. F. Meyer, M. Pesaro, and R. Scheffold, *Angew. Chem.*, **76**, 393 (1964).
21a. M. S. Kharasch, G. Stampa, and W. Nudenberg, *Science*, **116**, 309 (1952).
21b. K. Schmid and H. Schmid, *Helv. Chim. Acta*, **36**, 687 (1953).
21c. D. P. Kelly and J. T. Pinhey, *Tetrahedron Lett.*, **1964**, 3427.
21d. D. W. Boykin, Jr., and R. E. Lutz, *J. Amer. Chem. Soc.*, **86**, 5046 (1964).
21e. D. E. McGregor, M. G. Vinje, and R. S. McDaniel, *Can. J. Chem.*, **43**, 1417 (1965).
21f. F. L. Bach and J. C. Barclay, Abstracts, 150th National Meeting of the American Chemical Society, Atlantic City, N. J., Sept. 12–17, 1965, p. 9S.
21g. J. Wiemann, Nguyen Thoai, and F. Weisbuch, *Tetrahedron Lett.*, **1965**, 2983.
21h. J. Hill, *Chem. Commun.*, **1966**, 260.
21i. D. P. Kelly, J. T. Pinhey, and R. D. G. Rigby, *Tetrahedron Lett.*, **1966**, 5953.
21j. P. Scribe, M. R. Monot, and J. Wiemann, *Tetrahedron Lett.*, **1967**, 5157.
21k. W. O. Godtfredsen, W. v. Daehne, and S. Vangedal, *Experientia*, **23**, 280 (1967).
21l. M. K. M. Dirania and J. Hill, *J. Chem. Soc., C*, **1968**, 1311.
21m. G. Koga, N. Kikuchi, and N. Koga, *Bull. Chem. Soc. Japan*, **41**, 745 (1968).
21n. J. T. Pinhey and K. Schaffner, *Aust. J. Chem.*, **21**, 2265 (1968).
21o. H. J. Hageman, *Chem. Commun.*, **1968**, 401.
22. H. I. Joschek and S. I. Miller, *J. Amer. Chem. Soc.*, **88**, 3269 (1966).
23a. W. Kirmse and M. Buschhoff, *Angew. Chem.*, **77**, 681 (1964).
23b. J. E. Baldwin and L. E. Walker, *J. Amer. Chem. Soc.*, **88**, 4191 (1966).
24. J. E. Baldwin and L. E. Walker, *J. Amer. Chem. Soc.*, **88**, 3769 (1966).
25. H. Michio, F. Takeshi, Y. Odaira, and S. Tsutsumi, *Nippon Kagaku Zasshi*, **88**, 1091 (1967); *Chem. Abstr.*, **69**, 35607 (1968).
26a. F. D. Chattaway and K. J. Orton, *Proc. Chem. Soc.*, **18**, 200 (1902).
26b. J. J. Blanksma, *J. Chem. Soc.*, **82** II, 646 (1902); *Rec. Trav. Chim.*, **21**, 366 (1902).
26c. J. H. Mathews and R. V. Williamson, *J. Amer. Chem. Soc.*, **45**, 2574 (1923).
26d. C. W. Porter and P. Wilbur, *J. Amer. Chem. Soc.*, **49**, 2145 (1927).
26e. F. W. Hodges, *J. Chem. Soc.*, **1933**, 241.
26f. K. N. Ayad, C. Beard, R. F. Garwood, and W. J. Hickinbottom, *J. Chem. Soc.*, **1957**, 2981.

26g. J. Coulson, G. H. Williams, and K. M. Johnston, *J. Chem. Soc.*, *B*, **1967**, 174.
27. D. V. Banthorpe and J. A. Thomas, *J. Chem. Soc.*, **1965**, 7158.
28. J. I. G. Cadogan and W. R. Foster, *J. Chem. Soc.*, **1961**, 3076.
29. H. Kobsa, *J. Org. Chem.*, **27**, 2293 (1962).
30. G. M. Coppinger and E. R. Bell, *J. Phys. Chem.*, **70**, 3479 (1966).
31. A. A. Lamola and G. S. Hammond, *J. Chem. Phys.*, **43**, 2129 (1965).
32. J. G. Calvert and J. N. Pitts, Jr., *Photochemistry*, Wiley, New York, 1966: (a) p. 298; (b) p. 301; (c) p. 242; (d) pp. 515–520; (e) p. 382.
33. M. R. Sandner and D. J. Trecker, *J. Amer. Chem. Soc.*, **89**, 5725 (1967).
34. M. R. Sandner, E. Hedaya, and D. J. Trecker, *J. Amer. Chem. Soc.*, **90**, 7249 (1968).
35. A. J. Fry, R. S. H. Liu, and G. S. Hammond, *J. Amer. Chem. Soc.*, **88**, 4781 (1966).
36. H. E. Zimmerman, H. G. Dürr, R. S. Givens, and R. G. Lewis, *J. Amer. Chem. Soc.*, **89**, 1863 (1967).
37. R. A. Finnegan and J. J. Mattice, *Tetrahedron*, **21**, 1015 (1965).
38a. M. Rausch, M. Vogel, and H. Rosenberg, *J. Org. Chem.*, **22**, 903 (1957).
38b. R. E. Bozak, *Chem. Ind.* (London), **1969**, 24.
39. D. A. Plank, *Tetrahedron Lett.*, **1968**, 5423.
40. J. C. Anderson and C. B. Reese, *J. Chem. Soc.*, **1963**, 1781.
41. D. Belluš, K. Schaffner, and J. Hoigné, *Helv. Chim. Acta*, **51**, 1980 (1968).
42. C. L. Braun, S. Kato, and S. Lipsky, *J. Chem. Phys.*, **39**, 1645 (1963).
43. N. J. Turro, *Molecular Photochemistry*, Benjamin, New York-Amsterdam, 1965: (a) pp. 44–79; (b) p. 149.
44. R. M. Hochstrasser and G. B. Porter, *Quart. Rev.*, **14**, 146 (1960).
45. H. H. Jaffé and A. L. Miller, *J. Chem. Educ.*, **43**, 469 (1966).
46. A. Davis and J. H. Golden, *J. Chem. Soc.*, *B*, **1968**, 425.
47. T. J. Stone and W. A. Waters, *J. Chem. Soc.*, **1964**, 213.
48. N. M. Atherton, E. J. Land, and G. Porter, *Trans. Faraday Soc.*, **59**, 818 (1963).
49. J. A. McRae and M. C. R. Symons, *J. Chem. Soc.*, *B*, **1968**, 428.
50. J. S. Humphrey. Private communication, 1968.
51. L. Schutte and E. Havinga, *Tetrahedron*, **23**, 2281 (1967).
52. D. A. Plank, Thesis, Purdue University, 1966.
53. B. A. M. Oude Alink, Thesis, Leiden, 1966.
54. C. D. Gutsche and B. A. M. Oude Alink, *J. Amer. Chem. Soc.*, **90**, 5855 (1968).
55. G. Quinkert, *Angew. Chem., Int. Ed.*, **4**, 211 (1965).
56. J. Kovář, M. Hudlický, and I. Ernest, *Preparative Reactions in Organic Chemistry. VIII. Molecular Rearrangements*, NČSAV, Prague, 1965, p. 989.
57a. Z. R. Grabowski, *Z. Phys. Chem.* (*Neue Folge*), **27**, 239 (1961).
57b. T. Latowski, E. Latowska, and M. Brudka, *Zeszyty Nauk. Mat., Fiz., Chem., Gdansk*, **4**, 95 (1964); *Chem. Abstr.*, **65**, 8710c (1966).
57c. J. T. Pinhey and R. D. G. Rigby, *Tetrahedron Lett.*, **1969**, 1267.
57d. *Ibid.*, **1969**, 1271.
58a. E. J. Baum and J. N. Pitts, Jr., *J. Phys. Chem.*, **70**, 2066 (1966).
58b. A. Basinski and E. Latowska, *Rocz. Chem.*, **40**, 1747 (1966).
58c. E. Latowska and T. Latowski, *Rocz. Chem.*, **40**, 1977 (1966).
59. P. Sláma, D. Belluš, and P. Hrdlovič, *Collect. Czech. Chem. Commun.*, **33**, 3752 (1968).
60. J. S. Bradshaw, E. L. Loveridge, and L. White, *J. Org. Chem.*, **33**, 4127 (1968).
61a. D. H. R. Barton, Y. L. Chow, A. Cox, and G. W. Kirby, *Tetrahedron Lett.*, **1962**, 1055.

61b. Ibid., *J. Chem. Soc.*, **1965**, 3571.
62. R. A. Finnegan and D. Knutson, *J. Amer. Chem. Soc.*, **89**, 1970 (1967).
63. W. H. Horspool and P. L. Pauson, *Chem. Commun.*, **1967**, 195.
64. R. A. Finnegan and D. Knutson, *Tetrahedron Lett.*, **1968**, 3429.
65. R. Hoffmann and R. B. Woodward, *Accounts Chem. Res.*, **1**, 17 (1968).
66. D. Seebach, *Fortschr. Chem. Forsch.*, **11**, 177 (1969).
67. J. A. Berson, *Accounts Chem. Res.*, **1**, 152 (1968).
68. R. S. Givens, *Tetrahedron Lett.*, **1969**, 663.
69a. H. H. Jaffé, *Chem. Rev.*, **53**, 191 (1953).
69b. C. D. Ritchie and W. F. Sager, *Progress in Physical Organic Chemistry*, Vol. 2., Wiley-Interscience, New York-London, 1964, p. 323.
70. R. A. Finnegan and D. Knutson, *J. Amer. Chem. Soc.*, **90**, 1670 (1968).
71. D. Belluš, P. Hrdlovič, and P. Sláma, *Collect. Czech. Chem. Commun.*, **33**, 2646 (1968).
72. D. C. Neckers, *Mechanistic Organic Photochemistry*, Reinhold, New York, 1967, p. 107.
73. D. V. Rao and V. Lamberti, *J. Org. Chem.*, **32**, 2896 (1967).
74. A. A. Lamola and L. J. Sharp, *J. Phys. Chem.*, **70**, 2634 (1966).
75. F. Jortner, *J. Polym. Sci.*, **37**, 199 (1959).
76. N. C. Yang, A. Shani, and G. R. Lenz, *J. Amer. Chem. Soc.*, **88**, 5369 (1966).
77. C. Pac and S. Tsutsumi, *Bull. Chem. Soc. Japan*, **37**, 1392 (1964).
78a. C. H. Kuo, R. D. Hoffsommer, H. L. Slates, D. Taub, and N. L. Wendler, *Chem. Ind.* (London), **1960**, 1627.
78b. D. Taub, C. H. Kuo, H. L. Slates, and N. L. Wendler, *Tetrahedron*, **19**, 1 (1963).
78c. M. R. Stoner, *Dissertation Abstr.*, **25**, 6243 (1965).
78d. G. M. Coppinger, French Pat. 1,379,593; *Chem. Abstr.*, **62**, 7946b (1965).
78e. C. D. Pande and B. Venkataramani, *Indian J. Technol.*, **4**, 342 (1966); *Chem. Abstr.*, **66**, 54824 (1967).
78f. J. M. Bruce and E. Cutts, *J. Chem. Soc.*, C, **1966**, 449.
78g. M. G. Kuzmin, J. A. Michejev, and L. N. Guseva, *Dokl. Akad. Nauk SSSR*, **176**, 368 (1967).
78h. T. Matsuura and Y. Kitaura, *Tetrahedron Lett.*, **1967**, 3311.
78i. H. Goeth, *Kunststoffe Plastics*, **15**, 96 (1968); *Chem. Abstr.*, **69**, 28246 (1968).
78j. C. D. Pande, B. N. Tripathi, and B. Venkataramani, *Indian J. Chem.*, **6**, 542 (1968).
79a. J. H. Chaudet, G. C. Newland, H. W. Patton, and J. W. Tamblyn, *SPE Trans.*, **1**, 26 (1961).
79b. G. C. Newland and J. W. Tamblyn, *J. Appl. Polym. Sci.*, **8**, 1949 (1964).
79c. M. Okawara, S. Tani, and E. Imoto, *Kogyo Kagaku Zasshi*, **68**, 223 (1965); *Chem. Abstr.*, **63**, 3068g (1965).
79d. S. B. Maerov, *J. Polym. Sci.*, Part A, **3**, 487 (1965).
79e. D. Belluš, P. Hrdlovič, and Z. Maňásek, *J. Polym. Sci.*, Part B, **4**, 1 (1966).
79f. D. Belluš, Z. Maňásek, P. Hrdlovič, and P. Sláma, *J. Polym. Sci.*, Part C, **16**, 267 (1967).
80. R. A. Finnegan and D. Knutson, *Chem. Ind.* (London), **1965**, 1837.
81. R. A. Finnegan and D. Knutson, *Chem. Commun.*, **1966**, 172.
82. O. Kling, E. Nikolaiski, and H. L. Schläfer, *Ber. Bunsenges. Phys. Chem.*, **67**, 883 (1963).
83. D. Belluš, Z. Maňásek, P. Hrdlovič, P. Sláma, and Ľ. Ďurišinová, IUPAC Int Symp. Macromol. Chem., Brussel-Louvain, 1967, Preprint 28; *J. Polym. Sci.* Part C, **22**(2), 629 (1969).

84. G. Porter and P. Suppan, *Trans. Faraday Soc.*, **61**, 1664 (1965).
85a. J. N. Pitts, Jr, H. W. Johnson, and T. Kuwana, *J. Phys. Chem.*, **66**, 2456 (1962).
85b. T. S. Godfrey, G. Porter, and P. Suppan, *Discussions Faraday Soc.*, **1965**, 194.
85c. E. J. O'Connell, Jr., *J. Amer. Chem. Soc.*, **90**, 6550 (1968).
86a. J. N. Pitts, Jr., L. D. Hess, E. J. Baum, A. E. Schuck, J. K. S. Wan, P. A. Leermakers, and G. Vesley, *Photochem. Photobiol.*, **4**, 305 (1965).
86b. E. J. Baum, J. K. S. Wan, and J. N. Pitts, Jr., *J. Amer. Chem. Soc.*, **88**, 2652 (1966).
86c. J. A. Barltrop and J. D. Coyle, *J. Amer. Chem. Soc.*, **90**, 6584 (1968).
87. H. I. Joschek and S. I. Miller, *J. Amer. Chem. Soc.*, **88**, 3273 (1966).
88. D. Elad, *Tetrahedron Lett.*, **1963**, 873.
89. D. Elad, D. V. Rao, and V. I. Stenberg, *J. Org. Chem.*, **30**, 3252 (1965).
90. B. S. Thyagarajan, N. Kharasch, H. B. Lewis, and W. Wolf, *Chem. Commun.*, **1967**, 614.
91. H. Shizuka and I. Tanaka, *Bull. Chem. Soc. Japan*, **41**, 2343 (1968).
92. G. S. Hammond, P. A. Leermakers, and N. J. Turro, *J. Amer. Chem. Soc.*, **83**, 2369 (1961).
93. J. Reisch and D. Niemeyer, *Tetrahedron Lett.*, **1968**, 3247.
94. H. Stobbe, *Chem. Ber.*, **58**, 2859 (1925).
95. O. L. Chapman and W. R. Adams, *J. Amer. Chem. Soc.*, **90**, 2333 (1968).
96. P. G. Cleveland and O. L. Chapman, *Chem. Commun.*, **1967**, 1064.
97. H. Obara and H. Takahashi, *Bull. Chem. Soc. Japan*, **40**, 1012 (1967).
98. A. Mustafa, *Chem. Rev.*, **51**, 1 (1952).
99. K. H. Grellmann and E. Tauer, *Tetrahedron Lett.*, **1967**, 1909.
100. B. H. Smith, *Bridged Aromatic Compounds*, Academic Press, New York-London, 1964, pp. 24–200.
101. M. Fischer, *Chem. Ber.*, **101**, 2669 (1968).
102a. M. Feldkimel-Gorodetsky and Y. Mazur, *Tetrahedron Lett.*, **1963**, 369.
102b. M. Gorodetsky and Y. Mazur, *Tetrahedron*, **22**, 3607 (1966).
103. C. L. McIntosh, *Can. J. Chem.*, **45**, 2267 (1967).
104. E. L. Eliel, *Stereochemistry of Carbon Compounds*, McGraw-Hill, New York, 1962, pp. 239–247.
105. A. Yogev, M. Gorodetsky, and Y. Mazur, *J. Amer. Chem. Soc.*, **86**, 5208 (1964).
106. G. Marsh, D. R. Kearns, and K. Schaffner, *Helv. Chim. Acta*, **51**, 1890 (1968).
107. D. Belluš, D. R. Kearns, and K. Schaffner, *Helv. Chim. Acta*, **52**, 971 (1969).
108. T. S. Cantrell, W. S. Haller, and J. C. Williams, *J. Org. Chem.*, **34**, 509, (1969).
109. K. Nozaki, Z. Yamaguti, T. Okada, R. Noyori, and M. Kawanisi, *Tetrahedron*, **23**, 3993 (1967).
110. C. L. McIntosh, P. deMayo, and R. W. Yip, *Tetrahedron Lett.*, **1967**, 37.
111a. H. Nozaki, Z. Yamaguti, and R. Noyori, *Tetrahedron Lett.*, **1965**, 37.
111b. J. T. Pinhey and K. Schaffner, *Aust. J. Chem.*, **21**, 1873 (1968).
112a. R. C. Cookson, A. G. Edwards, J. Hudec, and M. Kingsland, *Chem. Commun.*, **1965**, 98.
112b. H. U. Hostettler, *Tetrahedron Lett.*, **1965**, 1941.
113. Z. Horii, Y. Hori, and C. Ywata, *Chem. Commun.*, **1968**, 1424.
114. P. T. Izzo and A. S. Kende, *Tetrahedron Lett.*, **1966**, 5731.
115. N. C. Yang and G. R. Lenz, *Tetrahedron Lett.*, **1967**, 4897.
116. R. W. Hoffmann and R. R. Eicken, *Tetrahedron Lett.*, **1968**, 1759.
117. M. P. Cava and S. C. Havlicek, *Tetrahedron Lett.*, **1967**, 2625.
118. F. R. Stermitz, in *Organic Photochemistry* Vol. 1, O. L. Chapman, Ed., Dekker, New York, 1967, pp. 242–282.

Photoassociation in Aromatic Systems

BRIAN STEVENS, *Department of Chemistry, University of South Florida, Tampa, Florida 33620*

I.	Introduction.	162
II.	Reversible Photoassociation	164
	A. Fluor Concentration Dependence of the Fluorescence Spectrum.	164
	B. Dependence of the Fluorescence Spectrum on Pressure	172
	C. Dependence of the Fluorescence Spectrum on Quencher Concentration.	173
	D. Decay Characteristics of Molecular and Excimer (Exciplex) Fluorescence	178
III.	Excimer (Exciplex) Dissociation.	182
	A. Dependence of Fluorescence Spectrum on Temperature	182
	B. Temperature Dependence of Experimental Quenching Constants	188
	C. Fluorescence Quenching Efficiency	189
	D. Energy Transfer Efficiencies	192
IV.	Excimer (Exciplex) Binding Energies.	194
	A. The Excimer Configuration	194
	B. Exciton and Charge Resonance Concepts	195
	C. The Semiempirical LCAO MO Approximation	198
	D. Exciplex Binding Energies.	200
V.	Excimer (Exciplex) Relaxation	200
	A. Fluorescence Emission.	200
	B. Intersystem Crossing	203
	C. Internal Conversion	206
	D. Photodimerization (Photoaddition)	207
	E. Ionic Dissociation.	209
VI.	Photoassociation in Ordered Systems	211
	A. The Crystalline State	211
	B. The Polyphenyl Alkanes	213
	C. The Paracyclophanes	215
	D. DNA.	215
VII.	Alternative Excimer Formation Modes.	218
	A. Molecular Triplet–Triplet Annihilation.	218
	B. Ionic Doublet–Doublet Annihilation	219
	C. Dimer Cation Neutralization.	221
VIII.	Conclusions.	222
	References.	222

I. INTRODUCTION

The electronic relaxation of an aromatic molecule M in its lowest excited singlet state $^1M^*$ is well understood in terms of fluorescence emission, internal conversion to a spin-paired ground state, and intersystem-crossing to the triplet state manifold. In fluid media these processes compete with bimolecular (collisional) quenching which is promoted with varying degrees of collisional efficiency by an increase in concentration of either the fluorescent species itself or of a foreign quenching species Q; the different types of quenching processes are usually described with reference to some property of the quenching species (Table I).

Insofar as the absorption spectrum of the fluor-quencher system is independent of its composition, the quenching process reflects an interaction of the quenching species with the potentially fluorescent molecule $^1M^*$, thus self-quenching may be interpreted quantitatively in terms of reversible photoassociation

$$^1M^* + M \rightleftharpoons {}^1M_2^*$$

and subsequent relaxation of the excimer $^1M_2^*$. By analogy with molecular relaxation routes, excimer relaxation may be expected to involve fluorescence emission and internal conversion to a dissociated ground state, together with intersystem crossing to the triplet excimer. These processes are summarized in Scheme 1 which includes photoassociation in the triplet manifold and formation of the stable photodimer M_2 from the excimer singlet state only; the intervention of a triplet excimer intermediate in photodimerization would require at least one intersystem crossing and is the less probable in consequence.

TABLE I
Bimolecular Quenching Processes

Quencher[a]	Quenching process	Efficiency
M	Self	Variable
O_2, NO	Paramagnetic	High
RBr, ϕBr, Xe	External heavy atom	Variable
RNH_2, ϕNH_2	Charge transfer	Variable
M'	Energy transfer	High

[a] R ≡ alkyl group. ϕ ≡ aryl group. M' denotes aromatic hydrocarbon of lower excited singlet state energy than M.

SCHEME 1. Homomolecular Photoassociation and Excimer Relaxation.

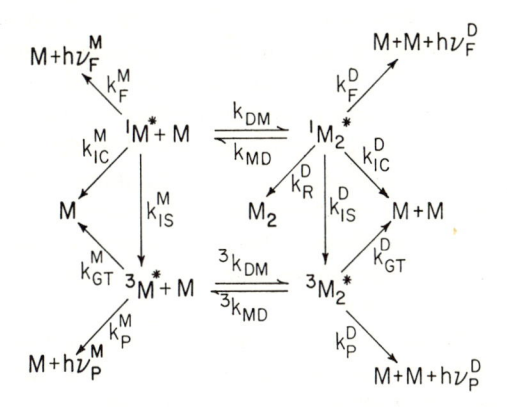

In reviewing the evidence for, and consequences of, photoassociation in aromatic systems it is of interest to examine the extent to which these processes constitute a particular example of a more general quenching scheme (Q ≠ M) based on the reversible photoassociation of fluor M and quencher Q together with relaxation of the exciplex $^1MQ^*$. The general Scheme 2 is more complex than Scheme 1 since exciplex dissociation may generate either molecular component in an electronically excited state thereby accommodating electronic energy transfer to Q as a quenching process.

Rate constants for the various processes follow the nomenclature of Birks et al.[1] insofar as this is possible; however, in view of the possible participation

SCHEME 2. Heteromolecular Photoassociation and Exciplex Relaxation.

of four electronically excited species the unimolecular constants are distinguished by the superscript M (molecule), D (excimer), C (exciplex), and Q (quencher), while the subscripts refer to fluorescence emission (F), internal conversion (IC), intersystem crossing from the excited singlet state (IS), phosphorescence emission (P), and nonradiative transition from triplet to ground state (GT); other rate constants are labeled according to excited states involved, e.g., $^3k_{QC}$ refers to dissociation of triplet exciplex to triplet quencher. Despite its complexity, this nomenclature is preferred to that based on the indiscriminate numbering of the 24 processes involved.

It is emphasized that the terms *excimer*[2] and *exciplex*[3,4] are reserved here for homomolecular and heteromolecular excited double molecules formed *after* the act of light absorption by one component in a process of *photoassociation*, in the absence of spectroscopic or cryoscopic evidence for molecular association in the ground state. Recent findings indicate that excimer (or exciplex) formation may also result from triplet–triplet annihilation,[5,6] cation-anion combination[7] (doublet–doublet-annihilation), and electron capture by the (relatively stable) dimer (or complex) cation[8]; these processes are discussed in Section VII.

II. REVERSIBLE PHOTOASSOCIATION

A. Fluor Concentration Dependence of the Fluorescence Spectrum

Direct evidence for photoassociation of aromatic hydrocarbons in solution is afforded by the appearance of a structureless emission band, at longer wavelengths than the molecular fluorescence spectrum, as the solute concentration is increased; the molecular fluorescence undergoes a corresponding reduction in intensity as shown in Figure 1. The absence of permanent chemical change is confirmed by the invariance of the absorption spectrum under these conditions and the restoration of the molecular emission spectrum on dilution.

This phenomenon was first reported for pyrene in benzene by Förster and Kasper[9] who identified the origin of the long wave emission component as the radiative relaxation of an excited double molecule or excimer $^1M_2^*$ formed after the act of light absorption by one of its molecular components (in a process of photoassociation). The origin of the emission bands is shown schematically in Figure 2. Table II lists other compounds which exhibit similar behavior, while excimer fluorescence of the compounds in Table III is not observed following singlet excitation in solution.

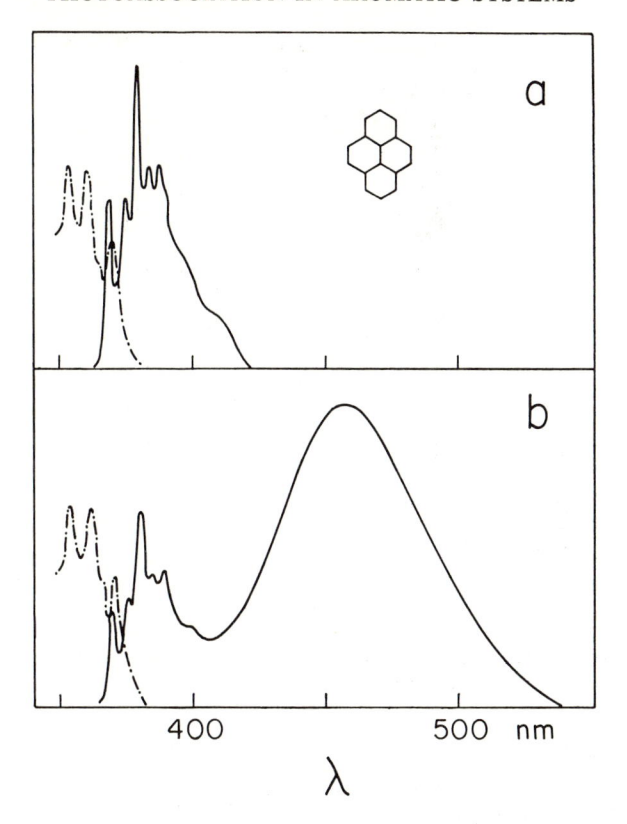

Fig. 1. Absorption (—·—) and fluorescence (———) spectra of pyrene in ethanol at 20°C; (a) $10^{-4}M$; (b) $10^{-2}M$.

The quantum yields of molecular and excimer fluorescence are related to the appropriate rate constants (Scheme 1) as follows: the photostationary conditions

$$\frac{d[^1M^*]}{dt} = I_a + k_{MD}[^1M_2^*] - (k_F^M + k_{IC}^M + k_{IS}^M + k_{DM}[M])[^1M^*] = 0$$

$$\frac{d[^1M^*]}{dt} = k_{DM}[M][^1M^*] - (k_F^D + k_{IC}^D + k_{IS}^D + k_R^D + k_{MD})[^1M_2^*] = 0$$

yield the following expressions for the photostationary concentrations of

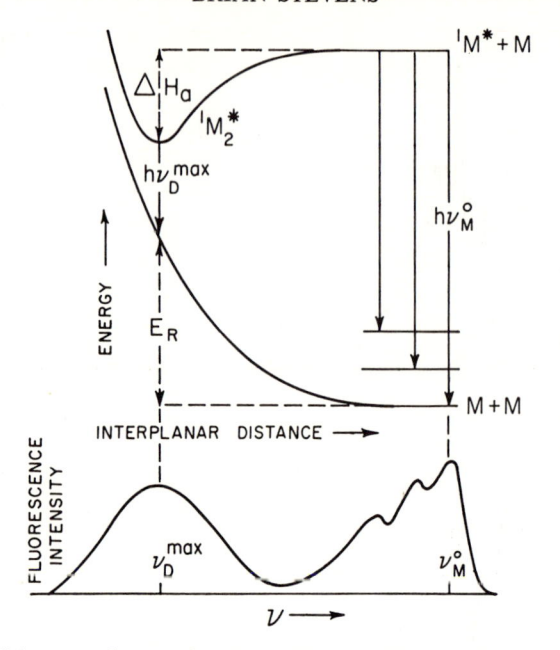

Fig. 2. Potential energy diagram for photoassociation illustrating origin of excimer and molecular fluorescence bands.

emitting species:

$$[^1M^*] = \frac{I_a(k_F^D + k_{IC}^D + k_{IS}^D + k_R^D + k_{MD})}{(k_F^M + k_{IC}^M + k_{IS}^M)(k_F^D + k_{IC}^D + k_{IS}^D + k_R^D + k_{MD}) + k_{DM}[M](k_F^D + k_{IC}^D + k_{IS}^D + k_R^D)}$$

$$[^1M_2^*] = \frac{I_a k_{DM}[M]}{(k_F^M + k_{IC}^M + k_{IS}^M)(k_F^D + k_{IC}^D + k_{IS}^D + k_R^D + k_{MD}) + k_{DM}[M](k_F^D + k_{IC}^D + k_{IS}^D + k_R^D)}$$

where I_a denotes the rate of light absorption. The quantum yields of molecular and excimer fluorescence are accordingly expressed by Eqs. (1) and (2)

$$\gamma_D^M = k_F^M[^1M^*]/I_a = q_M/(1 + [M]/[M]_{1/2}) \tag{1}$$

and

$$\gamma_F^D = k_F^D[^1M_2^*]/I_a = q_D/(1 + [M]_{1/2}/[M]) \tag{2}$$

where $q_M = [k_F^M/(k_F^M + k_{IC}^M + k_{IS}^M)]$ is the limiting yield of molecular fluorescence at infinite dilution, $q_D = [k_F^D/(k_F^D + k_{IC}^D + k_{IS}^D + k_R^D)]$ denotes the limiting yield of excimer fluorescence at infinite concentration, and

$$[M]_{1/2} = \frac{(k_F^M + k_{IC}^M + k_{IS}^M)(k_F^D + k_{IC}^D + k_{IS}^D + k_R^D + k_{MD})}{k_{DM}(k_F^D + k_{IC}^D + k_{IS}^D + k_R^D)} \tag{3}$$

TABLE II
Compounds Exhibiting Excimer Fluorescence in Solutions at Room Temperature[a]

Compound	Excited state[b]	$(\bar{\nu}_M^0 - \bar{\nu}_D^{max})$, cm^{-1}	ΔL_{ab}, cm^{-1} [c]
Benzene	^1L_b	5400	9000
Naphthalene	^1L_b	5200	3700
Acenaphthene	^1L_b	5900	3300
Anthracene (9-methyl)	^1L_a	7400	—
Pyrene	^1L_b	6100	3100
1,2-Benzpyrene	^1L_b	(4200)	4400
3,4-Benzpyrene	^1L_b	5300	2700
1,2,4,5-Dibenzpyrene	^1L_b	4600	3100
1,2-Benzanthracene	^1L_b	6300	1900
1,2,3,4-Dibenzanthracene	^1L_b	5800	2000
Perylene	^1L_a	5500	—
1,12-Benzperylene	^1L_b	4200	1200
Anthanthrene	^1L_a	5600	—
Cholanthrene	^1L_b	6100	900
2,5-Diphenyloxazole[d]	—	6300	—

[a] Does not include the many derivatives of these compounds which exhibit similar behavior, cf. Ref. 10.
[b] In notation of Platt.
[c] Energy separation of ^1L_a and ^1L_b states when latter is lowest state.
[d] Ref. 12.

TABLE III
Compounds not Exhibiting Excimer Fluorescence in Solution[a]

Compound	Lowest state	ΔL_{ab}, cm^{-1}	Remarks
Anthracene A	^1L_a	—	Form photodimers with $k_R^D \gg k_F^D$
Naphthacene	^1L_a	—	
Pentacene	^1L_a	—	
9,10-Diphenyl A	^1L_a	—	Photoassociation sterically-hindered $k_{DM} = 0$
9,10-Dimethyl-1,2-benzanthracene	^1L_a	—	
Rubrene	^1L_a	—	
Azulene[b]	^1L_b	13900	$\Delta L_{ab} \geq$ stabilization energy of CI between CR and exciton state of ^1L_a origin
Phenanthrene	^1L_b	5100	
Triphenylene	^1L_b	5800	
Picene	^1L_b	3900	
3,4-Benzphenanthrene	^1L_b	4900	
1,2,5,6-Dibenz A	^1L_b	3200	
1,2,7,8-Dibenz A	^1L_b	3200	
3,4,5,6-Dibenzphenanthrene	^1L_b	5100	
Chrysene	^1L_b	2700	$[M]_{max} \ll 1/k_{DM}\tau_F^M$

[a] Cf. J. B. Birks and L. G. Christophorou, *Nature* (London), **197**, 1064 (1963); *Proc. Roy. Soc., Ser A*, **277**, 571 (1964).
[b] B. Stevens and J. T. Dubois, *Trans. Faraday Soc.*, **62**, 1525 (1966).

is a half-quenching concentration at which

$$\gamma_F^M([M]_{1/2}) = q_M/2$$

and

$$\gamma_F^D([M]_{1/2}) = q_D/2$$

The measurement of relative yields γ_F^M/q_M and γ_F^D/q_D as a function of fluor concentration [M] at constant temperature provides a value for $[M]_{1/2}$ at this temperature, as shown in Figure 3; this is related to the rate constant for photoassociation k_{DM} and the lifetimes τ_F^M and τ_F^D of the emitting species at the same temperature by Eq. (3) in the form

$$[M]_{1/2} = (1 + k_{MD}\tau_F^D)/k_{DM}\tau_F^M \tag{4}$$

where

$$\tau_F^M = 1/(k_F^M + k_{IC}^M + k_{IS}^M) \tag{5}$$

and

$$\tau_F^D = 1/(k_F^D + k_{IC}^D + k_{IS}^D + k_R^D) \tag{6}$$

If it can be shown independently that $k_{MD} \ll 1/\tau_F^D$ at the prevailing temperature (Section III.A), then a value for τ_F^M at infinite dilution may be used to compute k_{DM} directly from $[M]_{1/2}$; in all other cases the evaluation of k_{DM} requires an examination of the fluorescence decay parameters as described in Section II.D below.

Values for k_{DM} obtained in this way are compared in Table IV with the diffusion-limited rate constant k_D computed from the relationship[10]

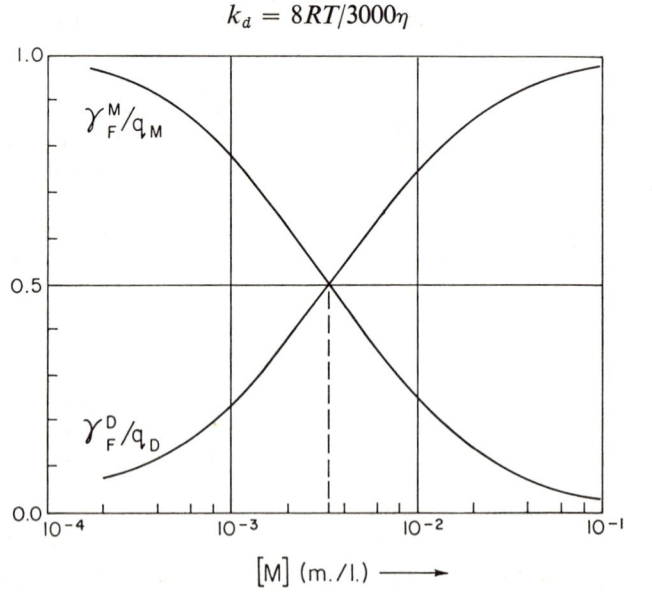

$$k_d = 8RT/3000\eta \tag{7}$$

Fig. 3. Schematic variation of relative molecular (γ_F^M/q_M) and excimer (γ_F^D/q_D) fluorescence quantum yields with solute concentration according to Eqs. (1) and (2).

TABLE IV

Rate Constants (k_{DM}) and Encounter Efficiencies (k_{DM}/k_d) for Photoassociation

M	Solvent	T, °K	k_{DM}, $10^9 M^{-1} \sec^{-1}$	k_d,[a] $10^9 M^{-1} \sec^{-1}$	k_{DM}/k_d	Ref.
Benzene	Methylcyclohexane	195	0.92	0.74	1.2	140
Benzene	Benzene	293	78			37
Toluene	Toluene	293	51			37
p-Xylene	p-Xylene	293	49			
Mesitylene	Mesitylene	293	39			37
1-Methylnaphthalene	Ethanol	293	3.2	5.5	0.58	38
2-Methylnaphthalene	Ethanol	293	2.0	5.5	0.36	38
1-Methoxynaphthalene	Toluene	293	0.034	11	0.003	46
2-Methoxynaphthalene	Toluene	293	0.11	11	0.01	46
1,6-Dimethylnaphthalene	Ethanol	293	0.4	5.5	0.07	38
1,2-Benzanthracene	Cyclohexane	293	2.8	6.7	0.43	42
5-Methylbenzanthracene	Cyclohexane	293	4.1	6.7	0.61	42
6-Methylbenzanthracene	Cyclohexane	293	3.0	6.7	0.45	42
10-Methylbenzanthracene	Cyclohexane	293	5.4	6.7	0.81	42
Pyrene	Benzene	293	11	10	1.1	47
Pyrene	Acetone	293	14	19	0.74	43
Pyrene	Ethanol	293	7.0	5.5	1.3	43
Pyrene	Cyclohexane	293	6.7	6.7		
Pyrene	Cyclohexane	300	7.5	7.7		
Pyrene	Cyclohexane	310	8.9	9.4	~1.0	1
Pyrene	Cyclohexane	320	10	11		
Pyrene	Cyclohexane	330	12	13		
Pyrene	Cyclohexane	340	14	15		
3-Cyanopyrene	Benzene	293	10	10	1.0	47
3-Chloropyrene	Benzene	293	9	10	0.9	47
3-Bromopyrene	Benzene	293	7	10	0.7	47
4-Methylpyrene	Benzene	293	10	10	1.0	47

[a] From Eq. (7).

for a solvent of viscosity η at temperature T; the encounter probability for photoassociation k_{DM}/k_d is virtually unity for pyrene in cyclohexane over an appreciable temperature range whereas for other molecules of lower symmetry a steric factor may be operative.

The absence of an excimer band ($\gamma_F^D = 0$) in concentrated solutions of low viscosity is not necessarily an indication that photoassociation does not take place; thus an examination of Eq. (2) shows that $\gamma_F^D = 0$ if either (a) excimer relaxation is predominantly nonradiative (or $k_R^D \gg k_F^D$) i.e., $q_D \simeq 0$; or (b) $[M]_{1/2} \gg [M]$, i.e., $k_{DM} \ll k_{MD}\tau_F^D/\tau_F^M[M]$, a condition established by rapid excimer dissociation at the prevailing temperature (k_{MD} large, cf. Section III.A) or in the absence of photoassociation ($k_{DM} = 0$).

These alternatives may be distinguished by examining the concentration dependence of the molecular fluorescence yield given by Eq. (1); the observation of self-quenching invalidates condition (b) and provides strong evidence for photoassociation in systems where $\gamma_F^D \simeq q_D \simeq 0$. These criteria are applied to the behavior of anthracene and certain of its derivatives in Table V, from which it is concluded that photoassociation of 9,10-diphenylanthracene (as in 9,10-dimethyl-1,2-benzanthracene and rubrene) is prohibited by steric hindrance of the bulky substituents (Table III).

A possible exception to the use of self-quenching of molecular fluorescence as a photoassociation criterion is provided by azulene which emits[13,14] from the second excited singlet state $^1M^{**}$; in this case self-quenching may involve energy transfer[15] to the lowest excited singlet state $^1M^*$ which has a negligible fluorescence yield. Moreover the absence of self-quenching does not eliminate the possibility of photoassociation if the excimer dissociation probability $k_{MD}\tau_F^D$ vastly exceeds the photoassociation probability $k_{DM}\tau_F^M[M]$ at the prevailing temperature; this situation may be resolved by temperature-dependent studies of the emission spectrum and the self-quenching parameters (Section III.B).

Since the excimer fluorescence yield γ_F^D increases with fluor concentration at a given temperature (Eq. 2), this should have its maximum value for the

TABLE V
Application of Photoassociation Criteria to Anthracene Derivatives

Compound	Self-quenching	Excimer fluorescence	Photo-association	Remarks
Anthracene	Yes	No	Yes	$q_D = 0$
9-Methylanthracene	Yes	Yes	Yes	$q_D > 0$
9,10-Diphenylanthracene	No	No	No	$[M]_{1/2} \gg [M]$, $k_{DM} = 0$

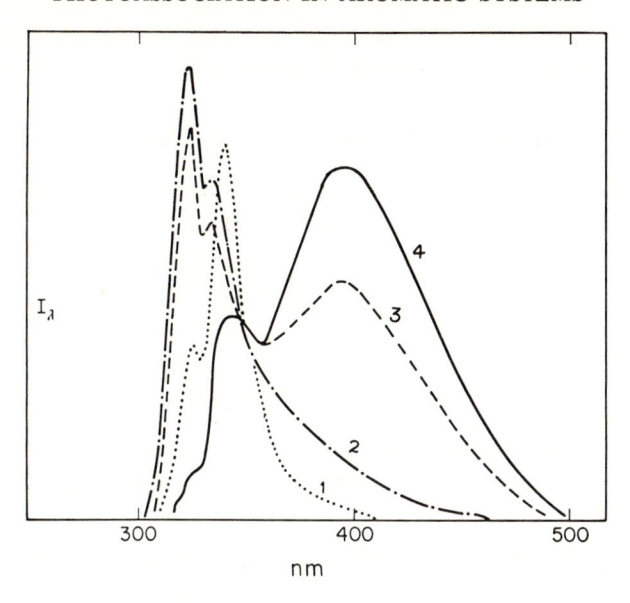

Fig. 4. Fluorescence spectra of 1-fluoronaphthalene,[18] (*1*) solid at 236°K; pure liquid at 356°K, (*2*), 295°K (*3*), and 260°K (*4*).

pure liquid fluor provided the intrinsic yield q_D has a finite value at the fluor melting point. This has been confirmed for a number of low melting aromatic hydrocarbons, notably derivatives of benzene[16,17] and naphthalene[18,19] (see Fig. 4) while even at their respective melting point of 151 and 179°C the fluorescence spectra of liquid pyrene and 1,2-benzpyrene consists solely of the corresponding excimer band.[19] The absence of spectroscopic evidence for excited states of higher aggregation in these systems may be rationalized from the viewpoint of resonance stabilization since the excimer and excited molecular singlet state are no longer degenerate while the ground state excimer configuration that might be expected to interact with the excimer has considerable destabilization energy (Fig. 2 and Table IX) and by definition does not exist[2] in fluid media under normal conditions.

Evidence for photoassociation in the triplet manifold is at present inconclusive. Although Hoytink et al.[20] have reported excimer phosphorescence from cooled ethanolic solutions of phenanthrene and naphthalene, concentration and temperature-dependent studies of the emission characteristics must be extended in order to distinguish photoassociation of the triplet state from intersystem crossing of the singlet excimer and possible triple–triplet annihilation. Certainly the decay constant of the molecular triplet state in fluid media is relatively insensitive to solute concentration[21] although this may be a consequence of rapid dissociation of the weakly-bound triplet excimer to regenerate the original components.

B. Dependence of the Fluorescence Spectrum on Pressure

In the absence of photodimerization ($k_R^D = 0$) the ratio of excimer/molecular fluorescence quantum yields is given by (Eqs. 1–3)

$$\frac{\gamma_F^D}{\gamma_F^M} = \frac{q_D}{q_M} \frac{[M]}{[M]_{\frac{1}{2}}} = \frac{k_F^D k_{DM}[M]}{k_F^M(k_F^D + k_{IC}^D + k_{IS}^D + k_{MD})} \quad (8)$$

which reduces to

$$\gamma_F^D/\gamma_F^M \simeq (k_F^D/k_F^M) K_a [M]$$

under conditions such that $k_{MD} \gg (k_F^D + k_{IC}^D + k_{IS}^D)$. Insofar as the quotient k_F^D/k_F^M is insensitive to changes in pressure, the pressure-dependence of the experimental quantity

$$R = (\gamma_F^D/\gamma_F^M)_P [M]_0/(\gamma_F^D/\gamma_F^M)_0 [M]_P = (\gamma_F^D/\gamma_F^M)_P \rho_0/(\gamma_F^D/\gamma_F^M)_P \rho_P = (K_a)_P/(K_a)_0$$

reflects the variation of photoassociation equilibrium constant $K_a = k_{DM}/k_{MD}$

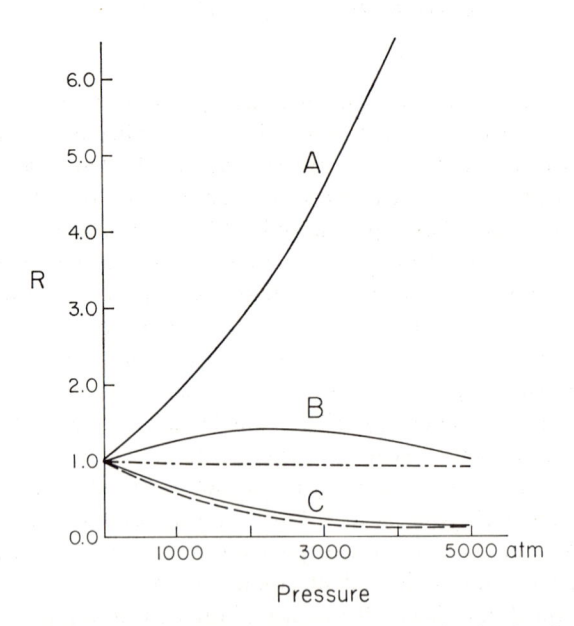

Fig. 5. Effect of pressure on relative yield R of excimer/molecular fluorescence for (A), 6-dimethylnaphthalene; (B) 1,2-benzanthracene and (C) pyrene Curves —·— and ——— describe pressure-dependence of relative solvent density ρ_0/ρ_P and viscosity η_0/η_P (after Seidel and Selinger[23]).

with pressure P at the prevailing temperature;[22] ρ denotes here the solution density, and the zero subscript refers to atmospheric pressure. Seidel and Selinger[23] report a value of $-20 \pm 4 \text{ cm}^3 \text{ mole}^{-1}$ for the volume change ΔV accompanying the photoassociation of 1,6-dimethylnaphthalene at 20°C from the data in Figure 5 expressed in the form

$$(d \ln R/dP)_T = (d \ln K/dP)_T = -\Delta V/RT$$

This represents a contraction of 10% of the volume of two unexcited molecules.

In the limit of negligible excimer dissociation at the prevailing temperature ($k_{MD}\tau_F^D \ll 1$), Eqs. (7) and (8) reduce to

$$\gamma_D^D/\gamma_F^M \simeq q_D k_{DM}[M]/k_F^M \simeq 8RTq_D[M]/3000\eta k_F^M$$

whence

$$R \simeq \eta_0/\eta_P$$

if q_D/k_F^M is assumed to be pressure-invariant. Under these conditions the reduction in excimer/molecular fluorescence yield ratio with increasing pressure closely follows the simultaneous decrease in relative solvent viscosity η_0/η as illustrated by the behavior of pyrene in toluene at 20°C.

The corrected yield ratio R for 1,2-benzanthracene in toluene at the same temperature increases with pressure up to 2500 atm but decreases at still higher pressures; this is attributed to a transition between the limiting conditions described above as excimer dissociation is suppressed at higher pressures. The limiting slope at lower pressures is consistent with a value of $-\Delta V \geq 6 \pm 2 \text{ cm}^3 \text{ mole}^{-1}$ for this molecule.[22,23]

C. Dependence of the Fluorescence Spectrum on Quencher Concentration

Weller and coworkers[24] and Mataga et al.[25] have found that the quenching of aromatic hydrocarbon molecular fluorescence by aromatic amines Q in nonpolar solvents is accompanied by the appearance of a structureless emission band some 5000 cm^{-1} to the red of the molecular fluorescence origin as shown in Figure 6. In the absence of any corresponding change in the absorption spectrum these authors assign the structureless emission to the fluorescence of an excited charge transfer complex (exciplex) formed after the act of light absorption (by either component) in a process of photoassociation:

$$^1M^* + Q \longrightarrow {}^1M^{\pm}Q^{\mp}$$

TABLE VI
Dipole Moments from Solvent Effect on Exciplex Fluorescence Maxima[a]

Solvent	$\dfrac{f-f'}{2}$	$C_6H_5N(C_2H_5)_2$[b] Anthracene*	$C_6H_5N(C_2H_5)_2$[b] Pyrene*	$C_6H_5N(C_2H_5)_2$*[b] Biphenyl	p-$C_6H_4(CN)_2$ Pyrene*[b]	Pyrene* Pyrene
Methylcyclohexane	0.106	21100	22200	27100	—	21000
Toluene	0.117	20450	21300	25400	22800	20900
Diethylether	0.256	20100	21150	25150	22400	21050
Chlorobenzene	0.261	19750	20500	23500	21250	21050
Ethylacetate	0.293	19250	20000	22900	21200	21000
Dimethoxyethane	0.304	18850	19600	23000	20450	21000
Dichloroethane	0.324	18750	19600	22900	20200	21050
Acetone	0.374	18600	19300	22300	—	21050
$\mu°$, Debye		10	11	13.5	12	0.0
$-\Delta H_a$,[d] kcal/mole		8.5	6.5	6.0	3.0	

[a] Units of cm^{-1} (Beens, Knibbe, and Weller[28]).
[b] Electron donor.
[c] From Eq. (15).
[d] From Eq. (27).
Asterisk denotes primarily excited species.

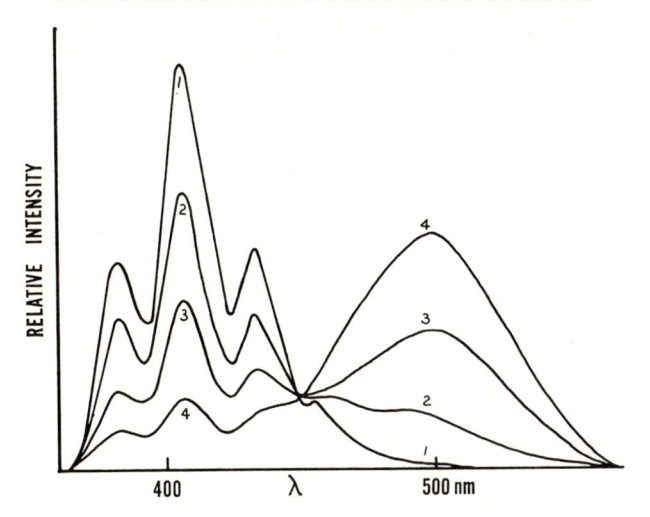

Fig. 6. Fluorescence spectra of anthracene ($3 \times 10^{-4}M$) in toluene in presence of diethylaniline at concentrations $0.000M$ (*1*); $0.005M$ (*2*); $0.025M$ (*3*); $0.100M$ (*4*) (after Weller[24]).

The dependence of molecular and exciplex fluorescence yields, γ_F^M and γ_F^C, on quencher concentration are described by Eqs. (9) and (10)

$$\gamma_F^M = \frac{q_M}{(1 + [Q]/[Q]_{1/2})} \tag{9}$$

$$\gamma_F^C = \frac{q_C}{(1 + [Q]_{1/2}/[Q])} \tag{10}$$

analogous to Eqs. (1) and (2), with

$$[Q]_{1/2} = \frac{(k_F^M + k_{IC}^M + k_{IS}^M)(k_F^C + k_{IC}^C + k_{IS}^C + k_R^C + k_{MC})}{k_{CM}(k_F^C + k_{IC}^C + k_{IS}^C + k_R^C)} \tag{11}$$

$$= (1 + k_{MC}\tau_C^F)/k_{CM}\tau_M^F \tag{12}$$

for those systems where $E_{^1M^*} < E_{^1Q^*}$ and $k_{QC} \ll k_{MC}$ (Scheme 2). The origin of the different emission bands is illustrated schematically in Figure 2 with $M \equiv Q$ and $^1M_2^* \equiv {}^1M^{\pm}Q^{\mp}$, and examples of fluorescent exciplexes are given in Tables VI and VII.

The direction of electron transfer in the exciplex, determined by the inequality

$$(IP + EA)_{acceptor} > (IP + EA)_{donor}$$

where IP and EA denote ionization potential and electron affinity, respectively, has been established as follows.[26] If exciplex solvation energy and

TABLE VII
Limiting Photoassociation Equilibrium Constants K_a for Exciplex MQ from $[Q]_{1/2}$ in Hexane at Room Temperature (Q = diethylaniline)[a]

M	$\bar{\nu}_c$, cm^{-1}	$[Q]_{1/2}$, M	τ_F^M, nsec	τ_F^C, nsec	K_a, M^{-1} [b]
Anthracene	21150	0.0157	4.65	105[c]	> 1400
Pyrene	22250	0.0034	300	130[d]	> 130
Perylene	19000	0.18	5.05	37[e]	> 40

[a] Weller.[24]
[b] From Eq. (16).
[c] Knibbe, Röllig, Schafer, and Weller.[31]
[d] Mataga, Okada, and Yamamoto.[32]
[e] Ware and Richter.[34]

Franck-Condon terms are neglected, the exciplex (charge-transfer) energy relative to the unexcited configuration energy is given approximately by

$$E_{CT} \simeq IP_{donor} - EA_{acceptor} - C - \frac{2\mu^2}{\sigma^3}\left\{\frac{\epsilon - 1}{2\epsilon + 1} - \frac{n^2 - 1}{4n^2 + 2}\right\} \quad (13)$$

where C is the coulombic stabilization energy of the anion-cation pair in the exciplex configuration and the final term represents the stabilization energy of the resultant dipole of moment μ occupying a cavity of radius σ in a solvent of dielectric constant ϵ and refractive index n.[27] For exciplexes of similar configuration in the same solvent it follows that

$$E_{CT}^{max} = h\nu_{CT}^{max} \simeq IP_{donor} - EA_{acceptor} + \text{constant} \quad (14)$$

where ν_{CT}^{max} is the frequency of the exciplex fluorescence peak. For a series of 26 fluorescent hydrocarbons M in liquid N,N'-diethylaniline Q, Knibbe, Rehm, and Weller[26] find that $h\nu_{CT}^\circ$ is expressed as

$$h\nu_{CT}^{max} \text{ (eV)} = 1.17 - 0.65 E_{M^-/M}$$

and conclude that insofar as the half-wave reduction potential $E_{M^-/M}$ and EA_M are both a measure of the lowest vacant orbital energy of the fluor M this acts as electron acceptor in the exciplex M^-Q^+ formed with this quencher.
For a given quencher-fluor (donor-acceptor) system, Eq. (13) in the form

$$\bar{\nu}_{CT}^{max} \text{ (cm}^{-1}) \simeq \text{constant} - 2\mu^2(f - f'/2)/hc\sigma^3 \quad (15)$$

provides a quantitative expression for the observed red shift of the exciplex emission peak with solvent polarity, where $f = (\epsilon - 1)/(2\epsilon + 1)$ and $f' = (n^2 - 1)/(2n^2 + 1)$. From appropriate plots of the data in Table VI, Weller and coworkers [28] obtain the values tabulated for the exciplex dipole moment μ on the assumption that the cavity radius σ has a value of 5 Å. Whereas the (excimer) complex of identical (pyrene) molecules has zero dipole moment, the magnitude of μ obtained for the exciplexes listed confirms the pronounced charge-transfer character of these states in the direction indicated.

In the absence of detailed analyses of fluorescence decay curves (Section II.D), limiting values for the photoassociation equilibrium constants K_a may be obtained from the experimental quantities $[Q]_{1/2}$, τ_F^M, and τ_F^C which are related by the expression (cf. Eq. 12)

$$K_a = \frac{k_{CM}}{k_{MC}} \gtrsim \frac{k_{CM}\tau_F^C}{1 + k_{MC}\tau_F^C} = \frac{\tau_F^C}{\tau_F^M [Q]_{1/2}} \tag{16}$$

As shown in Table VII, the lower limits of K_a vary considerably with exciplex composition, but are of the order of those obtained for excimer formation.

The charge-transfer character of the exciplex introduces the possibility of higher states of aggregation of fluor and quenching species. Thus fluorescence of the exciplex (in this case M^+Q^-) of naphthalene (M) and 1,4-dicyanobenzene (Q) in toluene is gradually replaced by a second structureless emission band at still lower frequencies as the naphthalene concentration is increased, which is attributed [29] to the formation and radiative relaxation of the triple complex $^1M_2Q^*$. The solvent polarity dependence of this additional band maximum parallels that of the exciplex band, confirming the dipolar character of the emitting species which is assigned the asymmetrical configuration MM^+Q^-. A value of 1.1 V (vs. SCE) for the polarographic oxidation potential E_{MM/MM^+} of the naphthalene dimer is estimated from the peak separation of exciplex and triple complex emission.

Change-transfer complexes of solute-alcohol stoichiometry 1:2 have been reported by Walker, Bednar, and Lumry[3] for indole and certain methyl derivatives (M) in mixtures of associating solvents n-butanol and methanol (Q) with n-pentane; these authors introduced the term exciplex to describe the emitter of the red-shifted structureless fluorescence band which increases in intensity with the alcohol content of the mixed solvent. The shift of the exciplex band to longer wavelengths as the solvent polarity is increased, described by Eq. (15), confirms the dipolar nature of the complex that must have the structure $M^+Q_2^-$. No emission corresponding to the 1:1 complex is observed in these systems which indicates (but does not prove) that the photoassociation involves the alcohol dimer. The complex stoichiometry $M^+Q_n^-$ determined from (Eqs. 9, 10, and 12)

$$\frac{\gamma_F^C}{\gamma_F^M} = \frac{q_D}{k_F^M} \frac{k_{CM}[Q]^n}{(1 + k_{MC}\tau_F^C)} = K'[Q]^n \tag{17}$$

is consistent with the value of $n = 1$ for nonassociating solvents diethyl ether, dioxane, and acetonitrile.

D. Decay Characteristics of Molecular and Excimer (Exciplex) Fluorescence

In order to compute absolute rate constant values from photostationary parameters such as γ_F^M, γ_F^D, and $[M]_{1/2}$, it is necessary to obtain independent measurements of the over-all decay constants $1/\tau_F^M$ and $1/\tau_F^D$ of the emitting species (cf. Eqs. 4 and 12). In the absence of photoassociation ($[M] \ll [M]_{1/2}$) the appropriate relaxation equation for the 2-state system

$$-d[^1M^*]_t/dt = (k_F^M + k_{IS}^M + k_{IC}^M)[^1M^*]_t = [^1M^*]_t/\tau_F^M$$

is readily integrated to yield

$$\ln [^1M^*]_t = \ln [^1M^*]_0 - t/\tau_F^M$$

or, since the instrumental response function

$$f^M(t) = \alpha [^1M^*]_t/\tau_F^M$$

is related to the temporal population of emitting states by an instrumental constant α, the instrumental response following a δ-function excitation pulse decays exponentially with time according to

$$f_t^M = f_o^M \exp(-t/\tau_F^M)$$

and provides a value for the decay constant $1/\tau_F^M$ of the emitting species.

The introduction of reversible photoassociation at higher fluor (or quencher) concentrations invalidates the analysis of fluorescence decay curves of either emitting species in terms of a single decay constant since these species do not relax independently; accordingly the decay curves each represent the sum of two exponential components and the corresponding decay constants λ_1 and λ_2 are related to the photoassociation rate constants k_{DM} and k_{MD} and to the lifetimes τ_F^M and τ_F^D of both emitting species. The treatment given below follows that of Birks, Dyson, and Munro.[1]

On the assumption that a δ-function excitation pulse produces a population $[^1M^*]_0$ of excited molecules at $t = 0$, the appropriate equations

$$-d[^1M^*]/dt = \left(\sum_i k_i^M + k_{DM}[M]\right)[^1M^*] - k_{MD}[^1M_2^*]$$

$$-d[^1M_2^*]/dt = \left(\sum_i k_i^D + k_{MD}\right)[^1M_2^*] - k_{DM}[M][^1M^*]$$

may be solved subject to the conditions $[^1M^*] = [^1M^*]_0$ and $[^1M_2^*]_0 = 0$ at $t = 0$ to obtain the temporal distribution of emitting states given by

$$[^1M^*]/[^1M^*]_0 = \{(\lambda_2 - X)e^{-\lambda_1 t} + (X - \lambda_1)e^{-\lambda_2 t}\}/(\lambda_2 - \lambda_1) \quad (18)$$

$$[^1M_2^*]/[^1M^*]_0 = k_{DM}[M]\{e^{-\lambda_1 t} - e^{-\lambda_2 t}\}/(\lambda_2 - \lambda_1) \quad (19)$$

where

$$\lambda_{1,2} = \tfrac{1}{2}[X + Y \mp \{(Y - X)^2 + 4k_{MD}k_{DM}[M]\}^{1/2}] \quad (20)$$

$$X = \sum_i k_i^M + k_{DM}[M] = 1/\tau_F^M + k_{DM}[M]$$

$$Y = \sum_i k_i^D + k_{MD} = 1/\tau_F^D + k_{MD}$$

Under these conditions the instrumental response curves (cf. Fig. 7) are expressed by

$$f^M(t) = \{f_0^M(\lambda_2 - X)/(\lambda_2 - \lambda_1)\}\{e^{-\lambda_1 t} + Ae^{-\lambda_2 t}\} \quad (21)$$

where

$$A = (X - \lambda_1)/(\lambda_2 - X)$$

and

$$f^D(t) = \{f_{max}^D \lambda_2/(\lambda_2 - \lambda_1)\}\{e^{-\lambda_1 (t - t_{max})} - e^{-(\lambda_2 t - \lambda_1 t_{max})}\} \quad (22)$$

where f_{max}^D is the maximum excimer fluorescence response exhibited at time t_{max} given by

$$t_{max} = [\ln(\lambda_1/\lambda_2)]/(\lambda_1 - \lambda_2) \quad (23)$$

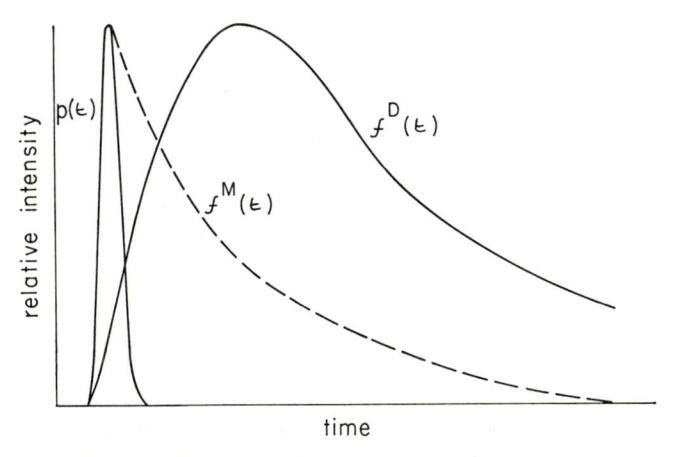

Fig. 7. Schematic molecular $f^M(t)$ and excimer $f^D(t)$ fluorescence response curves following pulsed-flash excitation $p(t)$ (after Birks, Dyson, and Munro[1]).

For a finite excitation pulse width and decay constant λ_p (Fig. 7) the observed fluorescence response function $f'(t)$ is expressed as a superposition integral of the (true) δ-excitation response function $f(t)$ and the over-all time function of source and detector $p(t)$ according to

$$f'(t) = \int_0^t f(t') p(t - t') \, dt'$$

where $p(t) = 0$ at $t \leq 0$. If λ_p is very much larger than either λ_1 or λ_2, as is often the case for the pulsed-flash sampling fluorimeter, it can be shown that $f'(t)$ approximates closely to $f(t)$ when $t \gtrsim 2/\lambda_p$. In this case λ_1 is obtained as $-d \ln f^M/dt$ at low solute concentrations where $\lambda_1 \to X$ and $A \to 0$ (Eqs. 20 and 21), and as $-d \ln f^D(t)/dt (t \to \infty)$ at higher concentrations where $\lambda_2 \gg \lambda_1$ (Eq. 27); λ_2 is most conveniently computed from λ_1 and the peak separation $t_{\max}$ of the excitation pulse and the excimer response maximum according to Eq. (23).

The phase-shift fluorimeter on the other hand provides measurements of the phase angle ϕ and degree of modulation m relative to those of the excitation signal modulated at an angular frequency ω; however, these parameters are also related to the decay constants of both emitting species by λ_1 and λ_2 which are readily computed from the relationships

$$\phi_D = \phi_1 + \phi_2 \qquad \phi_n = \tan^{-1}(\omega/\lambda_n)$$

$$m_D = m_1 m_2 \qquad m_n = \lambda_n/(\lambda_n^2 + \omega^2)^{1/2}$$

where the subscript D refers to excimer (or exciplex) emission.

From measurements of λ_1 and λ_2 as a function of fluor (or quencher) concentration, photoassociation constants k_{MD} and k_{DM}, and lifetimes τ_F^M and τ_F^D of the emitting species may be evaluated using the relationships (Eq. 20)

$$\left. \begin{array}{l} \lambda_1 = 1/\tau_F^M \\ \lambda_2 = 1/\tau_F^D + k_{MD} \end{array} \right\} [M] = 0$$

$$\left. \begin{array}{l} \lambda_1 \to 1/\tau_F^D \\ \lambda_2 \to 1/\tau_F^M + k_{MD} + k_{DM}[M] \end{array} \right\} [M] \to \infty$$

$$\left. \begin{array}{l} \lambda_1 + \lambda_2 = 1/\tau_F^M + 1/\tau_F^D + k_{MD} + k_{DM}[M] \\ d(\lambda_1 + \lambda_2)/d[M] = k_{DM} \end{array} \right\} \text{at all } [M]$$

Values of k_{DM} obtained in this way are listed in Table IV.

It should be noted that excimer dissociation (process MD) is responsible for a nonexponential decay of molecular fluorescence which affords a criterion of photoassociation under conditions ($k_{MD}\tau_F^D \gg 1$) where neither self-quenching of molecular fluorescence nor excimer emission is observed; such behavior has been reported for solutions of naphthalene at moderate temperatures.[33] Alternatively, the exponential decay of molecular fluorescence over several half-lives provides a good indication that excimer dissociation is relatively insignificant ($k_{MD}\tau_F^D \ll 1$) or that excited molecules undergo independent relaxation; in this case k_{DM} may be obtained directly from the measured decay constant at different fluor (or quencher) concentrations according to (Eqs. 20 and 21)

$$-d\ln f^M/dt = \lambda_1 = X = 1/\tau_F^M + k_{DM}[M] \qquad (24)$$

The effect of exciplex dissociation (process MC) on the over-all kinetics of molecular fluorescence decay has been examined by Ware and Richter[34] for the system perylene–dimethylaniline in solvents with dielectric constants (ϵ) varying from 2.3 to 37. In low dielectric media ($\epsilon = 2.3$–4) the perylene fluorescence response may be fitted to a two-component exponential curve and exciplex emission is also observed, whereas in more polar solvents ($\epsilon > 12$) exciplex fluorescence is absent (at ambient temperatures) and the molecular fluorescence decays exponentially. These observations are consistent with both an increase in exciplex stability toward molecular dissociation with solvent polarity (Eq. 13) and the increased probability of dissociation into solvated ions

$$M^-Q^+ \to M_s^- \cdots Q_s^+$$

under the same conditions reported by Weller (Section V.E). Values for k_{CM} in polar media obtained from Eq. (24) (with $k_{CM} \equiv k_{DM}$), and listed in Table VIII, are in fair agreement with diffusion-limited expectation values computed from Eq. (7) but show no correlation with dielectric constant.

TABLE VIII

Photoassociation Rate Constants (k_{CM}) and Encounter Efficiencies (k_{CM}/k_d) for Perylene and Dimethylaniline (Ware and Richter[34])

Solvent	T, °K	ϵ	k_{CM}, 10^9 M^{-1} sec^{-1}	k_d, 10^9 M^{-1} sec^{-1}	k_{CM}/k_d
Pyridine	298	12.0	5.05	7.55	0.67
n-Propanol	298	20.1	4.67	3.32	1.4
Acetone	298	20.7	12.6	22.1	0.57
Ethanol	298	24.3	6.51	6.15	1.06
Ethanol	309	24.3	7.93	7.38	1.07
Methanol	298	32.6	10.5	12.3	0.85
Acetonitrile	298	36.7	13.0	19.5	0.72
Dimethylacetamide	298	37.8	5.45	7.22	0.75

The application[34] of time-dependent diffusion theory to the quenching constant $k_{CM}\tau_F^M$ for the perylene–dimethylaniline system in acetonitrile, which exhibits a pronounced dependence on quencher concentration, provides a value of 6 Å for the quenching encounter diameter as the adjustable parameter which considerably exceeds that of 3 Å computed for the same system in benzene. This accords with the conclusions of Leonhardt and Weller[35] that electron transfer may occur with direct contact of donor and acceptor in polar media (Section V.E).

III. EXCIMER (EXCIPLEX) DISSOCIATION

$$^1M_2^*(^1MQ^*) \xrightarrow{k_{MD(MC)}} {}^1M^* + M(Q)$$

A. Dependence of Fluorescence Spectrum on Temperature

This is illustrated by the fluorescence spectrum of acenaphthene in Figure 8. In the absence of photochemical change ($k_R^D = k_R^C = 0$) the relative quantum yields of excimer/molecular fluorescence is given by Eq. (8)

$$\frac{\gamma_F^D}{\gamma_F^M} = \frac{k_F^D k_{DM}[M]}{k_F^M(k_F^D + k_{IC}^D + k_{IS}^D + k_{MD})} = \frac{k_F^D k_{DM}[M]\tau_F^D}{k_F^M(1 + k_{MD}\tau_F^D)} \qquad (8)$$

in which the fluorescence decay constants k_F^M and k_F^D are assumed to be temperature independent.†

In the low temperature limit defined by the condition $k_{MD}\tau_F^D \ll 1$, Eq. (8) reduces with Eq. (7) to

$$(\gamma_F^D/\gamma_F^M) \simeq (8RTk_F^D[M]\tau_F^D/3000k_F^M\eta_0) \exp(-E_d/RT) \qquad (25)$$

where the temperature-dependence of solvent viscosity is expressed as

$$\eta = \eta_0 \exp(E_d/RT)$$

Accordingly the increase in excimer/molecular fluorescence yield ratio with temperature in this region (Fig. 9) reflects the corresponding increase in encounter frequency $k_{DM}[M]$ of excited and unexcited molecules if $d([M]\tau_F^D)/dT \simeq 0$, and the temperature coefficient is related to the activation energy for viscous flow of the solvent E_d.

The increase in molecular fluorescence yield, and concomitant reduction in yield of excimer fluorescence with temperature at higher temperatures, illustrated in Figure 9, affords direct evidence for excimer dissociation, and under

† This assumption, on which the subsequent treatment is based, is not necessarily valid for k_F^D (cf. page 201), and is limited by the dependence of solvent refractive index on temperature.[36]

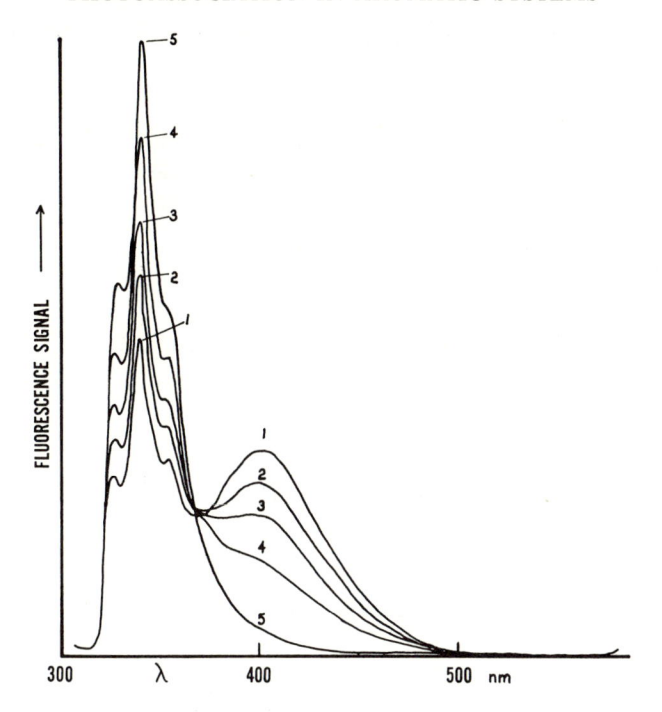

Fig. 8. Fluorescence spectrum of acenaphthene[11] (0.2M) in toluene at 211°K (*1*), 217°K (*2*), 226°K (*3*), 240°K (*4*), and 284°K (*5*).

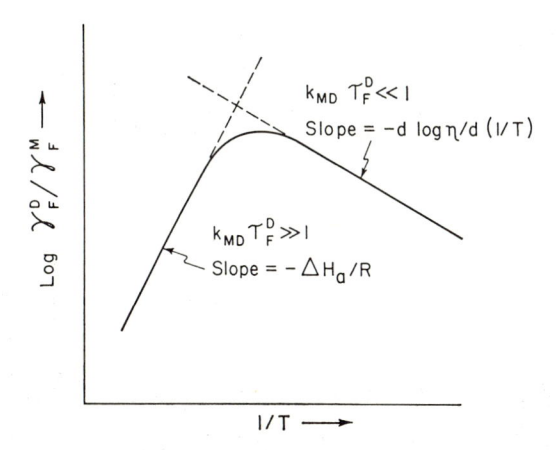

Fig. 9. Schematic variation of relative excimer/molecular fluorescence yield (γ_F^D/γ_F^M) with temperature (cf. Eqs. 25 and 26).

TABLE IX

Enthalpies, Entropies, and Equilibrium Constants for Photoassociation of Aromatic Molecules M

M	Solvent	T, °K	K_a, M^{-1}	$-\Delta H_a$, kcal mole^{-1}	E_R, kcal mole^{-1}	ΔS_a, cal mole^{-1} °C^{-1}	Ref.
Benzene	Benzene	293	0.12	5.1	11.3	−20.3	37
Benzene	Methylcyclohexane	298	0.25	6.2	11.7		140
Benzene	Methylcyclohexane	195	52				140
Toluene	Toluene	293	0.05	6.7	7.1		37
Toluene	Methylcyclohexane	298	0.19	5.1			140
Toluene	Methylcyclohexane	195	18				140
Ethylbenzene	Methylcyclohexane	195	17	4.8			140
Cumene	Methylcyclohexane	195	8.1	4.6			140
o-Xylene	Methylcyclohexane	195	4.5	4.6			140
m-Xylene	Methylcyclohexane	195	8.1	4.8			140
p-Xylene	Methylcyclohexane	195	7.5	6.0			140
p-Xylene	p-Xylene	293	0.020	≥2.5	≤7.9	≤−13.8	37
Mesitylene	Methylcyclohexane	195	1.4	3.9			140
Mesitylene	Mesitylene	293	0.078	2.8	8.1	−15.6	37
1,2,3-Trimethylbenzene	Methylcyclohexane	195	3.6	3.9			140
1,2,4-Trimethylbenzene	Methylcyclohexane	195	2.7	3.7			140
Phenol	Phenol	281–313		4.5	11.6	−25	17
Anisol	Anisole	239–297		5.1	10.7	−22	17
Naphthalene	Ethanol	173–293		6.2	12.5		38
Naphthalene	n-Heptane	214–293		5.8	12.9		39
1-Fluoronaphthalene	Toluene	222–295		5.8	10.7	−21.8	4, 11
1-Methylnaphthalene	Ethanol	173–293		4.7	14.0	−10.0	38
1-Methylnaphthalene	n-Heptane	223–293		6.9	11.8		39
2-Methylnaphthalene	Ethanol	173–293		5.1	12.9	−12.0	38
2-Methylnaphthalene	n-Heptane	248–293		6.9	11.1		39
2-Methylnaphthalene	Ethyl ether	204–273		5.8	12.9	−21.0	11
1,6-Dimethylnaphthalene	Ethanol	173–293		4.6	9.4	−12.5	38
1,6-Dimethylnaphthalene	n-Heptane	293	1.94	6.2	7.8	−19.8	39, 40

1,8-Dimethylnaphthalene	n-Heptane	233–293	3.2	14.8		39
α-Methoxynaphthalene	Toluene	170–390	5.2		−4.4	46
β-Methoxynaphthalene	Toluene	170–380	8.0		−5.8	46
Acenaphthene	Toluene	203–234	4.9		−19.0	11
9-Methylanthracene	Benzene	240–340	4.6	11.9		41
9-Methylanthracene	Chloroform	240–340	5.3	15.1		41
9,10-Dimethylanthracene	Benzene	240–340	4.4	14.2		41
9,10-Dimethylanthracene	Chloroform	240–340	3.9	12.9		41
9,10-Dimethylanthracene	Ethyl ether	240–340	3.9	13.6		41
9,10-Dimethylanthracene	Ethyl ether	249–265		~16.0		41
1,2-Benzanthracene	Cyclohexane	293	6.0	11.4	−17.4	11
1,2-Benzanthracene	Cyclohexane	293	28			42
5-Methylbenzanthracene	Cyclohexane	293	140			42
6-Methylbenzanthracene	Cyclohecane	293	70			42
10-Methylbenzanthracene	Cyclohexane	293	150			42
Pyrene	Acetone	293	1170	8.1		43
Pyrene	Cyclohexane	293	1030	7.7		43
Pyrene	Ethanol	293	1000	7.9	−17.6	43
Pyrene	Ethanol	333–358		7.9	−18.5	11
Pyrene	Nonane		10.9	6.2		44
Pyrene	Hexadecane		10.1	7.0		44
Pyrene	Paraffin oil	298–573	11.0	6.0	−20	45

TABLE X

Frequency Factors A_{MD} (sec^{-1}) for Excimer Dissociation[a]

M = Solvent	Pyrene Paraffin oil	1-Methyl-naphthalene Ethanol	2-Methyl-naphthalene Ethanol	1,6-Dimethyl-naphthalene Ethanol	1-Methoxy-naphthalene Toluene	2-Methoxy-naphthalene Toluene
A_{MD}	9×10^{17}	1.1×10^{14}	2.4×10^{14}	0.1×10^{14}	1.2×10^{11}	3.4×10^{11}
Ref.	45	38	38	38	46	46

[a] From Eq. (28).

the limiting condition $k_{MD}\tau_F^D \gg 1$ is described quantitatively by Eq. (8) in the reduced form

$$\frac{\gamma_F^D}{\gamma_F^M} \simeq \frac{k_F^D}{k_F^M}\frac{k_{DM}[M]}{k_{MD}} = \frac{k_F^D}{k_F^M} K_a[M]$$

$$\simeq (k_F^D[M]/k_F^M) \exp(\Delta S_a/R) \exp(-\Delta H_a/RT) \tag{26}$$

The variation of log (γ_F^D/γ_F^M) with reciprocal temperature over the whole temperature range is shown schematically in Figure 9 where the slope of the data lines in the limiting high and low temperature regions provide values for the photoassociation enthalpy ΔH_a and activation energy for solute diffusion in accordance with Eqs. (26) and (25), respectively. Independent estimates of the relative radiative decay constants $k_F^D/k_F^M = q_D\tau_F^M/q_M\tau_F^D$ at one temperature are necessary to obtain values for the photoassociation equilibrium constant K_a and entropy of photoassociation ΔS_a from the experimental yield ratio at high temperatures[11] (Eq. 26). The accuracy of the data obtained in this way is improved by compensation for the temperature coefficient of [M] or of solvent density which is usually neglected.

Measurements of $[M]_{1/2}$ over the limiting temperature regions provides values for the same parameters[38,45,46] from Eq. (4) in the form

$$[M]_{1/2} = \frac{3000\eta_0}{8RT\tau_F^M} \exp(E_d/RT) + \frac{\tau_F^D}{\tau_F^M} \exp(-\Delta S_a/R) \exp(\Delta H_a/RT)$$

although this procedure strictly requires the independent measurement of the decay constants $1/\tau_F^M$ and $1/\tau_F^D$ over the appropriate range of temperatures.

At a given temperature K_a may be evaluated directly as the quotient k_{DM}/k_{MD} from analyses of the fluorescence decay curves[16] as described in Section II.D. If ΔH_a is also available for the same system, then ΔS_a may be computed[39-43] from the relationship

$$\Delta S_a = R \ln K_a + \Delta H_a/T$$

The thermodynamic quantities listed in Table IX together with destabilization energies E_R of the unexcited excimer configuration computed as (Fig. 2)

$$E_R = h(\nu_M^0 - \nu_D^{max}) + \Delta H_a \tag{27}$$

support the following conclusions:

(a) K_a is independent of solvent at a given temperature†; it is interesting to note in this respect that in at least two cases examined (toluene[37] and 1,6-dimethylnaphthalene[39]) a similar value for K_a is found for the pure liquid fluor.

† Hirayama and Lipsky[140] find, however, that K_a varies with solute concentration in the case of benzene and some alkyl derivatives.

(b) ΔH_a (and therefore ΔS_a) is also independent of solvent (if values reported by the same observer are compared).

(c) Except in the case of pyrene, the ground state destabilization energy is largely responsible for the observed red shift of the excimer band maximum at ν_D^{max} relative to the molecular $0''$–$0'$ fluorescence band frequency ν_M^0.

(d) Entropies of photoassociation (apart from values reported by Selinger[38,46]) are on the order of -20 cal °C^{-1} mole^{-1} which is the expectation value for a molecular association process; significantly lower values of $-\Delta S_a$ may be attributed to a reduction in photoassociation encounter probability (Table IV).

In solution the over-all process of excimer dissociation includes escape of the dissociation products from the solvent cage, and the corresponding rate constant expressed as

$$k_{MD} = A_{MD} \exp\left[(-E_d + \Delta H_a)/RT\right] \qquad (28)$$

varies with the activation energy E_d for viscous flow of the solvent.[44] A solvent dependence of the frequency factor A_{MD} may also be responsible for the wide range of values listed in Table X, in which case excimer dissociation could be complex in the kinetic sense.

Weller[24] has estimated enthalpies of exciplex formation from the energy separation $\nu_M^0 - \nu_C^{max}$ of the molecular $0''$–$0'$ and exciplex fluorescence maximum using the appropriate form of Eq. (27) with E_R assumed to have the value found for pyrene; despite the doubtful validity of this approximation the values listed for ΔH_a in Table VI are sufficiently low to permit exciplex dissociation during its radiative lifetime and the total emission spectrum of these systems may be expected to vary with temperature in the manner described above for one-component systems. This has recently been confirmed by Knibbe, Rehm, and Weller[30] who obtain the enthalpies and entropies of photoassociation of the donor-acceptor pairs listed in Table XI. From a detailed analysis of the fluorescence decay curves for the perylene–diethylaniline system in benzene, Ware and Richter[34] find that

$$k_{MC} = 2 \times 10^{15} \exp(-10200/RT) \text{ sec}^{-1}$$

from which it follows that with

$$h\nu_M^0 - h\nu_C^{max} = 9.6 \text{ kcal mole}^{-1}$$

$$-\Delta H_a = 10.2 - E_d = 8.1 \text{ kcal mole}^{-1}$$

and

$$E_R = 9.6 - 8.1 = 1.5 \text{ kcal mole}^{-1}$$

(cf. Eqs. 27 and 28). These values of ΔH_a and E_R are in fair agreement with

TABLE XI
Enthalpies and Entropies of Heteromolecular Photoassociation of Donor (D) and Acceptor (A) in Hexane Solution[30]

A	D	$-\Delta H_a$, kcal mole^{-1}	$-\Delta S_a$, cal mole^{-1} °C^{-1}	E_R,[a] kcal mole^{-1}
Anthracene*	Diethylaniline	10.2	18.7	5.5
Pyrene*	Diethylaniline	8.4	18.3	5.0
1,2-Benzanthracene*	Diethylaniline	8.3	19.7	4.4
Perylene*	Diethylaniline	6.6	15.6	3.0
Naphthalene*	Triethylamine	7.2	—	15.2
Anthracene*	Triethylamine	4.1	—	13.6
1,2-Benzanthracene*	Triethylamine	2.7	—	15.4
Biphenyl	TMPD*	10.2	18.0	10.5
Biphenyl	Diethylaniline*	7.3	17.3	8.0

[a] Ground state destabilization energy (cf. Fig. 2).
Asterisk denotes primarily excited species.

the data in Table XI for this system, which are the more reliable, and the frequency factor $A_{MC} = 2 \times 10^{15}$ sec^{-1} lies within the range of values reported for A_{MD} (Table X).

B. Temperature Dependence of Experimental Quenching Constants

The concentration dependence of molecular fluorescence yield may be expressed by Eqs. (1) and (4) in the form

$$\frac{1}{\gamma_F^M} = \frac{1}{q_M}\left\{1 + \frac{k_{DM}\tau_F^M[M]}{1 + k_{MD}\tau_F^D}\right\} \quad (29)$$

Excimer dissociation (process MD) therefore reduces the Stern-Volmer self-quenching constant from its expectation value $K_M^0 = k_{DM}\tau_F^M$ to

$$K_M = d(q_M/\gamma_F^M)/d[M] = k_{DM}\tau_F^M/(1 + k_{MD}\tau_F^D) \quad (30)$$

with

$$d\ln(1/K_M)/d(1/T) \simeq d\ln k_{MD}/d(1/T) = (\Delta H_a - E_d)/R \quad (31)$$

Just as photoassociation may be inferred from the observation of fluorescence self-quenching, so may the negative temperature coefficient of K_M be adopted as a criterion of excimer dissociation in the absence of excimer fluorescence. The application of Eq. (31) to data obtained for the self-quenching of anthracene in the vapor phase leads to a value of $-\Delta H_a \simeq 7.6$

kcal mole^{-1} in this case[48] (where $E_d = 0$) which is of the magnitude obtained for other compounds in solution (Table IX).

It should be noted that experimental verification of the inequality

$$K_M < 8RT\tau_F^M/3000\eta$$

at a single temperature in solution is not necessarily diagnostic of excimer dissociation since a reduction in the photoassociation rate constant $k_{DM} = 8pRT/3000\eta$ with $p < 1$ due to steric factors could be responsible for the same effect (Table IV). However, since the encounter reaction probability p should not exhibit a negative temperature dependence these alternatives are resolved by measurements of K_M at different temperatures.

Equations (29)–(31) may also be applied with appropriate modification to the physical quenching of molecular fluorescence by any quencher Q, in which case a negative temperature coefficient of the quenching constant

$$K_Q = d(q_M/\gamma_F^M)/d[Q] = k_{CM}\tau_F^M/(1 + k_{MC}\tau_F^C) \tag{32}$$

affords a criterion of exciplex dissociation in the absence of spectroscopic evidence for photoassociation. This has been demonstrated for the perylene–diethylaniline complex by Ware and Richter[34] who use Eq. (31) to obtain a value of 10.2 kcal mole^{-1} for $\Delta H_a - E_d$ for this system in benzene.

C. Fluorescence Quenching Efficiency

The very wide range of quenching efficiencies exhibited by different quenchers under the same conditions (cf. Table I) can be rationalized in terms of reversible photoassociation and provides indirect evidence for this phenomenon. Thus if K_Q is significantly less than $k_d \tau_F^M$ it may be assumed that $k_{MC}\tau_F^C \gg 1$ and Eq. (32) reduces to

$$K_Q \simeq k_{CM}\tau_F^M/k_{MC}\tau_F^C$$

$$\simeq (\tau_F^M/\tau_F^C) \exp(\Delta S_a/R) \exp(-\Delta H_a/RT) \tag{33}$$

If the exciplex M^-Q^+ is treated as a pure charge-transfer state with (cf. Eq. 13)

$$-\Delta H_a \sim h\nu_M^D - IP_Q + EA_M + C + 2\mu^2[(f - f')/2]/\sigma^3 \tag{34}$$

then the quenching constant measured for the same fluor M in the same solvent should increase with ionization potential of the quencher Q according to the approximate expression

$$\ln K_Q \sim \text{constant} - IP_Q$$

insofar as the exciplex properties τ_F^C, μ, and C are independent of quencher.

This is qualitatively illustrated by the data[49] in Table XII which confirms the direction of electron transfer in these systems.

Although no such correlation has been established for fluorescence quenching by conjugated dienes,[50] the wide range of quenching constants obtained experimentally[51] may be explained in terms of exciplex stability if this depends on the π-orbital overlap of the interacting species. The steric requirements in this case appear to be very critical.

It is emphasized that the measured quenching constant depends on the relative frequencies of exciplex dissociation and exciplex relaxation expressed as the quantity $k_{MC}\tau_F^C$; thus a change in properties of the quenching molecule may influence both the exciplex stability and its lifetime. This behavior is illustrated by the quenching of molecular pyrene fluorescence in benzene by halogenated molecules, which it is suggested[52] increase spin-orbital coupling in the exciplex and thereby promote intersystem crossing to the triplet exciplex manifold. If this is the dominant exciplex relaxation process (no radiative relaxation is observed), then

$$1/\tau_F^C \sim k_{IS}^C \propto \left(\sum_i \xi_i\right)^2$$

where ξ_i is the spin-orbit coupling parameter of the ith atom.[53] In this approximation the measured quenching rate constant $k_Q = K_Q/\tau_F^M < k_d$ is given by (Eqs. 33 and 34)

$$\log\left[k_Q\bigg/\left(\sum_i \xi_i\right)^2\right] \sim \text{constant} + (\text{EA}_A - \text{IP}_D)/RT \tag{35}$$

for the same fluor/solvent system if the exciplex configuration is independent of Q (i.e., $\Delta C = \Delta \mu = 0$). The linear variation of $\log [k_Q/(\sum_i \xi_i)^2]$ with half-wave reduction potential ($\sim \text{EA}_A$) of the quencher shown in Figure 10 is consistent with the interpretation given and indicates a direction of electron transfer from fluor to quencher in the exciplex.

It has recently been shown[54] that oxygen quenching of potentially fluorescent molecules involves a collision-induced intersystem crossing from the excited singlet state, possibly via the formation of a charge transfer complex[55] $M^+O_2^-$. Although the multiplicity of the quencher in this case prevents a description of the quenching sequence in terms of Scheme 2, the high efficiency of the over-all process is consistent with the formation of a stable complex (with respect to dissociation) of very short lifetime. Variations of k_Q by a factor of ~ 3 for different fluorescent species[56] may, however, reflect differences in complex stability at the prevailing temperature, and a temperature-dependent dissociation frequency would account for the much lower collisional quenching efficiencies of oxygen at higher temperatures in the vapor phase.[57]

TABLE XII

Variation of Experimental Rate Constant k_Q with Ionization Potential of Quencher of Acridine Fluorescence in Aqueous NaOH (0.03M) at 25°C (cf. Weller[49])

Quencher	NH$_3$	CH$_3$NH$_2$	i-C$_3$H$_7$NH$_2$	n-C$_4$H$_9$NH$_2$	C$_6$H$_5$CH$_2$NH$_2$	(CH$_3$)$_2$NH	(CH$_3$)$_3$N	(C$_2$H$_5$)$_3$N
$10^{-8}k_Q$, M^{-1} sec^{-1}	0.28	1.8	1.8	4.8	6.2	16	19	30
IP_Q, eV	10.16	8.97	8.72	8.70	7.56	8.24	7.82	7.56

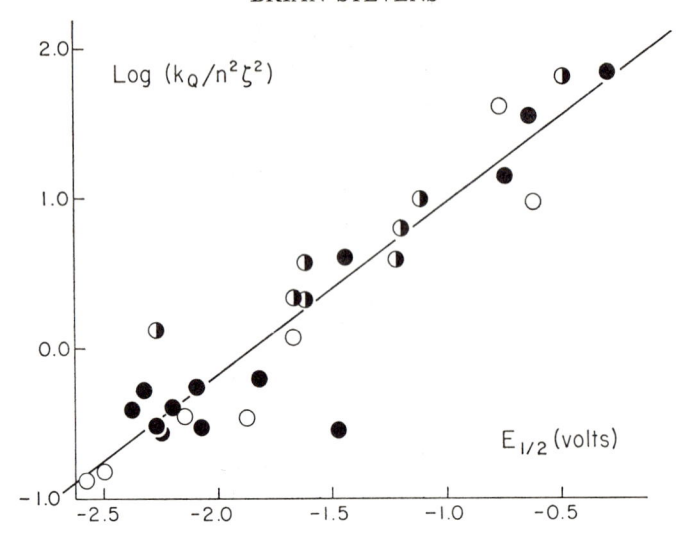

Fig. 10. Plot of data for external heavy-atom quenching of pyrene molecular fluorescence in benzene at 20°C according to Eq. (35); polarographic half-wave reduction potential $E_{1/2}$ is taken as measure of electron affinity of quenching species containing chlorine (○), bromine (●), or iodine (◐) atoms (Thomaz and Stevens[52]).

D. Energy Transfer Efficiencies

In the absence of long-range dipole-induced dipole interaction of fluor and quencher, as evidenced by the negligible overlap of fluor emission and quencher absorption spectra,[58] the diffusional energy transfer process

$$^1M^* + Q \xrightarrow{k_t} M + {}^1Q^*$$

may be accommodated by the general quenching Scheme 2 if photoassociation is followed by the alternative mode of exciplex dissociation represented by

$$^1M^* + Q \underset{k_{MC}}{\overset{k_{CM}}{\rightleftarrows}} {}^1MQ^* \xrightarrow{k_{QC}} M + {}^1Q^*$$

The transfer rate constant k_t, obtained from measured transfer efficiencies expressed as

$$\frac{\gamma_F^Q}{q_Q} = \frac{k_t[Q]\tau_F^M}{1 + k_t[Q]\tau_F^M} \tag{36}$$

where γ_F^Q is the observed yield of quencher fluorescence produced by selective excitation of M, is related to the photoassociation and exciplex dissociation constants by

$$k_t = k_{CM}\{k_{QC}/(k_{QC} + k_{MC} + \sum_i k_i^C)\} \tag{37}$$

If therefore the transfer process is diffusion-limited ($k_t \simeq k_d$), as found for a number of (energy) donor-acceptor systems by Dubois and coworkers,[59] it must be concluded that $k_{QC} \gg (k_{MC} + \sum_i k_i^C)$ with $k_{CM} \simeq k_d$. This is in contrast to the findings for triplet energy transfer where the measured rate constant ^3k_t falls below the diffusion-limited value (Eq. 7) as the energy separation $^3\Delta E_{MQ}$ of donor ($^3M^*$) and acceptor ($^3Q^*$) triplet states is reduced.[60] With $^3k_{QC} \gg k_P^C + k_{GT}^C = 1/\tau_P^C$ the appropriate form of Eq. (37)

$$^3k_t = {}^3k_{CM}\{^3k_{QC}/(^3k_{QC} + {}^3k_{MC} + 1/\tau_P^C)\}$$

reduces to

$$^3k_t \simeq {}^3k_{CM}/(1 + {}^3k_{MC}/{}^3k_{QC}) \simeq k_d/(1 + \exp(-{}^3\Delta E_{MQ}/RT))$$

which is in qualitative agreement with observation. However, more extensive measurements of $^3k_t(\tau_P^Q, T)$ are required to distinguish this mechanism from that involving (endothermic) transfer in the reverse direction[61,62] following the separation of M and $^3Q^*$.

Indirect evidence for the formation of a complex intermediate in diffusional energy transfer processes may be provided by the measurement of k_t from both donor quenching and acceptor sensitization at low temperatures where $k_{QC} \sim \sum_i k_i^C = 1/\tau^C$; in this case exciplex relaxation should reduce the quencher sensitization efficiency but leave the donor fluorescence quenching constant unchanged.

In photoassociating solvents (e.g., benzene and its alkyl derivatives) the transfer of electronic excitation energy from solvent to fluorescent solute is characterized[37,63] by rate constants k_t which are approximately twice those computed from the less approximate form of Eq. (7):

$$k_t = k_d = 2\pi N(D_M + D_Q)R_{MQ}/1000$$

using independently-measured values for the diffusion coefficients D and a critical (resonance) transfer distance R_{MQ} estimated[58] from spectroscopic properties of the solvent (M) and solute (Q). This is accommodated[63] by the introduction of an additional diffusion coefficient Λ to describe the migration of excitation energy through the solvent which, it is suggested, contributes substantially to the over-all solvent-solute transfer process with

$$k_t = 2\pi N(D_M + D_Q + \Lambda)R_{MQ}/1000 \tag{38}$$

Birks and coworkers[37] have provided a quantitative description of this solvent energy migration based on the successive formation and dissociation of the solvent excimer, i.e.,

$$^1M_a^* + M_b \underset{k_{MD/2}}{\overset{k_{DM}[M]}{\rightleftarrows}} {}^1M_{ab}^* \xrightarrow{k_{MD/2}} M_a + {}^1M_b^*$$

TABLE XIII
Solvent Energy Migration Parameters at 293°K[a]

	Benzene	Toluene	p-Xylene	Mesitylene
[M], mole liter^{-1}	11.2	9.7	8.1	7.2
K_a, liter mole^{-1}	0.12	0.055	0.02	0.018
k_{DM}, liter mole^{-1} sec^{-1}	7.8×10^{10}	5.1×10^{10}	4.9×10^{10}	3.9×10^{10}
Λ, cm^2 sec^{-1}	6.8×10^{-5}	7.5×10^{-5}	9.9×10^{-5}	7.3×10^{-5}
δ	0.57	0.37	0.14	0.11
$\bar{a}$, Å	3.3	3.6	4.2	4.3

[a] From transfer to solute PPO.[37]

which provides a mean transit time $\bar{t}$ for energy transfer between adjacent solvent molecules given by

$$\bar{t} = 1/k_{DM}[M] + 1/k_{MD} = (1 + K_a[M])/k_{DM}[M] \quad (39)$$

This is related by elementary diffusion theory to the migration coefficient Λ and the root-mean-square energy-displacement $\bar{a}$ according to

$$\Lambda = \bar{a}^2/6\bar{t} \quad (40)$$

whence from Eqs. (39) and (40)

$$\Lambda = \bar{a}^2 k_{DM}[M]/6(1 + K_a[M]) = \bar{a}^2 \, \delta k_{MD}/6 \quad (41)$$

where δ is the fraction of excited solvent molecules present in the photo-associated form. Λ is estimated from experimental measurements of k_t (Eq. 36) using Eq. (38) and average values of D and R for molecular and excimer diffusion and transfer to the solute. The corresponding values of $\bar{a}$ are listed in Table XIII, together with the quantities required to evaluate these from Eq. 41, for the solvents quoted with 2,5-diphenyloxazole as solute.[37] The energy displacement increases with alkyl substitution of the solvent as expected for a photoassociation mechanism.

IV. EXCIMER (EXCIPLEX) BINDING ENERGIES

A. The Excimer Configuration

Theoretical estimates of the excimer stabilization energy $W = -\Delta H_a$ (Table IX) have been made on the assumption of a symmetrical sandwich structure of one planar identical molecule exactly superimposed on the other; this configuration is based on the observation of characteristic emission bands exhibited by rigid systems (e.g., 4,4′-paracyclophane, crystalline pyrene) in which the molecular components are suitably oriented prior to excitation at an interplanar distance R of ~ 3.5 Å.

Chandross and Ferguson[64] find that the absorption spectra of dimers, produced[65] by photolytic cleavage of photodimers of anthracene and monoderivatives in a rigid methylcyclohexane glass at 77°K, are consistent with a symmetrical sandwich configuration; these dimers also emit the characteristic excimer fluorescence. On the other hand, it is necessary to assume a 60° rotation of one component about the intermolecular axis of the 9,10-dichloroanthracene dimer (as in the crystalline compound) to account for the observed resonance splittings of both absorption bands.[64]

B. Exciton and Charge Resonance Concepts

In the excimer configuration the excited molecular states ^1L_a and ^1L_b are each split into nondegenerate exciton states as shown schematically in Figure 11. The first-order perturbation energy ΔE_{exc} at the equilibrium interplanar separation R_0 is given by[66]

$$\Delta E_{\text{exc}} = m^2/R_0^3 \tag{42}$$

where the transition moment m (e.s.u) is obtained from the experimental oscillator strength f of the appropriate molecular transition and the relationship

$$f = 4.703 \times 10^{29} \bar{\nu} |m^2|$$

for an average transition energy $\bar{\nu}$ (cm^{-1}). Owing to the larger $^1L_a \leftarrow ^1A$ oscillator strength, the exciton stabilization energy is greater for excimer states of ^1L_a origin than for those of ^1L_b parentage. However, for molecules of lowest ^1L_b states (Fig. 11) the energy separation ΔL_{ab} of molecular ^1L_a and ^1L_b states must be subtracted from the ^1L_a exciton state energy $\Delta E_{\text{exc}}(^1L_a)$ to obtain the excimer binding energy $W_{\text{exc}}(^1L_a)$ measured relative to the lowest excited molecular state ^1L_b; i.e., if $^1L_a > ^1L_b$

$$W_{\text{exc}}(^1L_a) = m^2(^1L_a)/R_0^3 - \Delta L_{ab} \tag{43a}$$

$$W_{\text{exc}}(^1L_b) = m^2(^1L_b)/R_0^3 \tag{43b}$$

whereas for $^1L_b > ^1L_a$

$$W_{\text{exc}}(^1L_a) = m^2(^1L_a)/R_0^3 \tag{44}$$

In the absence of experimental values for R_0, these are computed from Eqs. (43) and (44) with $W = -\Delta H_a$ (Table X) and are given for a number of molecules in columns 2 and 3 of Table XIV. Since these are significantly

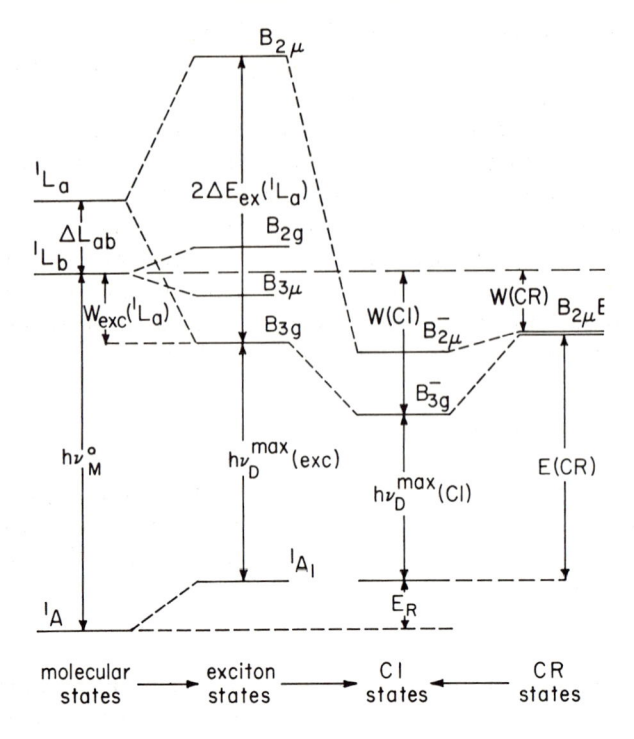

Fig. 11. Energy diagram illustrating origin and configuration interaction of lowest exciton and charge resonance states of point group D_{2h}.

TABLE XIV
Equilibrium Interplanar Distances R_0 (Å) Calculated for D_{2h} Excimer Configuration

Molecule	Lowest excited state	Exciton (^1L_b)[a]	Exciton (^1L_a)[b]	CI (exc CR)[c]	LCAO MO[d]
Naphthalene	^1L_b	0.9	2.5	3.2	3.0
Anthracene	^1L_a		2.7	3.2	3.5
Pyrene	^1L_b	0.7	2.5	3.3	—
Perylene	^1L_a		4.1	3.6	4.0

[a] Equation (43b).
[b] Equations (43a) and (44).
[c] Reference 68.
[d] Reference 73.

lower than the expected value 3.0 Å $< R_0 <$ 3.5 Å, it is concluded[67] that the excimer binding energy cannot be explained solely in terms of ^1L exciton states; however, in certain cases states of ^1B origin may contribute to the observed excimer stability.[71]

The charge resonance states (CR) represented by

$$M^+M^- \longleftrightarrow M^-M^+$$

arise from the interaction of two degenerate charge transfer states produced by the promotion of an electron from the highest-filled orbital of one molecular component to the lowest vacant orbital of the other. The energy of the lowest CR state relative to the molecular ground state is given by

$$E(CR) = IP - EA - C - \Delta$$

where C is the coulomb attraction energy of the two ions at the equilibrium separation and the interaction energy Δ of zeroth order charge transfer states M^+M^- and M^-M^+ is sufficiently small[67] to be neglected in this approximation. The CR binding energy $W(CR)$, relative to that of the lowest molecular ^1L state, is therefore obtained as (Fig. 11)

$$W(CR) = E(^1L) - E(CR) - E_R$$

whence

$$h\nu_D^{max} = E(^1L) - W(CR) - E_R \simeq IP - EA - C$$

The linear dependence of $h\nu_D^{max}$ on $IP - EA$ has been demonstrated[67] for a limited number of photoassociating molecules (including benzene) indicating the approximate constancy of the coulomb energy term C. However, the coefficient $h\nu_D^{max}/(IP - EA)$ is appreciably less than unity and the evaluation of C using a point-charge approximation at $R_0 = 3$ Å leads to values for $h\nu_D^{max}$ which are 1–2 eV greater than those observed.[67]

Azumi, Armstrong, and McGlynn[68] have subsequently treated the configuration interaction (CI) of zeroth-order exciton and charge resonance states in a four electron approximation; a similar approach is adopted by Murrell and Tanaka[69] and by Hoytink and Konijnenberg.[70] Group-theoretical considerations of molecules and (symmetrical) excimers of point group D_{2h} show[68] that the exciton states of ^1L_a origin and the degenerate charge resonance states transform identically as B_{3g} and B_{2u} (Fig. 11) and differently from exciton states of ^1L_b origin; accordingly CR and exciton states of ^1L_a origin should exhibit pronounced configuration interaction. Energies E of the resulting CI states B_{3g}^+, B_{3g}^-, B_{2u}^+ and B_{2u}^- are obtained[68] relative to the ground state energy $E_0 (= E_R)$ as roots of the secular equations

$$\begin{vmatrix} (H_{aa} - E_0) - E & (H_{ab} - S_{ab}E_0) - S_{ab}E \\ (H_{ab} - S_{ab}E_0) - S_{ab}E & (H_{bb} - E_0) - E \end{vmatrix} = 0$$

with, e.g.,

$$H_{aa}(B_{3g}) = E(^1L_a) - m^2(^1L_a)/R^3$$

$$H_{bb}(B_{3g}) = H_{bb}(B_{2u}) = \text{IP} - \text{EA} - C$$

$$H_{ab} = \langle(\text{exc})|\mathscr{H}|(\text{CR})\rangle$$

where the total Hamiltonian $\mathscr{H}$ includes the core Hamiltonian and the sum of electrostatic π-electron repulsion terms. The overlap integral S_{ab} is expressed in terms of carbon 2p atomic orbital overlap for corresponding atoms in the two molecules. Of the adjustable parameters R and effective nuclear charge Z in the Slater orbital exponent, these authors assign $Z = 3.18$ for all integrals (cf. Refs. 69 and 70) to compute $E(R)$ for each CI state. For the lowest of these (B_{3g}^-), $E(R)$ corresponds to the observed value of $h\nu_D^{\max}$ at the values of R_0 listed in column 4 of Table XIV, which are of the expected magnitude. At the equilibrium separation given for naphthalene the computed mixing coefficients show a 60% contribution of the exciton state to the final configuration.

In the case of benzene (point group D_{6h}) excimer states arise from configuration interaction of the 8-fold degenerate CR states and exciton states of both ^1L_a and ^1L_b origin. Owing to the very large separation of ^1L_a and ^1L_b states in this molecule, the lowest exciton state is of ^1L_b character and contributes to the lowest excimer state after configuration interaction.[68]

For the symmetric (D_{2h}) excimer the dipole transition to the ground state is polarized parallel to the principal molecular axes. However since the transition is symmetry forbidden in this configuration, a slight rotation or displacement of one molecule relative to the other is necessary to induce dipole-allowed character to excimer fluorescence. This does not produce a significant change in the emission frequency, but the lowering of excimer symmetry introduces some contribution of in-plane polarization together with the possibility of CI between CR states and exciton states of L_b origin.[68]

C. The Semiempirical LCAO MO Approximation

Although the orbital overlap induced mixing of zeroth-order CR and exciton states provides a quantitative description of excimer binding energies, the predicted asymmetric splitting of excimer π-electron states is not confirmed by analyses of the dissociated photodimer spectrum of anthracene derivatives reported by Chandross and Ferguson.[64] An alternative description [72–74] of the exciner energy levels is based on a semiempirical LCAO MO "supermolecule" approximation in which all π-electrons of the two interacting molecules are treated using Hückel dimer orbitals and taking full account of electron interaction; the approach adopted by Chandra and Lim[73] is outlined below.

For a molecule containing $2m$ π-electrons in orbitals $1, 2, 3, \ldots, m$, the Hückel dimer orbitals are designated $\phi_1, \phi_2, \phi_3, \ldots, \phi_{2m}$ where ϕ_{2k} and ϕ_{2k-1} are both of k molecular orbital origin split by an energy equal to twice the intermolecular resonance integral $\beta' < 0$, i.e.,

$$E_{2k} = E_k - \beta'; \quad E_{2k-1} = E_k + \beta'$$

The four lowest singlet excited states of the dimer described by the configurational wave-functions

$$\chi_1 = \phi_{2m}\phi_{2m+1}$$
$$\chi_2 = \phi_{2m}\phi_{2m+2}$$
$$\chi_3 = \phi_{2m-1}\phi_{2m+1}$$
$$\chi_4 = \phi_{2m-1}\phi_{2m+2}$$

are symmetric (χ_2 and χ_3) or antisymmetric (χ_1 and χ_4) with respect to reflection in the plane of symmetry perpendicular to the intermolecular axis. Configurational interaction states of the same symmetry removes the degeneracy of χ_2 and χ_3 and leads to the four CI excited dimer singlet states

$$\sigma = (\chi_1 + \lambda\chi_4)/(1 + \lambda^2)^{1/2}$$
$$\rho = (\chi_1 + \lambda'\chi_4)/(1 + \lambda'^2)^{1/2}$$
$$\gamma = (\chi_2 + \chi_3)/2^{1/2}$$
$$\delta = (\chi_2 - \chi_3)/2^{1/2}$$

where λ and λ' denote mixing coefficients in the lower and higher states, respectively. The excitation energy of the lowest excited dimer state σ is shown to be an approximately linear function of the molecular ^1L_a (Clar p band) excitation energy expressed as

$$E(\sigma) \simeq [(1 + \lambda)^2/2(1 + \lambda^2)]E(^1L_a)$$

and confirmed by the experimental data for 14 molecules with[73]

$$\lambda = \frac{2\beta' - (4\beta'^2 + \langle\chi_1|\mathcal{H}|\chi_4\rangle^2)^{1/2}}{\langle\chi_1|\mathcal{H}|\chi_4\rangle} \simeq 0.23$$

Matrix elements $\langle\chi_1|\mathcal{H}|\chi_4\rangle$ and resonance integrals β' are evaluated as a function of interplanar separation R to obtain the values of R_0 given in Table XIV (column 5) consistent with the spectroscopic data.

An extension of this treatment to the excimer triplet levels[73] leads to the approximate expression

$$E(^1\sigma) - E(^3\sigma) \simeq 0.73\{[E(^1L_a) - E(^3L_a)] - 0.4\} \text{ eV}$$

for intermolecular coulombic repulsion integrals of 0.2 eV at an interplanar separation of 3.0–3.5 Å; this predicts a phosphorescence maximum for 2,2'-paracyclophane at 450 mµ compared with the observed peak at 480 mµ.

D. Exciplex Binding Energies

The interaction of nondegenerate molecular or charge-transfer states is insufficient to describe the stability of photoassociation products of molecules with different electronic energy levels, ionization potentials, and electron affinities. On the other hand, treatments[26,28] of the exciplex as a pure charge-transfer state afford a quantitative description of the shift in fluorescence peak with solvent polarity and with electron affinity of the (fluorescent) donor in the same quencher-solvent system (Eq. 13); moreover, estimated values for the dipole moment of the emitting species (Table VI) confirm its pronounced charge-transfer character.

The phosphorescence and fluorescence spectra of aromatic hydrocarbons with electron acceptors in low-temperature glasses are well-documented[75-77] and have been reviewed[78]; the observed red shift in emission with reduced ionization potential of the donor parallels the change in charge transfer absorption spectrum and confirms the charge-transfer character of the emitting species. However, the observation of distinct fluorescence and phosphorescence bands with characteristic decay constants requires the interaction of locally-excited (molecular) configurations[79] to remove the degeneracy of "singlet" and "triplet" charge transfer states. The extent of the perturbation may be expected[80] to vary with the position of molecular states relative to the charge transfer energy at the prevailing donor-acceptor separation and is reflected by the correspondence of the observed emission spectrum with those of donor and acceptor alone.

V. EXCIMER (EXCIPLEX) RELAXATION

A. Fluorescence Emission $^1M_2^*(^1MQ^*) \xrightarrow{k_F^{D(C)}} M + M(Q) + h\nu_F^{D(C)}$

In the absence of the reverse absorption the radiative transition probability $k_F^{D(C)}$ is evaluated from the intrinsic excimer (exciplex) quantum yield of fluorescence $q_{D(C)}$ and the decay constant $1/\tau_F^{D(C)} = \sum_i k_i^{D(C)}$ measured under the same conditions, which are related by

$$k_F^{D(C)} = q_{D(C)}(k_F^{D(C)} + k_{IS}^{D(C)} + k_{IC}^{D(C)} + k_R^{D(C)}) = q_{D(C)}/\tau_F^{D(C)}$$

From Eqs. (2) and (10) the intrinsic yields $q_{D(C)}$ are equal to the measured fluorescence yields $\gamma_F^{D(C)}$ at infinite fluor (or quencher) concentrations; a sufficient condition that $[M] \gg [M]_{1/2}$ (or $[Q] \gg [Q]_{1/2}$) is established by the complete quenching of molecular fluorescence (Eqs. 1 and 9), as in pure liquids at moderate temperatures or for solutions of fluor in liquid quenchers. Alternatively the intrinsic yields may be computed from the measured yields at the half-quenching concentration $[M]_{1/2}$ or $[Q]_{1/2}$ (Eqs. 2 and 10), or, following Hirayama and Lipsky,[140] from linear plots of γ_F^M against γ_F^D which yield q_M and q_D as intercepts according to the relationship

$$\gamma_F^M/q_M + \gamma_F^D/q_D = 1$$

Lifetimes $\tau_F^{D(C)}$ are available from analyses of fluorescence decay curves as described in Section II.D where, to a good approximation, $1/\tau_F^{D(C)}$ is given as the experimental parameter λ_1 describing terminal decay for a system exhibiting excimer (exciplex) fluorescence only.

Excimer fluorescence parameters obtained in this way are listed in Table XV together with the corresponding molecular fluorescence decay constants k_F^M. To the extent that excimer fluorescence originates from configurationally-interacted states of ^1L_a origin, the relatively forbidden nature of this electric dipole transition must be a consequence of the symmetrical (sandwich) configuration of the emitting state from which the transition moment integral to a totally symmetric ground state is strictly zero. Chandra and Lim[73] conclude that thermal excitation of torsional oscillation of the molecular excimer components, described by the appropriate angular potential function, can effect a change of 5° in the angle θ about the naphthalene excimer interplanar axis at room temperature and compute the excimer oscillator strength as

$$f^D = f^M(^1L_a)\left[\frac{\nu_D^{max}}{\nu_M(^1L_a)}\right]\left(1 + \frac{2\lambda}{1 + \lambda^2}\right)(1 - \cos\theta)$$

where λ denotes the configurational interaction coefficient (p. 199) and f^M and ν_M refer to the oscillator strength and average frequency of the molecular $^1A \leftarrow {}^1L_a$ transition computed from the appropriate absorption band. With $\lambda(\theta)$ equal to 0.137 at $\theta = 5°$ this equation yields a radiative lifetime $1/k_F^D = 2.35 \times 10^{-6}$ sec compared with the experimental[38] value of 0.9×10^{-6} sec. This thermal origin of dipole-allowed character of excimer fluorescence is supported by the findings of Hirayama and Lipsky[140] that q_D for benzene and toluene exhibits a positive temperature coefficient which these authors attribute to a thermal destruction of excimer symmetry and a corresponding increase in k_F^D with temperature.

The absence of vibrational structure in the excimer fluorescence spectrum is attributed to a purely repulsive interplanar potential of the final state

TABLE XV
Excimer Fluorescence Parameters

M	Solvent	T, °K	q_D	τ_F^D, sec	k_F^D, sec^{-1}	k_F^M, sec^{-1}	Ref.
Benzene	Cyclohexane	293	0.025	1.2×10^{-8}	2.1×10^6	4.2×10^6	16
Toluene	Cyclohexane	293	0.053	1.6×10^{-8}	3.3×10^6	9.8×10^6	16
Naphthalene	Toluene	295	0.32	3.8×10^{-7}	0.9×10^6	2.3×10^6	38
1-Methoxynaphthalene	Toluene	298	0.24	3.0×10^{-7}	0.8×10^6	2.8×10^7	46
2-Methoxynaphthalene	Toluene	298	0.31	2.2×10^{-7}	1.4×10^6	2.8×10^7	46
1-Methylnaphthalene	Liquid	295	0.34	3.0×10^{-7}	1.1×10^6	2.9×10^6	38
2-Methylnaphthalene	Liquid	295	0.25	2.2×10^{-7}	1.1×10^6	3.4×10^6	38
1,6-Dimethylnaphthalene	Liquid	295	0.36	2.8×10^{-7}	1.3×10^6	3.6×10^6	38
1,6-Dimethylnaphthalene	Liquid	293			2.2×10^6		39, 40
1,6-Dimethylnaphthalene	Heptane	293	0.09	1.6×10^{-7}	1.4×10^6	5×10^6	39, 40
1,2-Benzanthracene	Cyclohexane	293	0.47	4.1×10^{-8}	11.3×10^6	4.2×10^6	42
5-Methylbenzanthracene	Cyclohexane	293	0.49	9.3×10^{-8}	5.3×10^6	4.7×10^6	42
6-Methylbenzanthracene	Cyclohexane	293	0.33	3.8×10^{-8}	8.5×10^6	2.5×10^6	42
10-Methylbenzanthracene	Cyclohexane	293	0.65	9.8×10^{-8}	6.6×10^6	4.6×10^6	42
Pyrene	Cyclohexane	293	0.75	6.7×10^{-8}	12×10^6	1.5×10^6	43
Pyrene	Paraffin oil	293			9×10^6	3.6×10^6	45

(Fig. 2). Rice and coworkers[75] have presented a theoretical computation of the benzene excimer spectrum using Boltzmann-weighted Franck-Condon overlap integrals over an excimer Morse potential function and a ground state delta function peaked at the classical turning point of an exponential repulsive interaction. This provides a qualitative interpretation of the observed red-shift of ν_D^{max} and reduction in half-intensity band width at lower temperatures but is rather more symmetrical about ν_D^{max} than the observed spectrum at ambient temperature.

The data reported for exciplex fluorescence are less extensive; however, it is interesting to note that τ_F^C for the anthracene–diethylaniline system (105 nsec[31]) and the perylene–diethylaniline system (37 nsec[34]) exceed the reciprocal molecular ($^1L_a \rightarrow {}^1A$) radiative decay constants by a factor of ~ 10. It is unlikely that the forbidden nature of the exciplex fluorescence in these cases originates in the high symmetry of the emitting species.

Under the (fluid) conditions necessary to promote photoassociation, the excimer fluorescence is rotationally depolarized; however, measurements of pyrene crystal fluorescence have confirmed[81] the predicted direction of polarization along the major axis.

B. *Intersystem Crossing* $^1M_2^*(^1MQ^*) \xrightarrow{k_{IS}^{D(C)}} {}^3M_2^*(^3MQ^*)$

While the observation of excimer phosphorescence[20,82] confirms the existence of this triplet state, it may not be regarded as conclusive evidence for intersystem crossing unless photoassociation in the triplet manifold (Scheme 1) and triplet–triplet annihilation can be eliminated as contributing processes. By analogy with molecular electronic relaxation however, intersystem crossing from $^1M_2^*$ and $^1MQ^*$ is to be expected and its inclusion in the general photoassociation scheme accommodates bimolecular quenching of molecular fluorescence by species containing atoms of high nuclear charge. Indirect evidence for intersystem crossing is also provided by the kinetic considerations outlined below.

In the absence of permanent chemical change ($k_R^D = 0$) the intrinsic yield of excimer fluorescence is expressed by

$$q_D = k_F^D/(k_F^D + k_{IS}^D + k_{IC}^D) = k_F^D/(k_F^D + k_i^D)$$

where k_i^D is the rate constant of undefined nonradiative excimer relaxation, the presence of which is confirmed by values of $q_D < 1$ (Table XV). The temperature-dependence of q_D and of

$$\tau_F^D = 1/(k_F^D + k_i^D)$$

has in most cases been reproduced [38,40,43,45,46,83] by the inclusion of a single temperature-dependent rate parameter k_i^D in the form

$$k_i^D = A_i^D \exp(-E_i^D/RT) \tag{45}$$

and provides values for the constants A_i^D and E_i^D listed in Table XVI. It is concluded that, since the computed frequency factor A_i^D is some orders of magnitude lower than characteristic values in the range 10^{12}–10^{14} sec^{-1} for unimolecular reactions, this describes a spin-intercombination process or $k_i^D \equiv k_{IS}^D$. A similar temperature dependence, reported [84-86] for certain molecular intersystem crossing constants k_{IS}^M, has been attributed to the production of a triplet state of higher energy than the participating singlet state. If the activation energies E_i^D listed in Table XVI have the same significance then the energy $E(^3M_2^{**})$ of the excimer triplet state produced in this process is given by

$$E(^3M_2^{**}) = h\nu_M^0 + \Delta H_a + E_i^D \tag{46}$$

The further assumption that $^3M_2^{**}$ is degenerate with the correlating molecular triplet state $^3M^{**}$ provides an estimate of the energy $E(^3M^{**})$ of this state in the region $E(^1M^*) > E(^3M^{**}) > E(^3M^*)$ which may be spectroscopically inaccessible. Double intersystem crossing to different molecular triplet states of naphthalene [87] is also apparently exhibited by the excimer of 1,6-dimethylnaphthalene [40] in which the nonradiative process is characterized by a rate constant k_i^D which is the sum of temperature-dependent and temperature-independent terms. The value of the latter is also consistent with a spin-prohibited process (Table XVI).

TABLE XVI
Nonradiative Excimer Relaxation Parameters[a]

M	Solvent	A_i^D, 10^8 sec^{-1}	E_i^D, kcal	$E(^3M_2^{**})$,[b] cm^{-1}	Ref.
1-Methylnaphthalene	Ethanol	25	4.0	30,600	38
2-Methylnaphthalene	Ethanol	78	4.1		3
1,6-Dimethylnaphthalene	Ethanol	4	3.2		38
1,6-Dimethylnaphthalene	n-Heptane[c]	60	3.7		40
Pyrene	Acetone	5	2.3		43
Pyrene	Cyclohexane	2	2.3		43
Pyrene	Ethanol	4	2.3	24,400	43
Pyrene	Ethanol	0.29	1.15		83

[a] From Eq. (45).
[b] From Eq. (46).
[c] k_i^D includes a temperature-independent term of 5.1×10^6 sec^{-1}.

Intersystem crossing followed by dissociation of the excimer triplet state in the manner indicated

$$^1M_2^* \xrightarrow{k_{IS}^D} {}^3M_2^* \xrightarrow{{}^3k_{MD}} {}^3M^* + M$$

provides an additional route for population of the molecular triplet state, the concentration-dependent triplet state formation efficiency being expressed (with ${}^3k_{MD} \gg k_{GT}^D + k_P^D$) as

$$\gamma_{IS} = (k_{IS}^M[{}^1M^*] + k_{IS}^D[{}^1M_2^*])/I_a$$
$$= (\gamma_{IS}^M[M]_{\frac{1}{2}} + \gamma_{IS}^D[M])/([M] + [M]_{\frac{1}{2}})$$

(cf. Eqs. 1 and 2) where $\gamma_{IS}^{M(D)} = k_{IS}^{M(D)}/(k_{IS}^{M(D)} + k_F^{M(D)} + k_{IC}^{M(D)})$. This behavior has been reported for pyrene in ethanol at 20°C by Medinger and Wilkinson[88] who measured the instantaneous population of the molecular triplet state by flash kinetic spectrophotometry. Although these authors treat the over-all process as one of dissociative intersystem crossing

$$^1M_2^* \longrightarrow {}^0M^* + M$$

their results are consistent with an excimer intersystem crossing yield γ_{IS}^D of 0.12 and a rate constant $k_{IS}^D = \gamma_{IS}^D \tau_F^D = 2.3 \times 10^6$ sec^{-1}. The rapid dissociation of the triplet excimer would account for the failure to detect this state in absorption with the time resolution available.

Direct evidence for exciplex intersystem crossing is provided by the observation[77] of both fluorescence and phosphorescence of this species following excitation in the charge transfer absorption band of aromatic hydrocarbon-acceptor systems in rigid low temperature glasses. Although this excitation mode strictly precludes application of the term exciplex, there is little doubt that the emitting species are identical with the products of heteromolecular photoassociation in fluid media.

As in the excimer case, dissociative intersystem crossing of the exciplex

$$^1MQ^* \xrightarrow{k_{IS}^C} {}^3MQ^* \xrightarrow{{}^3k_{MC}} {}^3M^* + Q$$

provides an additional excitation sequence for the molecular triplet state which is not only dependent on quencher concentration but which should also be promoted by spin-orbital coupling due to heavy atoms in the quenching molecule. Elegant studies by Wilkinson and coworkers[89] have shown that the external heavy atom quenching of molecular singlet states is accompanied by a quantitative increase in population of the molecular triplet state which requires that ${}^3k_{MC} \gg k_{GT}^C + k_P^C$ (Scheme 2) and affords direct evidence for the postulated quenching sequence; however, the observation of a reduction in the sum of molecular fluorescence and triplet state yields at lower temperatures where ${}^3k_{MC} \simeq k_{GT}^C + k_P^C$ would provide confirmation of the triplet exciplex intermediate.

Intersystem crossing from the exciplex singlet state (whether dissociative or not) is essential to one quantitative description[52] of external heavy atom quenching of molecular fluorescence where the variation in observed quenching constant of four orders of magnitude reflects a competition between singlet exciplex dissociation and intersystem crossing. In terms of this theory[52] an increase in the nuclear charge of halogen atoms in the quenching molecule not only increases its electron affinity, and thereby the exciplex stability toward dissociative regeneration of the molecular singlet state, but also promotes spin-orbital coupling and intersystem crossing in the complex (cf. Section III.C).

C. Internal Conversion $^1M_2^*(^1MQ^*) \xrightarrow{k_{IC}^{D(C)}} M + M(Q)$

For the compounds listed in Table XVI excimer fluorescence competes with a single nonradiative relaxation process which is assigned as an intersystem crossing on the basis of its relatively low frequency factor; accordingly there is no evidence for internal conversion in the examples cited. On the other hand Medinger and Wilkinson[88] find that it is necessary to include this process to account for their observation that the intersystem crossing yield ($\gamma_{IS}^D = 0.12$) and fluorescence yield ($\gamma_F^D = 0.75$) do not sum to unity, and assign a value of $k_{IC}^D = 6.3 \times 10^6$ sec^{-1} for this system. A finite nonradiative probability (k_{GT}^D) for the triplet excimer (cf. Section V.B) in the competitive sequence

$$^1M_2^* \longrightarrow {}^3M_2^* \xrightarrow{k_{GT}^D} M + M$$
$$\phantom{^1M_2^* \longrightarrow {}^3M_2^*} \xrightarrow{^3k_{MD}} {}^3M + M$$

would resolve this discrepancy and would appear as an over-all internal conversion of $^1M_2^*$ in the absence of an intermediary triplet state $^3M_2^*$.

Perhaps the most convincing evidence for internal conversion is provided by the observations of Hammond et al.[51] who find that the quenching of molecular fluorescence by conjugated dienes is not accompanied by either an increase in yield of molecular triplet states ($k_{IS}^C = 0$), by any detectable photochemical change ($k_R^C = 0$), or by the appearance of a characteristic exciplex band ($k_F^C = 0$). Since the observed quenching constants, K_Q, vary by several orders of magnitude it may be assumed that reversible photoassociation is operative in these systems (Section III.C) in which case with (cf. Eq. 33)

$$K_Q = k_{CM}\tau_F^M(k_F^C + k_{IS}^C + k_{IC}^C + k_R^C)/k_{MC}$$

k_{IC}^C must be finite. Alternatively the absence of accompanying triplet state production could reflect a rapid nonradiative relaxation of the exciplex triplet state ($k_{GT}^C \gg {}^3k_{MC}$) following intersystem crossing ($k_{IS}^C > 0$). In view of the low binding energy expected for $^3MQ^*$ in these systems, this latter quenching sequence must be considered unlikely.

D. *Photodimerization* (*Photoaddition*) $\quad {}^1M_2^*({}^1MQ^*) \xrightarrow{k_R^{D(C)}} M_2(MQ)$

Derivatives of the linear polyacenes, naphthalene,[46] anthracene,[41,90,93] naphthacene,[91,92] and pentacene,[92] form stable photodimers M_2 when irradiated in concentrated O_2-free solution or (exceptionally) in the crystalline state.[94,95] Transannular σ-bonding of the molecular dimer components results in a folding of the aromatic planes about the bonded atoms and a reduced π-electron delocalization reflected in a shift of the absorption spectrum to much higher frequencies.[46,92,96]

A simultaneous increase in photodimerization yield and reduction in molecular fluorescence yield with increasing solute concentration provides convincing evidence that photodimerization involves the molecular singlet state,[90,93] and in certain systems, notably derivatives of naphthalene[46] and anthracene,[41] excimer fluorescence is exhibited simultaneously.

Birks and coworkers[41] have suggested that photodimerization and excimer formation are competitive photoassociation processes the course of which is

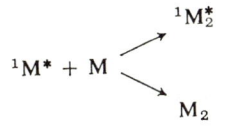

$$^1M^* + M \begin{array}{c} \nearrow {}^1M_2^* \\ \searrow M_2 \end{array}$$

dictated by the stereochemistry of the encounter product of (unsymmetrical) substituted molecules, e.g., excimer fluorescence and photodimerization result from antiparallel (*trans*) and parallel (*cis*) configurations of two 9-methylanthracene molecules, respectively. This is contrary to the findings of Calas, Lalande, and Mauvet[97] that dipole moments of di-9-cyanoanthracene and of di-9-bromoanthracene are consistent with the *trans* configuration; in the absence of steric factors *cis* and *trans* di-9-deuteroanthracene are formed in equal amounts.[98]

Stevens et al.[94,95] argue that photoassociation and dimerization are sequential processes

$$^1M^* + M \longrightarrow {}^1M_2^* \longrightarrow M_2$$

since if different bimolecular configurations are responsible for radiative and chemical relaxation these cannot occur simultaneously in the crystalline state where one configuration is preferred; however, the excimer fluorescence of crystalline 9-cyanoanthracene (and of one form of 1-chloroanthracene)[95] is reduced in intensity as photodimerization proceeds to completion.[94] This behavior introduces photodimerization as an additional criterion of photoassociation in systems which exhibit molecular fluorescence only, cf. Table XVII.

A comparison of the quantum yield of photodimerization given (in accordance with Scheme 1) by

$$\gamma_R^D = \left\{\frac{k_R^D}{k_F^D + k_{IS}^D + k_{IC}^D + k_R^D}\right\}\left\{\frac{[M]}{[M] + [M]_{1/2}}\right\} = q_R\left\{\frac{[M]}{[M] + [M]_{1/2}}\right\} \quad (47)$$

and the fluorescence yields

$$\gamma_F^D = q_D\left\{\frac{[M]}{[M] + [M]_{1/2}}\right\}$$

$$\gamma_F^M = q_M\left\{\frac{[M]_{1/2}}{[M] + [M]_{1/2}}\right\}$$

shows that the sum of these quantities is less than unity unless $q_R + q_D = q_M = 1$. Moreover the sum of γ_F^M and γ_R^D will be dependent on solute concentration unless (fortuitously) $q_D = q_M$, in agreement with observation.[93] From appropriate treatment of the experimental data $\gamma_R^D([M])$ values of q_R obtained for anthrancene[90] and 9-methylanthracene[93] are 0.9 and 0.22, respectively. The former value is consistent with the absence of excimer fluorescence in solutions of this compound at normal temperatures and a high intrinsic dimerization efficiency q_R must also be responsible for the concentration invariant fluorescence spectrum of naphthacene under the same conditions (cf. Table III). The observation[65] of anthracene excimer fluorescence from the photolyzed photodimer in rigid low temperature glasses indicates that rate constant k_R^D is temperature-dependent.

Recent attempts to correlate quantum yields of photodimerization of β-alkoxynaphthalenes with values of $[M]_{1/2}$ from spectroscopic data using Eq. (47) have not been entirely successful; thus Selinger et al.[46] conclude that

TABLE XVII
Photodimerization as a Criterion of Photoassociation

	Photodimerization	Excimer fluorescence	Photoassociation
Anthracene	Yes	No	Yes
9-Methylanthracene	Yes	Yes	Yes
9,10-Dimethylanthracene	No	Yes	Yes
9,10-Diphenylanthracene	No	No	No

photodimerization via excimer formation is not the only dimerization route. It should be noted that the photodimer is susceptible to thermal and photochemical decomposition into its molecular constituents, which if significant during the period of exposure, would be equivalent to an over-all excimer internal conversion process.

An example of the equivalent (photoaddition) reaction following heteromolecular photoassociation is provided by the photochemical addition of maleic anhydride to anthracene.[99] Livingston and coworkers[100] have shown that the anthracene triplet state is not involved in this reaction and that, in terms of Eq. (47) in the appropriate form, $q_R^C = 0.03$. However, if the excited complex $^1MQ^*$ formed directly by light absorption in the charge-transfer band is the reactive intermediate, this produces the adduct with a computed efficiency of 34%.

E. Ionic Dissociation $^1MQ^* \xrightarrow{k_I^C} M^- + Q^+$

The red shift of the exciplex band with increasing solvent polarity (Section II.C) is accompanied by a marked decrease in its intensity, until in highly polar solvents such as acetonitrile this band is no longer observed.[24,31,102] This behavior is illustrated in Figure 12 where the relative intensity of the

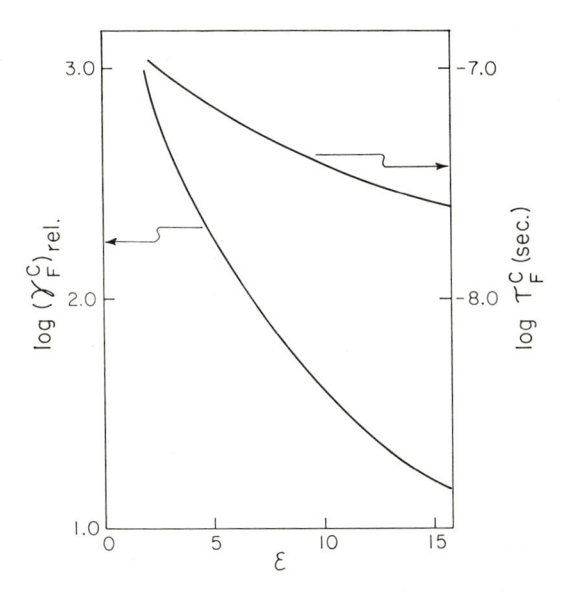

Fig. 12. Variation of lifetime τ_F^C and relative yield $(\gamma_F^C)_{rel}$ of anthracene–diethylaniline exciplex fluorescence with solvent dielectric constant ε (after Knibbe, Rollig, Schafer, and Weller[31]).

anthracene–diethylaniline exciplex band maximum, at the half-value concentration $[Q]_{1/2}$ (from Eq. 9) is plotted as a function of solvent dielectric constant ϵ. Since it is precisely under conditions of high solvent polarity that Weller et al.[35,101] have identified the hydrocarbon (perylene) radical anion M^- and the quencher (diethylamino) radical cation Q^+ by flash absorption spectroscopy, the reduction in exciplex fluorescence intensity is consistent with its dissociation to (partially solvated) ions in the manner indicated.[30]

Ionic dissociation is also responsible[31,102] for an increase in the measured fluorescence decay constant (Fig. 12) now given by

$$1/\tau_F^C = k_F^C + k_{IS}^C + k_{IC}^C + k_I^C$$

which, however, is insufficient to account for the observed reduction in fluorescence yield. Weller and coworkers[24,31] have suggested that the additional process that necessarily competes with exciplex formation involves the direct formation of a (solvated) ion-pair by electron transfer in the amended scheme

$$^1M^* + Q \underset{k_{IM}}{\overset{k_{CM}}{\rightleftarrows}} \begin{array}{c} {}^1(M^-Q^+) \\ \downarrow k_I^C \\ M^- + Q^+ \end{array}$$

Although this process (with rate constant k_{IM}) produces no additional quenching of molecular fluorescence, the quantum yield of exciplex fluorescence is reduced to

$$\gamma_F^C = \frac{q_D}{2}\left\{\frac{k_{CM}}{k_{CM} + k_{IM}}\right\} = \frac{k_F^C \tau_F^C k_{CM}}{2(k_{CM} + k_{IM})}$$

at the half-value concentration for molecular fluorescence quenching. γ_F^C therefore decreases with increase in solvent polarity due to the reduction in τ_F^C (by ionic dissociation) and a simultaneous increase in the electron transfer probability $k_{IM}(\epsilon)/[k_{CM} + k_{IM}(\epsilon)]$.

Energetic considerations based on the separation of solvated ions at the encounter distance a show that solvated ion-pair formation from $^1M^*$ is sufficiently exothermic in polar solvents to effectively prevent the production of excited singlet states $^1M^*$ by the reverse process. Table XVIII lists values for free energies ΔG_{IM} of ion-pair formation in acetonitrile estimated[24] from the oxidation and reduction potentials, E_{D/D^+} and $E_{A^-/A}$, of donor and acceptor using the relationship

$$\Delta G_{IM} = E_{D/D^+} - E_{A/A^-} - h\nu_M^0 - e^2/\epsilon a \tag{48}$$

a is evaluated as the encounter diameter from the molecular quenching rate

TABLE XVIII
Free Energies ΔG_{IM} of Ion-Pair Formation[a] with Diethylaniline ($E_{D/D+}$ = 0.76 V) in Acetonitrile (ϵ = 36.7) at Room Temperature (Weller[24])

M ($\equiv$ A)	[Q]½, mole liter^{-1}	k_{IM}, liter mole^{-1} sec^{-1}	$D_A + D_D$, cm^2 sec^{-1}	a, Å	$E_{A-/A}$, V	$-\Delta G_{IM}$, kcal mole^{-1}
Anthracene	0.0096	2.10 × 10^{10}	3.7 × 10^{-5}	7.5	−1.96	14
Pyrene	0.00021	1.76 × 10^{10}	3.7 × 10^{-5}	6.3	−2.10	12
Perylene	0.0081	2.00 × 10^{10}	3.5 × 10^{-5}	7.6	−1.65	11
Coronene	0.00017	1.96 × 10^{10}	3.3 × 10^{-5}	7.8	−2.04	5

[a] From excited singlet acceptor $^1M^*$, eq. (48).

constant k_{IM} (= k_Q in this solvent) and diffusion coefficients D_A and D_D to which it is related by

$$k_{IM} = 4\pi N(D_A + D_D)a/1000$$

VI. PHOTOASSOCIATION IN ORDERED SYSTEMS

A. The Crystalline State

The diffusional requirement for photoassociation is eliminated by suitable orientation of the molecular components prior to electronic excitation; this is the basis of an interpretation [103] of the gross spectral features of crystalline aromatic hydrocarbon fluorescence which is illustrated with reference to two limiting cases:

(a) Under normal conditions the catacondensed hydrocarbons of molecular formula $C_{4n+2}H_{2n+4}$ adopt a crystal lattice [104] shown schematically as Type A in Figure 13 in which the (rotational and translational) displacement of adjacent molecular planes to produce a symmetric sandwich configuration is prohibited by the interaction of neighboring molecules. These crystals exhibit a structured ("molecular") fluorescence spectrum red-shifted by ~100 cm^{-1} from the molecular spectrum observed in dilute solutions.

(b) The structureless fluorescence spectra of crystalline pericondensed hydrocarbons, pyrene, perylene, and 1,12-benzperylene, on the other hand, are red-shifted by ~6000 cm^{-1} from the 0″–0′ molecular fluorescence band, [103,105] and in the case of pyrene is virtually identical with the excimer band. As shown schematically in Figure 13, the molecular orientation in a

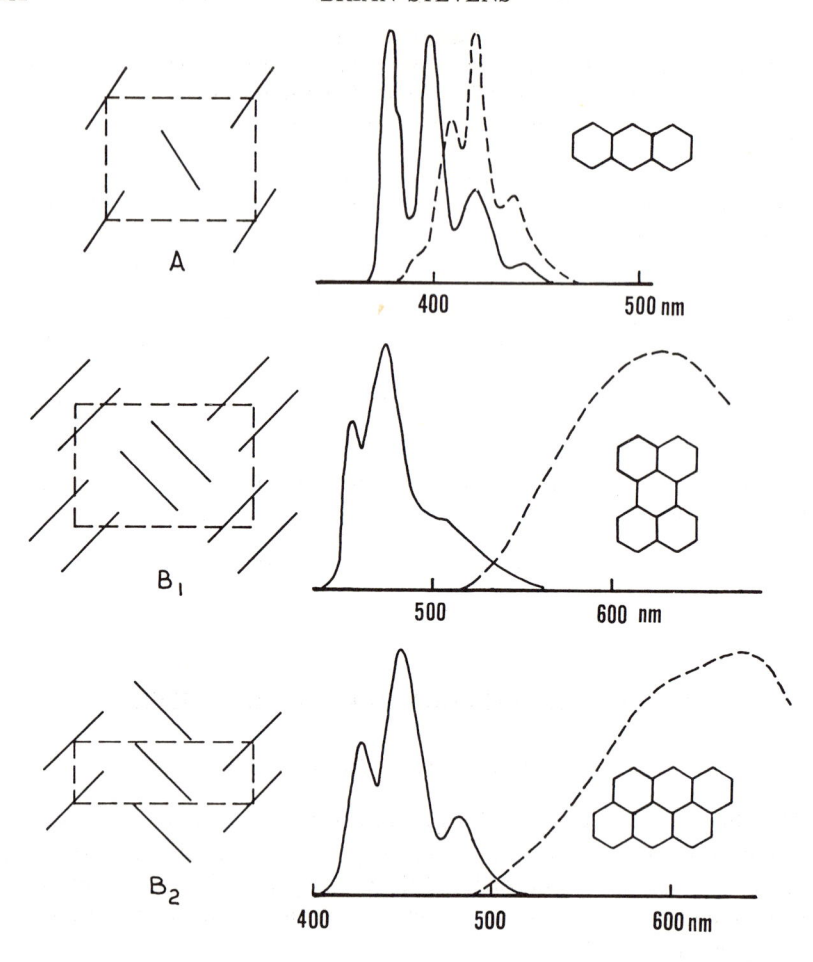

Fig. 13. Molecular orientation of catacondensed (*A*) and pericondensed (*B*₁, *B*₂) aromatic hydrocarbons in crystal lattice (schematic), with molecular (———) and crystal (— —) fluorescence spectra of representative molecules.[103]

Type B lattice preferred by these molecules is predisposed to excimer formation, although the absence of pronounced changes in the long-wave absorption spectrum indicates that the interplanar separation is reduced by photoassociation from the value [106] of ~ 3.5 Å in the unexcited configuration.

In a third type of crystal structure (Type B_2, Fig. 13) exhibited by the larger pericondensed hydrocarbons (e.g., coronene and ovalene), the characteristic excimer fluorescence band may originate from a small concentration of exciton traps produced by the relative translational displacement of adjacent molecules, or may be a consequence of the large splitting of molecular levels in these close-packed crystals.

On the basis of this correlation it has been suggested [103] that the crystal fluorescence spectrum is diagnostic of crystal structure which may vary with pressure, temperature, or the introduction of molecular substituent groups. Thus the catacondensed hydrocarbons, naphthalene,[107] anthracene,[107–109] chrysene,[109] and 1,2-benzanthracene[109] in the crystalline state exhibit an excimer-type spectrum at high pressures, while at very low temperatures the stable form of crystalline perylene exhibits a green, structured fluorescence [110] distinct from the orange-red structureless emission observed from the stable crystalline form at normal temperatures; in both cases X-ray analyses have confirmed the predicted change in crystal structure. The "molecular" fluorescence spectrum of crystalline anthracene is replaced by an excimer band [111] in crystalline 9-cyanoanthracene; this change in crystal structure, effected by molecular substitution, simultaneously promotes photodimerization in the solid state at the expense of excimer fluorescence intensity and indicates that photoassociation precedes photodimerization.[94] X-ray powder photographs confirm the different molecular packing of the blue (structured) and green (structureless) fluorescent modifications of crystalline 1-chloroanthracene [95]; whereas the former is photochemically stable, the latter also undergoes photodimerization to (however) a different solid modification of the same stereoisomer obtained from concentrated solutions. It therefore appears that the stereochemistry of the photodimer is determined by molecular geometry rather than crystal packing considerations although these are closely related.

An example of exciplex formation in the solid state may be afforded by perylene doped crystals of pyrene which emit a green structureless fluorescence in addition to the blue and orange-red excimer bands of pyrene and perylene, respectively. Hochstrasser[112] has shown that the energy of the emitting species is consistent with that of a charge transfer complex of pyrene and perylene molecules in a bimolecular unit of the pyrene lattice.

B. The Polyphenyl Alkanes

Following the initial observation [113] of excimer fluorescence from dissolved polystyrene, Hirayama [114] has reported a systematic survey of the fluorescence spectra of the di- and triphenylalkanes shown in Figure 14. In addition to the normal "molecular" spectrum of the phenyl group exhibited by all the molecules listed, an excimer band is exhibited by those systems in which the planar phenyl groups are separated by exactly three carbon atoms or a distance of 2.54 Å in the *trans* propane chain. Since there is no evidence of a

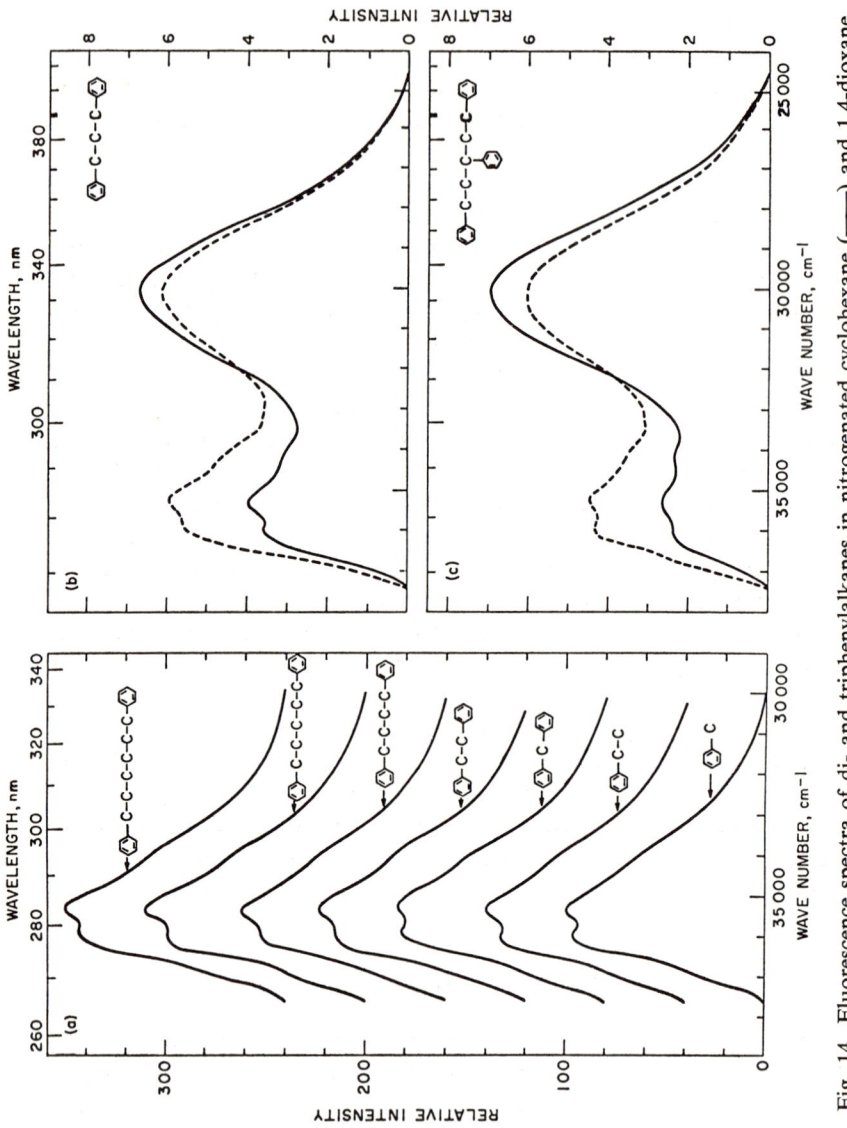

Fig. 14. Fluorescence spectra of di- and triphenylalkanes in nitrogenated cyclohexane (———) and 1,4-dioxane (– – –) solution at 0.01M phenyl constituent (reproduced with permission from Hirayama[114]).

corresponding change in absorption spectrum, and the computed photoassociation rate constant k_{DM} appears to vary with solvent viscosity, it is concluded that the parallel configuration of phenyl groups is established after the act of light absorption in a process of intramolecular photoassociation.[114] Rice and coworkers[115] have also observed a typical excimer band in polyvinylnaphthalene, while Klopffer reports excimer bands from poly-N-vinylcarbazole and 1,3-biscarbazolyl propane but not 1,4-biscarbazolyl butane.[139]

C. The Paracyclophanes

The paracyclophanes in which two benzene rings are held in a sandwich configuration by annular bridges of (n,n') methylene groups constitute an ideal series for an examination of spectral properties of dimers over a range of interplanar separations. In the lowest member of the series (2,2') the small interplanar separation of 2.75–3.09 Å results in a warping of the benzene rings which adopt a symmetrical boat configuration, and the appearance of additional bands in the absorption spectrum are a consequence of strong chromophoric interaction in the ground state. The broad structureless fluorescence spectrum peaking at 3500 Å follows direct excitation of the emitting state rather than a process of photoassociation.[115]

With an interplanar separation of 3.73 Å, 4,4'-paracyclophane is the lowest member of the series to exhibit an alkylbenzene absorption spectrum and the broad structureless fluorescence spectrum of this molecule with a peak intensity at 3400 Å is by definition an excimer band; further separation of the aromatic rings in 4,5' and 6,6'-paracyclophanes restores the fluorescence spectrum to that of the alkylbenzenes. These observations by Rice et al.[115] illustrate the critical nature of the interplanar separation in determining the extent of interaction between π-electron systems in the ground and excited configurations.

D. DNA

The close correspondence of the DNA absorption spectrum with that of a mixture of mononucleotides of the same composition illustrates the weak nature of the interactions between neighboring purine and pyrimidine bases guanine (G), cytosine (C), adenine (A), and thymine (T) at an interplanar separation of 3.36 Å in the unexcited double-helical configuration. On the other hand the structureless fluorescence band of (calf-thymus) DNA is redshifted by ~ 3500 cm^{-1} from the fluorescence spectral origin of the mononucleotides; it closely resembles the fluorescence spectrum of the dinocleotide ApT (and of poly dAT) and is accordingly identified[131] with the fluorescence of the adenine–thymine exciplex ¹AT*. That this photoassociation is a result of light absorption by either component is confirmed by recordings of the DNA fluorescence excitation spectrum which approaches the combined absorption spectrum of adenine and thymine mononucleotides.[132]

This exciplex configuration is largely determined by the relative orientation of adjacent pyrimidine bases which are rotated from the symmetrical "sandwich" configuration by an angle θ of 36° in the twin-stranded helical structure of DNA. Since the stabilization energy due to the interaction of transition dipoles at this angle should be reduced by a factor of $\cos \theta$, a red-shift of 3500 sec $\theta \sim 5900$ cm^{-1} might be expected for the exciplex band maximum originating from an unrotated configuration if the ground state destabilization energy is independent of θ. This is close to the characteristic red-shift observed for aromatic systems.

Although excimer (exciplex) fluorescence is also exhibited by most dinucleotides,[133] the observed phosphorescence from these systems, and from DNA, is characteristic of the lowest molecular triplet state. In the case of DNA at low temperatures this is identified[132] as the triplet state of thymine which, in the absence of molecular intersystem crossing, must be populated by intermolecular energy transfer in the triplet manifold or by intersystem crossing from the ¹AT* exciplex.[134]

The photodimerization of anthracene derivatives has been advanced as a criterion of photoassociation in these systems; it is therefore of interest to examine the extent to which similar behavior exhibited by the pyrimidine constituents of nucleic acids can be described in terms of the same reaction sequence.

In a fluid environment the photodimerization of thymine and its derivatives involves cycloaddition at the 5,6-double bond to form one or more of the four possible stereoisomers of the cyclobutane dimer shown

Irradiation of frozen aqueous thymine solutions produces the *cis* head-to-head (*chh*) dimer, the high quantum yield (0.5–1.0) being attributed to the preferred orientation of adjacent molecules in the microcrystalline thymine hydrate. The gradual isolation of substrate molecules in the photodimer matrix is associated[135] with the appearance and increase in intensity of molecular fluorescence as photodimerization proceeds; identical behavior exhibited by crystalline 9-cyanoanthracene and (one modification of) 1-chloroanthracene[111] is regarded as a strong indication that photodimerization involves the molecular singlet state in a process of photoassociation. Alternatively, the assignment of a triplet precursor of the thymine dimer requires that photoassociation leads to an increase in the over-all intersystem crossing yield by a factor of $\sim 10^4$ since $\gamma_{IS} \sim 0$ in the isolated molecule.[135]

As in the case of its aromatic analogs, the thymine dimer undergoes dissociation to its molecular components on exposure to radiation of shorter wavelengths[137]; in the absorption spectrum of the photodissociated *chh* dimer in a rigid glass the splitting (3600 cm^{-1}) of the molecular band at 270 nm into weak (out-of-phase) and stronger (in-phase) exciton components is consistent[136] in the point dipole approximation with an interplanar separation of 2.8 Å in this unexcited dimer configuration.

Perhaps the best-characterized lesion in DNA associated with uv inactivation and mutagenesis is that involving the intrastrand photodimerization of adjacent thymine residues; this lesion is almost wholly repaired by photodissociation of the dimers at shorter wavelengths in the photoreactivation process. Production of the *chh* dimer in this case, promoted by the configuration of adjacent molecules on the same sugar-phosphate strand, must however involve a rotational displacement of $\sim 36°$, following the reduction of ~ 0.6 Å in molecular separation.

Despite the attractive simplicity of a photoassociation mechanism for thymine dimerization, this cannot account for the observed dimer yields in dilute aqueous solution where the lifetime of the molecular singlet state is $\sim 10^{-11}$ sec and $k_{DM}[M]\tau_F^M \ll 1$; moreover, Lamola and Yamane[138] have reported that dimerization of thymine in DNA is photosensitized by acetophenone which selectively populates the pyrimidine triplet states. Spin considerations require that photodimerization in the triplet manifold is less likely to involve a concerted cycloaddition than the intermediate formation of free radicals (R) which would accommodate the significant rearrangement

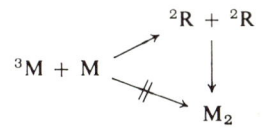

$$^3M + M \begin{array}{c} \nearrow \ ^2R + {}^2R \\ \searrow\!\!\!\!\!/\!\!\!\!\searrow \ \downarrow \\ M_2 \end{array}$$

of molecular components accompanying the dimerization of adjacent molecules in the helical configuration. It thus appears that the role of photoassociation in this over-all process may be limited to the increased population of molecular triplet states by intersystem crossing in the excimer.

VII. ALTERNATIVE EXCIMER FORMATION MODES

A. Molecular Triplet–Triplet Annihilation

A number of dissolved aromatic hydrocarbons exhibit both molecular and excimer bands in the delayed fluorescence spectrum [6,83,116,117] that originates [118] in the mutual annihilation of two molecular triplet states; if this can produce both the molecular singlet state and the excimer directly according to the scheme

$$^3M^* + {}^3M^* \xrightarrow{k_{MTT}} {}^1M^* + M \underset{k_{MD}}{\overset{k_{DM}}{\rightleftarrows}} {}^1M_2^* \xleftarrow{k_{DTT}} {}^3M^* + {}^3M^*$$

the relative yield of delayed excimer/molecular fluorescence is given by [83,121]

$$\Delta \gamma_D = \left\{\frac{\gamma_F^D}{\gamma_F^M}\right\}_D = \frac{k_F^D}{k_F^M} \left\{\frac{\alpha_D/\tau_F^M + (1 + \alpha_D)k_{DM}[M]}{1/\tau_F^D + \alpha_D k_{MD}}\right\} \qquad (49)$$

where $\alpha_D = k_{DTT}/k_{MTT}$. The linear dependence of $\Delta \gamma_D$ on solute concentration [M] is confirmed experimentally and extrapolation to zero concentration yields a finite value for $\Delta \gamma_D^0$ (and hence α_D) which requires that triplet–triplet annihilation produces $^1M^*$ and $^1M_2^*$ directly as shown.

Analyses [83,119–121] of the data in terms of Eq. (49) show that α_D varies both with temperature and the solute species; thus $\alpha_D \simeq 0.9$ for 1,2-benzanthracene and 5-methyl-1,2-benzanthracene in cyclohexane [121] at 20°C, but has a limiting value of 2.0 for pyrene [120] under the same conditions while the temperature variation in ethanol is given by the empirical expression [83]

$$\alpha_D \sim 2/(1 + 8 \times 10^7/k_d)$$

where k_d, the diffusion-limited rate constant at the prevailing temperature, is computed from Eq. (7).

Since, by the nature of this process, k_{DTT} is viscosity dependent, the temperature-dependence of k_{DTT}/k_{MTT} at low temperatures has been attributed to a viscosity-independent [116,117] component of k_{MTT} characteristic of a long range dipole–dipole interaction; however, this is subject to the objection that the transition $^3M^* \to {}^1M^*$ in the energy acceptor is spin-prohibited contrary to the requirement of an allowed acceptor transition for significant dipole–dipole induced resonance transfer.[58] The interpretation of these data, despite the varying degrees of ingenuity it has invoked,[122] is therefore still the subject of dispute. Most recently it has been suggested [123] that annihilation precedes photoassociation, i.e., $k_{DTT} = 0$, on the grounds that α_D appears to vary as the photoassociation encounter probability; the inequality $D_D/D_M > F_D/F_M$ is described semiquantitatively in terms of a random walk re-encounter probability of the annihilation products $^1M^*$ and M generated in close proximity, which establishes a higher local concentration of unexcited molecules in the vicinity of $^1M^*$.

It is interesting to note that excimer bands of phenanthrene[67] and of anthracene,[124] which have defied detection in the prompt fluorescence spectra even at low temperatures, have been observed in the delayed emission spectra of these compounds at $-75°K$. Presumably at the low temperatures necessary to observe these bands the high solvent viscosity completely suppresses photoassociation at the reduced concentration available, i.e., $k_{DM}[M] \ll 1/\tau_F^M$, whereas the reduced triplet–triplet annihilation rate constant $k_{M(D)TT}$ finds compensation in the longer lifetime and higher stationary concentration of the triplet state.

Parker and Joyce[125] have also observed a new band in the delayed emission spectrum of solutions of anthracene (A) and 9,10-diphenylanthracene (B) in ethanol at $-75°C$, which they attribute to the exciplex of these species formed in the process of mixed triplet–triplet annihilation

$$^3A^* + {}^3B^* \longrightarrow {}^1AB^*$$

B. Ionic Doublet–Doublet Annihilation

Chandross, Longworth, and Visco[126] have reported the observation of long-wave structureless emission bands in the vicinity of the electrodes during the ac electrolysis of anthracene, phenanthrene, perylene, and 3,4-benzpyrene in polar solvents such as acetonitrile and dimethylformamide; the similarity between the perylene band and the crystal fluorescence spectrum prompted the assignment of these bands to excimer fluorescence originating in the process

$$M^+ + M^- \longrightarrow {}^1M_2^*$$

A more detailed study of the electrochemiluminescence of 9,10-dimethylanthracene in dimethylformamide has been carried out by Parker and Short[127] who measured the ratio of excimer/molecular fluorescence yields $(\gamma_F^D/\gamma_F^M)_E$ in the chemiluminescent spectrum, and $(\gamma_F^D/\gamma_F^M)_D$ in the delayed emission spectrum, at different temperatures. The general scheme

$$M^+ + M^- \xrightarrow{k_{MII}} {}^1M^* + M \underset{k_{MD}}{\overset{k_{DM}[M]}{\rightleftarrows}} {}^1M_2^* \xleftarrow{k_{DII}} M^+ + M^-$$

leads to the relationship (cf. Eq. 49)

$$\Delta \gamma_E = \left\{\frac{\gamma_F^D}{\gamma_F^M}\right\}_E = \frac{k_F^D}{k_F^M} \left\{\frac{\alpha_E/\tau_F^M + (1+\alpha_E)k_{DM}[M]}{1/\tau_F^D + \alpha_E k_{MD}}\right\} \qquad (50)$$

where $\alpha_E = k_{DII}/k_{MII}$; the linear dependence of $\Delta\gamma_E$ on solute concentration is confirmed experimentally[127] and the finite value of $\Delta\gamma_E^0$ at zero concentration requires that $^1M^*$ and $^1M_2^*$ are both produced directly in the alternative doublet–doublet annihilation routes MII and DII if these are the only processes involved.

An alternative origin of electrochemiluminescence involving triplet–triplet annihilation following the production of triplet states in the process

$$M^+ + M^- \longrightarrow {}^3M^* + M$$

would lead to the expected relationship $\Delta\gamma_E \equiv \Delta\gamma_D$; since $\Delta\gamma_E^0$ is several times greater than $\Delta\gamma_D^0$ at lower temperatures and applied potentials it is concluded[127] that triplet–triplet annihilation does not contribute substantially to the production of electronically excited species under these conditions.

Weller[24] has shown that process MII is prohibited in polar solvents on energetic grounds (cf. Section V.E and Table XVIII), in which case the finite value of $\Delta\gamma_E^0$ must reflect the contributions of both doublet–doublet and triplet–triplet annihilations, i.e., with $k_{MII} = 0$

$$\alpha_E = \alpha_D(k_{DII} + k_{DTT})/k_{DTT} \tag{51}$$

If it is assumed that $k_{MD}\tau_F^D \ll 1$ at the lowest temperature (253°K) for which data are reported,[127] then from Eqs. (49), (50), and (51) it follows that

$$\Delta\gamma_E^0/\Delta\gamma_D^0 - 1 \approx \alpha_E/\alpha_D - 1 = k_{DII}/k_{DTT} > 1$$

which might be expected from the relative encounter frequencies of oppositely charged and uncharged species. At higher temperatures and applied potentials $\Delta\gamma_E^0$ approaches $\Delta\gamma_D^0$ (and $\alpha_E \to \alpha_D$) indicating an increased contribution of triplet–triplet annihilation to the electrochemiluminescence intensity under these conditions; however, the role of the relatively stable dimer cation M_2^+ (Section VII.C) in the production of emitting states,[8] e.g.,

$$M^+ + M \longrightarrow M_2^+ \xrightarrow{+e} \begin{array}{c} {}^1M^* + M \\ {}^1M_2^* \end{array}$$

has yet to be considered in these systems.

Exciplex chemiluminescence has been observed by Weller and Zachariasse[128] from the flow reactions of aromatic hydrocarbon anions M^- and Wurster's Blue Cations Q^+. Since the processes

$$M^- + Q^+ \longrightarrow \begin{array}{c} {}^1M^* + Q \\ M + {}^1Q^* \end{array}$$

are endothermic, the observation of molecular fluorescence is attributed to the triplet–triplet annihilations

$$^3M^* + {}^3M^* \longrightarrow {}^1M^* + M$$

$$^3Q^* + {}^3Q^* \longrightarrow {}^1Q^* + Q$$

following triplet state production by

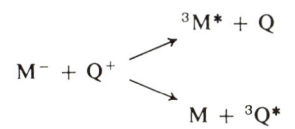

$$M^- + Q^+ \begin{array}{c} \nearrow {}^3M^* + Q \\ \searrow M + {}^3Q^* \end{array}$$

one or both of which may be energetically feasible. In the case of biphenyl, naphthalene, and triphenylene, the chemiluminescence spectrum includes the molecular fluorescence band of both M and Q (indicating the simultaneous production of $^3M^*$ and $^3Q^*$) together with the exciplex band which is believed to originate from the mixed triplet–triplet annihilation process

$$^3M^* + {}^3Q^* \rightarrow {}^1(M^-Q^+) \equiv {}^1MQ^*$$

C. Dimer Cation Neutralization

Simple MO considerations show that the dimer cation M_2^+ of an aromatic hydrocarbon M, with one less antibonding electron than the ground excimer configuration, should be stable with respect to its constituents $M^+ + M$; this is confirmed by esr[129] and optical absorption[8] studies of γ-irradiated solutions of benzene, naphthalene, and anthracene in low temperature glasses.

The appearance of the naphthalene excimer band in the thermoluminescence[130] or infrared stimulated luminescence[8] of irradiated naphthalene-saturated glasses is therefore attributed to the recombination of this dimer cation with released electrons

$$M_2^+ + e \longrightarrow {}^1M_2^*$$

which is consistent with a simultaneous reduction in intensity of the absorption-bands at 9600 and 17000 cm^{-1} assigned to M_2^+.

From the observed optical transition frequencies and appropriate photoassociation parameters Badger and Brocklehurst[8] estimate values in the range 10^3–10^6 liter mole^{-1} for the constant of the equilibrium

$$M^+ + M \rightleftharpoons M_2^+$$

Accordingly the formation and neutralization of dimer cations may be important processes in systems where molecular ions are generated.

VIII. CONCLUSIONS

Direct evidence for photoassociation is limited to those systems that exhibit either excimer (or exciplex) luminescence or excimer absorption which has recently been observed in benzene following pulse radiolysis.[141]

In the absence of spectroscopic evidence, however, photoassociation remains a logical primary process in the bimolecular reactions of electronically excited molecules where the excimer (or exciplex) represents a conceptually convenient origin of competitive secondary processes in the kinetic sequence of events. One of the more important of these processes is dissociative regeneration of the excited molecule which may effectively compete with radiative, nonradiative or chemical relaxation of the excimer (exciplex) to account for relatively low values and negative temperature coefficients of the measured bimolecular rate constants. This behavior perhaps provides the most convincing indirect evidence for photoassociation in these systems where it is anticipated that application of the recently developed laser excitation and spectro-analytical techniques[142] will furnish direct evidence of excimer absorption although the extent of photoassociation may be reduced by ground state depletion at the high light intensities employed.[143]

REFERENCES

1. J. B. Birks, D. J. Dyson, and I. H. Munro, *Proc. Roy. Soc.* (London), **A275**, 575 (1963).
2. B. Stevens and E. Hutton, *Nature*, **186**, 1045 (1960).
3. M. S. Walker, T. W. Bednar, and R. Lumry, *J. Chem. Phys.*, **45**, 3455 (1966); **47**, 1020 (1967).
4. There was some measure of agreement between participants at the Loyola International Conference on Molecular Luminescence, Chicago 1968, that this term be defined as stated although originally coined to describe complexes of 1.1 and 1.2 stoichiometry.[3] *Molecular Luminescence*, E. C. Lim, Ed., Benjamin, New York, 1969, p. 907; *Mol. Photochem.*, **1** 157 (1969).
5. J. Tanaka, C. Tanaka, E. Hutton, and B. Stevens, *Nature*, **198**, 1192 (1963).
6. C. A. Parker and C. G. Hatchard, *Trans. Faraday Soc.*, **59**, 284 (1963).
7. E. A. Chandross, J. W. Longworth, and R. E. Visco, *J. Amer. Chem. Soc.*, **87**, 14 (1965).
8. B. Badger, B. Brocklehurst, and R. D. Russell, *Chem. Phys. Lett.*, **1**, 122 (1967); B. Badger and B. Brocklehurst, *Nature*, **219**, 263 (1968).
9. Th. Förster and K. Kasper, *Z. Phys. Chem. N.F.*, **1**, 275 (1954); *Z. Elektrochem.*, **59**, 976 (1955).
10. J. B. Birks and L. G. Christophorou, *Proc. Roy. Soc.* (London), **A277**, 571 (1964).
11. B. Stevens and M. I. Ban, *Trans. Faraday Soc.*, **60**, 1515 (1964).
12. I. B. Berlman, *J. Chem. Phys.*, **34**, 1083 (1961).
13. M. Beer and H. C. Longuet-Higgins, *J. Chem. Phys.*, **23**, 1390 (1955); G. Viswanath and M. Kasha, *J. Chem. Phys.*, **24**, 574 (1956).

14. B. Stevens and J. T. Dubois, *Trans. Faraday Soc.*, **62**, 1525 (1966).
15. T. V. Ivanova, G. A. Mokeeva, and B. Y. Sveshnikov, *Opt. Spectrosc.*, **12**, 325 (1962).
16. J. B. Birks, C. L. Braga, and M. D. Lumb, *Proc. Roy. Soc.* (London) **A283**, 83 (1965).
17. S. S. Lehrer and G. D. Fasman, *J. Amer. Chem. Soc.*, **87**, 4687 (1965).
18. B. Stevens and T. Dickinson, *J. Chem. Soc.*, **1963**, 5492.
19. J. B. Birks and J. B. Aladekomo, *Spectrochim. Acta*, **20**, 15 (1964).
20. J. Langelaar, R. P. H. Rettschnick, A. M. F. Lamboy, and G. J. Hoytink, *Chem. Phys. Lett.*, **1**, 609 (1967).
21. G. Porter and M. W. Windsor, *Discussions Faraday Soc.*, **17**, 178 (1954); H. Linshitz, C. Steel, and J. Bell, *J. Phys. Chem.*, **66**, 2574 (1962).
22. Th. Förster, C. O. Lieber, H. P. Siedel, and A. Weller, *Z. Phys. Chem. N.F.*, **39**, 265 (1963).
23. H. P. Seidel and B. K. Selinger, *Aust. J. Chem.*, **18**, 977 (1965).
24. Cf. A. Weller, in *Fast Reactions and Primary Processes in Chemical Kinetics*, S. Claesson, Ed., Wiley-Interscience, New York, 1967, p. 413.
25. N. Mataga, T. Okada, and K. Ezumi, *Mol. Phys.*, **10**, 201, 203 (1966).
26. H. Knibbe, D. Rehm, and A. Weller, *Z. Phys. Chem. N.F.*, **56**, 95 (1967).
27. L. Onsager, *J. Amer. Chem. Soc.*, **58**, 1486 (1936); E. Lippert, W. Luder, and H. Boos, *Advan. Mol. Spectrosc.*, **1**, 443 (1962); N. Mataga, Y. Torihashi, and K. Ezumi, *Theor. Chim. Acta*, **2**, 158 (1964).
28. H. Beens, H. Knibbe, and A. Weller, *J. Chem. Phys.*, **47**, 1183 (1967).
29. H. Beens and A. Weller, *Chem. Phys. Lett.*, **2**, 140 (1968).
30. H. Knibbe, D. Rehm, and A. Weller, *Ber. Bunsenges. Phys. Chem.*, **73**, 839 (1969).
31. H. Knibbe, K. Röllig, F. P. Schafer, and A. Weller, *J. Chem. Phys.*, **47**, 1184 (1967).
32. N. Mataga, T. Okada, and N. Yamamoto, *Chem. Phys. Lett.*, **1**, 119 (1967).
33. I. B. Berlman and A. Weinreb, *Mol. Phys.*, **5**, 313 (1962).
34. W. R. Ware and H. P. Richter, *J. Chem. Phys.*, **48**, 1595 (1968).
35. H. Leonhardt and A. Weller, *Ber. Bunsenges. Phys. Chem.*, **67**, 791 (1963).
36. cf. W. R. Ware and B. A. Baldwin, *J. Chem. Phys.*, **43**, 1194 (1965).
37. J. B. Birks and J. C. Conte, *Proc. Roy. Soc.* (London), **A303**, 85 (1968).
38. B. K. Selinger, *Aust. J. Chem.*, **19**, 825 (1966).
39. J. B. Aladekomo and J. B. Birks, *Proc. Roy. Soc.* (London), **A284**, 551 (1965).
40. J. B. Birks and T. A. King, *Proc. Roy. Soc.* (London), **A291**, 244 (1966).
41. R. L. Barnes and J. B. Birks, *Proc. Roy. Soc.* (London), **A291**, 570 (1966).
42. J. B. Birks, D. J. Dyson, and T. A. King, *Proc. Roy. Soc.* (London), **A277**, 270 (1964).
43. J. B. Birks, M. D. Lumb, and I. H. Munro, *Proc. Roy. Soc.* (London), **A280**, 289 (1964).
44. Th. Förster, *Pure Appl. Chem.*, **7**, 73 (1963).
45. E. Döller and Th. Förster, *Z. Phys. Chem. N.F.*, **34**, 132 (1962).
46. P. Wilairat and B. K. Selinger, *Aust. J. Chem.*, **21**, 733 (1965).
47. E. Döller, *Z. Phys. Chem. N.F.*, **34**, 151 (1962).
48. K. H. Härdtl and A. Scharmann, *Z. Naturforsch. A*, **12**, 715 (1957); B. Stevens and P. J. McCartin, *Mol. Phys.*, **3**, 425 (1960).
49. A. Weller, in *Progress in Reaction Kinetics* Vol. 1, G. Porter, Ed., Pergamon, Oxford, 1961, p. 189.

50. G. S. Hammond, Private communication. L. M. Stephenson, D. G. Whitten, and G. S. Hammond, *The Chemistry of Ionisation and Excitation*, Taylor and Francis, London, 1967.
51. L. M. Stephenson, D. G. Whitten, G. F. Vesley, and G. S. Hammond, *J. Amer. Chem. Soc.*, **88**, 3665 (1966).
52. M. F. Thomaz and B. Stevens, in *Molecular Luminescence*, E. C. Lim, Ed., Benjamin, New York, 1969, p. 153.
53. D. S. McClure, *J. Chem. Phys.*, **17**, 905 (1949).
54. B. Stevens and B. E. Algar, *Chem. Phys. Lett.*, **1**, 58, 219 (1967); *J. Phys. Chem.*, **72**, 3468 (1968).
55. H. Tsubomura and R. S. Mulliken, *J. Amer. Chem. Soc.*, **82**, 5966 (1960).
56. B. Stevens and B. E. Algar, *J. Phys. Chem.*, **72**, 2582 (1968).
57. W. R. Ware and P. T. Cunningham, *J Chem. Phys.*, **43**, 3826 (1965).
58. Th. Förster, *Discussions Faraday Soc.*, **27**, 7 (1959).
59. J. T. Dubois and B. Stevens, in *Luminescence of Organic and Inorganic Materials*, H. P. Kallman and G. M. Spruch, Eds., Wiley, New York, 1962, p. 115; J. T. Dubois and M. Cox, *J. Chem. Phys.*, **38**, 2546 (1963); J. T. Dubois and R. L. Van Hemert, *J. Chem. Phys.*, **40**, 923 (1964).
60. G. Porter and F. Wilkinson, *Proc. Roy. Soc.* (London), **A264**, 1 (1961).
61. B. Stevens and M. S. Walker, *Proc. Chem. Soc.*, **26**, 109 (1964); *Proc. Roy. Soc.* (London), **A281**, 420 (1964).
62. K. Sandros and H. L. Backstrom, *Acta Chem. Scand.*, **18**, 2355 (1964).
63. R. Voltz, G. Laustriat, and A. Coche, *C. R. Acad. Sci., Paris*, **257**, 1473 (1963).
64. E. A. Chandross and J. Ferguson, *J. Chem. Phys.*, **45**, 397 (1966).
65. E. A. Chandross, *J. Chem. Phys.*, **43**, 4175 (1965).
66. G. J. Hoytink, *Z. Elektrochem.*, **64**, 156 (1960).
67. T. Azumi and S. P. McGlynn, *J. Chem. Phys.*, **41**, 3131 (1964).
68. T. Azumi, A. T. Armstrong, and S. P. McGlynn, *J. Chem. Phys.*, **41**, 3839 (1964); T. Azumi and S. P. McGlynn, *Ibid.*, **42**, 1675 (1965).
69. J. N. Murrell and J. Tanaka, *Mol. Phys.*, **4**, 363 (1964).
70. Cf. E. Konijnenberg, Thesis, Free University, Amsterdam, Holland, 1963.
71. J. B. Birks, *Chem. Phys. Lett.*, **1**, 304 (1967).
72. J. Koutecky and J. Paldus, *Collect. Czech. Chem. Commun.*, **27**, 599 (1962).
73. A. K. Chandra and E. C. Lim, *J. Chem. Phys.*, **48**, 2589 (1968).
74. M. T. Vala, I. H. Hillier, S. A. Rice, and J. Jortner, *J. Chem. Phys.*, **44**, 23 (1966).
75. L. Glass, I. H. Hillier, and S. A. Rice, *J. Chem. Phys.*, **45**, 3886 (1966).
76. S. P. McGlynn, J. D. Boggus, and E. Elder, *J. Chem. Phys.*, **32**, 357 (1960). J. Czekalla, G. Briegleb, W. Herre, and H. J. Vahlensieck, *Z. Elektrochem.*, **63**, 715 (1959); J. Czekalla and K. J. Mager, *Ibid.*, **66**, 65 (1962).
77. S. Iwata, J. Tanaka, and S. Nagakura, in *The Triplet State*, A. B. Zahlan, Ed., Cambridge Univ. Press, Cambridge, England, 1967, p. 433.
78. S. P. McGlynn, *Chem. Rev.*, **58**, 1113 (1958).
79. J. N. Murrell, *J. Amer. Chem. Soc.*, **79**, 4839 (1957).
80. R. S. Mulliken, *J. Chim. Phys.*, **61**, 20 (1963).
81. R. M. Hochstrasser and A. Malliàris, *J. Chem. Phys.*, **42**, 2243 (1965).
82. G. Castro and R. M. Hochstrasser, *J. Chem. Phys.*, **45**, 4532 (1966); E. C. Lim and S. K. Chakrabarti, *Mol. Phys.*, **13**, 293 (1967).
83. B. Stevens and M. I. Ban, *Mol. Cryst.*, **4**, 173 (1968).
84. B. Stevens, M. F. Thomaz, and J. Jones, *J. Chem. Phys.*, **46**, 405 (1967).
85. R. G. Bennett and P. J. McCartin, *J. Chem. Phys.*, **44**, 1969 (1966).

86. E. C. Lim, J. D. Laposa, and J. M. H. Ya, *J. Mol. Spec.*, **19**, 412 (1966).
87. B. Stevens and M. F. Thomaz, *Chem. Phys. Lett.*, **1**, 549 (1968).
88. T. Medinger and F. Wilkinson, *Trans. Faraday Soc.*, **62**, 1785 (1966).
89. T. Medinger and F. Wilkinson, *Trans. Faraday Soc.*, **61**, 620 (1965); A. R. Horrocks, A. Kearvell, K. Tickle, and F. Wilkinson, *Ibid.*, **62**, 3393 (1966).
90. E. J. Bowen and D. W. Tanner, *Trans. Faraday Soc.*, **51**, 475 (1955).
91. W. Koblitz and H. J. Schumacher, *Z. Phys. Chem.*, **B35**, 11 (1935); **B37**, 462 (1937).
92. J. B. Birks, J. H. Appleyard, and R. Pope, *Photochem. Photobiol*, **2**, 493 (1963).
93. A. S. Cherkasov and T. A. Vember, *Opt. Spectrosc.*, **4**, 319 (1959).
94. B. Stevens, T. Dickinson, and R. R. Sharpe, *Nature*, **204**, 876 (1964).
95. B. Stevens, R. R. Sharpe, and S. A. Emmons, *Photochem. Photobiol.*, **4**, 603 (1965).
96. C. A. Coulson, L. E. Orgel, W. Taylor, and J. Weiss, *J. Chem. Soc.*, 1955, 2961.
97. R. Calas, R. Lalande, and P. Mauret, *Bull. Soc. Chim.*, **1960**, 148; cf. D. E. Applequist, E. C. Friedrich, and M. T. Rogers, *J. Amer. Chem. Soc.*, **81**, 457 (1959).
98. H. Bouas-Laurent, Ph.D. Thesis, University of Bordeaux, France, 1964.
99. J. P. Simons, *Trans. Faraday Soc.*, **56**, 391 (1960).
100. R. Livingston, R. M. Go, and T. G. Truscott, *J. Phys. Chem.*, **70**, 1312 (1966).
101. H. Knibbe, D. Rehm, and A. Weller, *Ber. Bunsenges. Phys. Chem.*, **72**, 257 (1968).
102. N. Mataga, T. Okada, and N. Yamamoto, *Chem. Phys. Lett.*, **1**, 119, (1967).
103. B. Stevens, *Spectrochim. Acta*, **18**, 439 (1962).
104. J. M. Robertson, *Proc. Roy. Soc.* (London), **A207**, 101 (1951).
105. J. Ferguson, *J. Chem. Phys.*, **28**, 765 (1958).
106. J. M. Robertson and J. G. White, *J. Chem. Soc.*, **1947**, 358.
107. P. F. Jones and M. Nicol, *J. Chem. Phys.*, **43**, 3759 (1965).
108. J. Tanaka, T. Koda, S. Shionoya, and S. Minomura, *Bull. Chem. Soc. (Japan)*, **38**, 1559 (1966).
109. H. W. Offen, *J. Chem. Phys.*, **44**, 699 (1966).
110. J. Tanaka, *Bull. Chem. Soc.* (Japan), **36**, 1237 (1963).
111. B. Stevens and T. Dickinson, *Spectrochim. Acta*, **19**, 1865 (1963).
112. R. M. Hochstrasser, *J. Chem. Phys.*, **36**, 1099 (1962).
113. F. Hirayama, Thesis, University of Michigan, 1963.
114. F. Hirayama, *J. Chem. Phys.*, **42**, 3163 (1965).
115. M. T. Vala, J. Haebig, and S. A. Rice, *J. Chem. Phys.*, **43**, 886 (1965).
116. C. Tanaka, J. Tanaka, E. Hutton, and B. Stevens, *Nature*, **198**, 1192 (1963).
117. C. A. Parker, *Nature*, **200**, 331 (1963); *Spectrochim. Acta*, **19**, 989 (1963).
118. C. A. Parker and C. G. Hatchard, *Proc. Roy. Soc.* (London), **A269**, 574 (1962).
119. G. F. Moore and I. H. Munro, *Spectrochim. Acta*, **23**, 1291 (1967).
120. J. B. Birks, *J. Phys. Chem.*, **67**, 2199 (1963); **68**, 439 (1964).
121. J. B. Birks, G. F. Moore, and I. H. Munro, *Spectrochim. Acta*, **22**, 323 (1966).
122. E.g., J. B. Birks, *Phys. Lett.*, **24A**, 479 (1967).
123. B. Stevens, *Chem. Phys. Lett.*, **3**, 233 (1969).
124. C. A. Parker and T. Joyce, *Chem. Commun.*, **1967**, 744.
125. C. A. Parker and T. Joyce, *Chem. Commun.*, **1967**, 1138.
126. E. A. Chandross, J. W. Longworth, and R. E. Visco, *J. Amer. Chem. Soc.*, **87**, 3259 (1965).
127. C. A. Parker and G. D. Short, *Trans. Faraday Soc.*, **63**, 2618 (1967).
128. A. Weller and K. Zachariasse, *J. Chem. Phys.*, **46**, 4984 (1967).
129. I. C. Lewis and L. S. Singer, *J. Chem. Phys.*, **43**, 2712 (1965); O. W. Howarth and G. K. Fraenkel, *J. Amer. Chem. Soc.*, **88**, 4514 (1966).
130. B. Brocklehurst and R. D. Russell, *Nature*, **65**, 213 (1967).

131. J. Eisinger, M. Gúeron, R. G. Shulman, and T. Yamane, *Proc. Nat. Acad. Sci. U.S.*, **55**, 1015 (1966).
132. M. Gueron, R. G. Shulman, and J. Eisinger, in *The Triplet State*, A. B. Zahlan, Ed., Cambridge Univ. Press, 1967, p. 505.
133. J. Kondelka and L. Augenstein, *Photochem. Photobiol.*, **7**, 613 (1968).
134. Cf. J. Eisinger, *Photochem. Photobiol.*, **7**, 597 (1968).
135. Cf. A. A. Lamola, *Photochem. Photobiol.*, **7**, 619 (1968).
136. J. Eisinger and A. A. Lamola, *Biochem. Biophys. Res. Commun.*, **28**, 558 (1967).
137. Cf. R. B. Setlow, *Photochem. Photobiol.*, **7**, 643 (1968).
138. A. A. Lamola and T. Yamane, *Proc. Nat. Acad. Sci. U.S.*, **58**, 443 (1967).
139. W. Klopffer, *J. Chem. Phys.*, **50**, 2337 (1969); *Chem. Phys. Lett.*, **4**, 193 (1969).
140. F. Hirayama and S. Lipsky, *J. Chem. Phys.* **51**, 1939 (1969) *Molecular Luminescence*, E. C. Lim, Ed., Benjamin, New York, 1969, p. 237.
141. R. Cooper and J. K. Thomas, *J. Chem. Phys.*, **48**, 5097 (1968).
142. G. Porter and M. R. Topp, *Proc. Roy. Soc.* (London), **A315**, 163 (1970).
143. C. R. Goldschmidt, Y. Tomkiewicz, and I. B. Berlman, *Chem. Phys. Lett.*, **2**, 536 (1968).

Photochemistry in the Metallocenes

R. E. BOZAK, *Department of Chemistry, California State College, Hayward, Hayward, California 94542*

I. Nature of Electronic Absorption in Ferrocene, *bis*-Benzenechromium(I) and
    Derivatives 228
    A. Electronic Absorption Spectrum of Ferrocene 228
    B. Various Ferrocene Complexes 229
    C. Substituted Ferrocenes 231
    D. Other Metallocenes 231
II. Ferrocene in Photochemical Systems 234
    A. Quenching. 234
    B. Photosensitization 234
    C. Photolability 235
III. Other Metallocenes or Derivatives as Photochemical Systems 236
    A. Photochemical Transformations of Substituted Ferrocenes 236
    B. Ferrocene Derivatives as Protective Ultraviolet Absorbers 241
    C. Other Metallocenes 241
References 243

The discovery of ferrocene in 1951[1] heralded the beginning of several major new fields in organometallic chemistry. Of particular interest have been those species such as **1** and **2**, wherein two hydrocarbon rings are "sandwiched" onto a transition metal atom and a discrete, stable chemical compound results. These fall into the class which bears the name "metallocenes."[2]

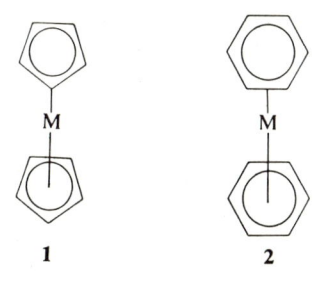

Another, although not so novel, field of vigorous research activity is organic photochemistry whose renaissance is witnessed by many articles, reviews,[3] and books.[4] However, while the photochemistry of transition metal coordination compounds has been collected and reported several times,[5,6] up to the present no survey has been available on photochemical investigation in the metallocenes. In this review we shall limit ourselves mainly to the photochemistry of the general class of compounds 1 and 2, or their derivatives.

I. NATURE OF ELECTRONIC ABSORPTION IN FERROCENE, bis-BENZENECHROMIUM(I) AND DERIVATIVES

Numerous spectral determinations have been made on this class of compounds. In some cases, very detailed spectra have been obtained on both vapor and solutions down to 180 mμ. In a few of the simple cases attempts have been made to assign the bands although, since a universally accepted picture of the bonding in the metallocenes has yet to arrive, the assignments perhaps should be regarded as preliminary.

A. Electronic Absorption Spectrum of Ferrocene

The ultraviolet and visible absorption spectra of ferrocene in 95% ethanol are characterized mainly by two bands. One is at 325 mμ ($\epsilon = 50$) and the other (which is somewhat broader) is centered at 440 mμ ($\epsilon = 90$).[7-9] High end absorption is observed ($\epsilon_{220} \sim 5000$, $\epsilon_{206} \sim 34{,}000$).

In aprotic solvents, such as isooctane or cyclohexane, the workers at Organic Laboratory Chemicals, J. T. Baker Chemical Co., report two bands: at 326 mμ ($\epsilon = 56$) and 437 mμ ($\epsilon = 96$) and high absorption at 202 mμ ($\epsilon = 54{,}000$).[8] Nesmeyanov et al. report[10] 325 mμ ($\epsilon = 51$), 440 mμ ($\epsilon = 87$), 265 mμ(sh) ($\epsilon = 1700$), 230 mμ(sh) ($\epsilon = 4400$), and 200 mμ ($\epsilon = 49{,}000$). More detailed investigations of the spectra of ferrocene include work by Scott and Becker[11] and Armstrong, Smith, Elder, and McGlynn.[12] The latter workers report the spectrum of ferrocene vapor. A weak band ($\epsilon = 8$) at 528 mμ in both protic and aprotic solvents seems to be of no practical value but may prove of theoretical importance in the future.[13,14]

The intriguing luminescence of ferrocene originally reported in 1962[11] is now believed to have stemmed from a spurious source. For instance, Tarr and Wiles[15] reported that "although emission (480–685 mμ) has been reported from ferrocene excited in its 324 mμ band, we were unable to detect any emission from ferrocene or benzoylferrocene, either solid or dissolved in 5:1 isopentane–methylcyclohexane glass at 77°K." Also, Guillory, Cook, and Scott[16] have noted that the emission most likely originated from an impurity, although it is noted by them that there still may be some evidence for a weak luminescence around 520 mμ. Independent observations may confirm this.

Most workers are in agreement about the general origin of the absorptions in ferrocene. In particular, the two long wavelength bands (we shall exclude the 528 mμ band from further discussion) arise due to transitions in the $3d$ electrons of the central iron atom. It has proven more challenging, however, to quantitatively assess the degree of ring-metal orbital mixing. For instance, Rosenblum says "The assignment of the 440 mμ band in ferrocene to a d-d type transition would appear plausible, but it seems likely that some mixing of ring orbitals in this transition must be assumed to account for the moderate effect of substituents on the band position."[17] According to Lundquist and Cais[9] the change in intensity observed when a —M(C$_5$H$_5$) is replaced by a —M(CO)$_3$ grouping indicates the 326 and 440 mμ maxima are associated with electronic transitions in the metal–carbon bonds between the rings and the iron atom. They suggest that these bands are characteristic of many organometallic systems and term them "MC-bands." Scott and Becker,[11] noting that the 440 mμ band maximum is, to their way of thinking, insensitive to substitution on the rings, assign its origin to a forbidden d-d transition between levels highly localized on iron. They ascribe the other band, 324 mμ, as arising from a forbidden transition unlike those normally assigned to a ligand field splitting of degenerate d levels and involving, to an unknown finite degree, the molecular orbitals of the rings. The charge transfer is deduced as being from the metal *to* the rings since the upper level is apparently mixed proportionately more with the ring orbitals and a bathochromic shift is observed with the introduction of electron withdrawing substituents.

The shorter wavelength bands (200–300 mμ) could be similar (but less forbidden) charge-transfer transitions or homoannular ring-localized transitions. For instance, Nesmeyanov[10] concludes, from an inspection of oscillator strengths, that the strong 200 mμ band is a $\pi \rightarrow \pi^*$ transition in the ring.

Removal of an electron from neutral ferrocene can yield a large number of different ferricenium salts. This is a dramatic visual transformation, due to the vivid blue color of the ferricenium cation in contrast to the pale orange of ferrocene solutions. The cation displays two maxima: 250 mμ ($\epsilon = 12,000$) and 617 mμ ($\epsilon = 340$).[18]

B. Various Ferrocene Complexes

Solutions of ferrocene in halogenated solvents at room temperature are reported to be stable in the dark, but upon exposure to light decomposition occurs at a relatively rapid, quite noticeable rate. Typically, a brown sludge is formed as the decomposition proceeds.

In 1957, Brand and Snedden,[19] noting that the uv spectrum of ferrocene is more intense in carbon tetrachloride than in hydrocarbon solvents and also that the 307 mµ band is not related to the 326 mµ ferrocene band in non-halogen solvents, assigned this new band to a dissociative charge-transfer species (Eq. 1) involving intermolecular charge-transfer in which the metal atom in ferrocene serves as the electron donor.

$$C_{10}H_{10}Fe + CCl_4 \xrightarrow{h\nu} C_{10}H_{10}Fe^{\oplus} + Cl^{\ominus} + \cdot CCl_3 \qquad (1)$$

Moreover, Brand and Snedden were also able to correlate the mass and the number of halogen substituents in the solvent with spectral changes. Even better correlation was found between the wavelength of the 307 mµ band and the half-wave potential of the halide—a measure of its tendency to act as an electron acceptor.

Since the spectrum of the ferricenium cation is so distinctively different from that of ferrocene in halogenated solvents, no ion must be formed as yet and the 307 mµ band must be due to a predissociative species probably close to that depicted as **3**.

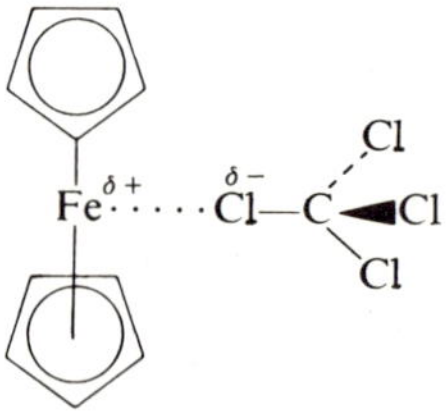

3

A stable ferrocene–TCNE adduct has been prepared and its structure, closely analogous to charge-transfer complexes of benzenoid systems has been determined to be that shown as **4**.[20] Whether a charge-transfer species involving the iron atom rather than the ring is possible has not been deter-

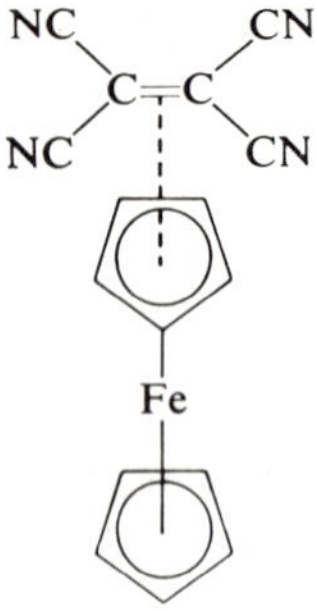

4

mined. The electronic absorption spectrum of 4 in KBr shows two strong bands near 975 and 1150 mμ.

Dannenberg and Richards[21] have reported evidence for a piperylene-ferrocene complex on the basis of (1) new absorptions between 220 and 300 mμ for piperylene/ferrocene mixtures in n-heptane and n-hexane, and (2) the appearance in the nmr spectrum of *trans*-piperylene of a new methyl doublet when ferrocene is present. Guillory, Cook, and Scott[16] have reported that, although they were unable to detect a complex using nmr, they did note small shifts in the uv spectrum of ferrocene when piperylene was added. This latter observation is consistent with a diene-ferrocene complex.

C. Substituted Ferrocenes

Several generalizations are made on the basis of such data as in Table I. For instance, the two rings tend to absorb independently of one another[10]; that is, no conjugation of one ring to another. Nesmeyanov has also noted that while intensities of bands in the 220–600 mμ region increased from ferrocene to monosubstituted ferrocenes to heteroannularly disubstituted ferrocenes, the opposite is true in the 180–220 mμ region.

While the band position in the spectra of acylmetallocenes is close to that of the parent, above 325 mμ its intensity is enhanced with the introduction of the —COR. (This is in contrast to the alkylmetallocenes, whose spectra are almost identical to the parent metallocene.) Below 300 mμ, however, the intense, broad end-absorption in the spectrum of the parent metallocene is replaced by two intense peaks (for acylferrocenes these appear near 230 and 270 mμ). These are believed to be associated in part with electronic transitions of the substituent with the ring, superimposed upon metal to ring charge-transfer bands.[7]

Also, conjugation of a delocalizing ferrocene group with the carbonyl should cause the intense $\pi \to \pi^*$ carbon–oxygen transition to move to longer wavelengths and "bury" the feeble $CO_{n \to \pi^*}$ band; hence, no such band is identifiable in acetylferrocene and the 267 mμ maximum most likely correlates with acetophenone's 239 mμ band.

D. Other Metallocenes

The long wavelength charge transfer bands (Lundquist and Cais' MC-bands[9]) noted earlier appear in various other cyclopentadienyl–metal derivatives. For instance, the electronic absorption spectra of ruthenocene has two bands: 321 mμ ($\epsilon = 240$) and 277 mμ ($\epsilon = 195$) in addition to high end absorption at 200 mμ.[7] In osmocene only one band appears, it being at 318 mμ ($\epsilon \approx 105$).[7]

TABLE I
Substituted Ferrocenes

Compound[a]	Electronic absorption bonds [λ_{max} (ϵ) or λ_{sh} (ϵ)]				Solvent	Reference
Ferrocene	440(87)	325(51)	243sh(2500)		200(49,000)	Isooctane	10
Ethylferrocene	438(95)	325(57)	245sh(3000)		203(42,000)	Isooctane	10
Benzylferrocene	440(136)	325(82)	254sh(3480)			Isooctane	15
1,1'-Dibenzylferrocene	438(110)	330(100)	250sh(6000)			95% EtOH	9
Phenylferrocene	447(330)	310sh(—)	278(10,600)	238(17,600)		95% EtOH	17
p-Nitrophenylferrocene	503(2580)	397(2640)	326(13,000)	280(—)		95% EtOH	17
p-Acetylphenylferrocene	466(1130)	376(2290)	304(16,280)	266(9850)		95% EtOH	17
1-Cyclopentenylferrocene	442(270)	325sh(424)	276(8400)	227sh(19,400)		Isooctane	22
Acetylferrocene	445(289)	318(1100)	267(5400)			Isooctane	22
Acetylferrocene		—	269(6500)	226(16,500)		95% EtOH	23
Benzoylferrocene	459(610)	355(1220)	273(6950)	240(20,400)		Isooctane	15
1,1'-Dibenzoylferrocene	470(700)	354(1970)	276(12,550)	247(22,300)		Isooctane	15
$\Phi COOCH_3$	444(180)	335(260)	304(860)	264(4100)	196(31,000)	Isooctane	10
$\Phi CH=N-\phi$	457(1270)	328(18,300)	264(19,300)			Isooctane	15
ΦNO_2	471(1310)	364(1780)	273(6560)	239(9630)		Isooctane	24
$\Phi N=N\Phi$	520(5200)	382(3600)	317(12,250)	268sh(7400)	252sh(8400)	Isooctane	24

[a] $\Phi = C_{10}H_9Fe$. This symbol shall be used for the ferrocenyl group throughout the rest of this review.

The electronic absorption spectrum of nickelocene has been recorded and analyzed in considerable detail by Scott and Becker.[11] They found band maxima at 1920, 3075, and 6920 Å and shoulders at 2700, 3450, 4400, and 5700 Å. The band at 1920 Å is believed to be an allowed transition which is designated as a N-V transition in ferrocene, but the intensity of this band is less than the N-V band in ferrocene. The shoulder at 2700 Å is denoted as a N-V or N-Q transition similar to the 2300 and 2600 Å shoulders in ferrocene. The band at 3075 Å is relatively intense and occurs at approximately the same wavelength as the intermolecular charge transfer band of ferrocene in carbon tetrachloride described by Brand and Snedden. The absorption spectrum of nickelocene in carbon tetrachloride shows no new bands other than those found in cyclohexane or ethanol.

Electronic absorption spectra of *bis*-benzenechromium (I) iodide (5) and its derivatives have been recorded and discussed by Yamada et al.[25,26] They noted a loss of the benzene rings' aromaticity in these compounds with the π electrons being drawn toward the metal and thereby conferring more a "cyclopentadienyl radical" character on the rings.

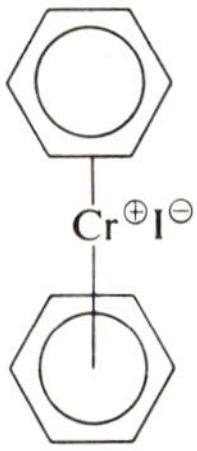

$$Cr^{\oplus}I^{\ominus}$$

5

Bands noted by these Japanese workers have been mainly attributed, however, by Scandola, Balzani and Carassiti[27] to iodide anion. Cr(benzene)$_2^{\oplus}$ cation most likely has a spectrum consisting of two bands between 200 and 400 mμ: 270 (ϵ = 6000) and 335 (ϵ = 6400) and high end absorption. Hence, "the benzene bands are not substantially shifted toward higher wavelengths,"[27] suggesting that in dibenzenechromium the rings retain substantial benzenoid character which is consistent with much of the spectroscopic data gathered in this system.

The carbonyl-containing metallocene, cymantrene[2] (6) shows one band,

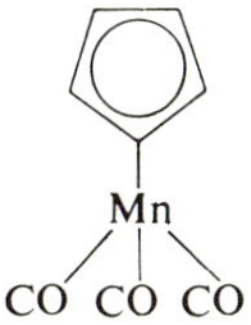

Mn
CO CO CO

6

λ_{max} 329 mµ (ϵ = 1000), a shoulder at 250 mµ ($\epsilon \sim$ 1400), and high end absorption. The 329 mµ band is the MC type mentioned earlier.[9]

II. FERROCENE IN PHOTOCHEMICAL SYSTEMS

A. Quenching

Up to the present, the most noteworthy photochemical property of the metallocenes has been triplet quenching. In 1966, Fry, Liu, and Hammond[28] published a study on the comparison of the rates of fast triplet quenching reactions. Ferrocene was found to be a superior quencher of species whose triplet energies range from 66.6 kcal/mole (triphenylene) down to anthracene at 42.6 kcal. The authors noted: ". . . the superiority of ferrocene as a quencher and the fact that acetylacetonates are more reactive than dipivaloylmethanates suggests that stereochemical factors are important in metal compounds functioning as triplet quenchers." They concluded that close contact between the triplet and the unsaturated ligands attached to the metal facilitates quenching. Since the unsaturated parts of ferrocene are so exposed, it is not hard to see why the quenching rates were as high as any ever observed in the Cal Tech laboratories.[29]

(The fast, relatively constant, rate of quenching lacks, at present, a convincing explanation. Perhaps all the metallic compounds found to be effective have very low-lying accessible states or the observed quenching does not always involve energy transfer as such. The latter possibility has been suggested.[30] Valentine[6] has considered the possibility that energy transfer involves the formation of metastable mixed excimers between the excited compound and the metal complex.)

No investigation has been published where substituted ferrocenes are used. Highly branched side chains should make the so-called unsaturated groups less labile for quenching since close contact would be impossible.

Ferrocene is marketed as one of the 24 compounds in the Photosensitizer and Quencher Kit wherein its efficient quenching capability is also noted.[8]

The photochemical reduction of benzophenone to benzopinacol in the presence of ferrocene[31] apparently relates to a high concentration of ketone so that sufficient unquenched benzophenone triplets are present and proceed to the dimer in the usual fashion.

B. Photosensitization

Photosensitization by ferrocene was first reported in 1964.[21,32] Richards noted that ferrocene catalyzes the *cis-trans* isomerization of various olefins (such as piperylene and stilbene), catalyzes the photodimerization of piperylene, and the internal cycloaddition of 1,5-cyclooctadiene. For instance, added ferrocene accelerates the photoisomerization of *trans*-piperylene (in benzene) by a factor of 5 and catalyzes the photodimerization of both piperylene and isoprene. In a later study, Guillory, Cook, and Scott[16] substantially confirmed this and concluded that the process goes, according to one of the possibilities suggested by Richards, via the diene-ferrocene complex, excited first to its singlet state which then intersystem-crosses to its triplet and finally dissociates to ground state ferrocene and triplet piperylene which can then isomerize. Scott, however, found that the rate of photoisomerization is low when *trans*-1,2-dimethylcyclohexane is substituted for benzene. The exact reason for the solvent difference has not been uncovered although a benzene/ferrocene complex may be a possibility.†

Richards further has presented evidence for an exception to the generalization that excitation to the lowest lying singlet is sufficient for photochemical reaction. He reports that even though the lowest lying singlet is at 450 mμ, light of 313 mμ or shorter is required to initiate the isomerization. He infers that the higher singlets do *not* rapidly decay to S_1, possibly because they involve excitation of the electrons associated to a great extent with the ligand while the lower states are localized on the metal and are, hence, "insulated."

Another phototransformation of what might be termed a loose ferrocene complex has been reported in the ferrocene-tetracyanoethylene case, where cyclohexane solutions of the components are irradiated through Vycor with a G. E. Sunlamp and the complex **4** is obtained.[33]

C. Photolability

The diamagnetic metallocenes are generally regarded as light-stable.[34,35] Under many different sets of conditions solutions of ferrocene are stable to actinic and ultraviolet irradiation. Indeed, the stability of excited ferrocene is generally thought to be very high as witnessed by the low G-values reported for its radiolysis.[36] With acid and molecular oxygen present, however, decomposition of ferrocene, via the ferricenium ion, occurs. With alkyl halides the photodecomposition already mentioned is well-known to take place,[19] having been observed by many workers, occasionally to their chagrin after a valuable sample had been stored in a halogenated solvent. The metallocenes containing the ligand carbon monoxide prefer decarbonylation, followed by donor substitution or metal-ring cleavage to intractable products (see below).

† This suggestion has been made by M. Rosenblum.

III. OTHER METALLOCENES OR DERIVATIVES AS PHOTOCHEMICAL SYSTEMS

A. Photochemical Transformations of Substituted Ferrocenes

The number of reported instances where substituted metallocenes undergo phototransformations cleanly to products is still relatively small. The first incident was a disclosure[37] by Finnegan and Hagen of a photochemical Fries rearrangement by Finnegan and Mattice.[38] They examined the photochemistry of both phenyl ferrocenoate (7) and *p*-tolyl ferrocenoate (8) at low molarities in cyclohexane. For instance, a solution of 0.82 g 7 in 750 ml cyclohexane was irradiated for 2 hr at 37° with a 450-W Hanovia lamp. The product was 0.12 g *p*-hydroxybenzoylferrocene (9) (no *o* was obtained).

The *p*-tolyl compound, however, was found to behave somewhat differently, yielding ca. 5% *p*-tolylferrocene and 2% ferrocenoic acid. The authors observe that this latter reaction might be in a class of photodecarboxylations observed elsewhere.[39]

Finnegan and Mattice do not go into detail as to the nature of the excited state, but they do imply that Kobsa's view[40] of the mechanism might apply. This would involve going through an excited state of the carbonyl to radicals 10 and 11 in a solvent cage (just after photodissociation so that radical recoupling follows efficiently). The observed preference of 10 for the *para* position of 11 to the exclusion of the *ortho* position is apparently without precedent.

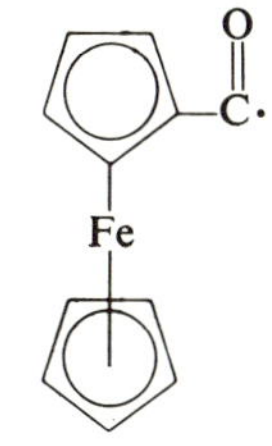

PHOTOCHEMISTRY IN THE METALLOCENES

[Structures 10 and 11]

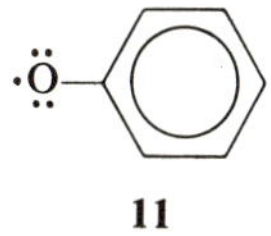

For the next set of photochemical transformations of substituted ferrocenes we must turn to the work of Nesmeyanov represented in the eight unbalanced Eqs. (2–9).

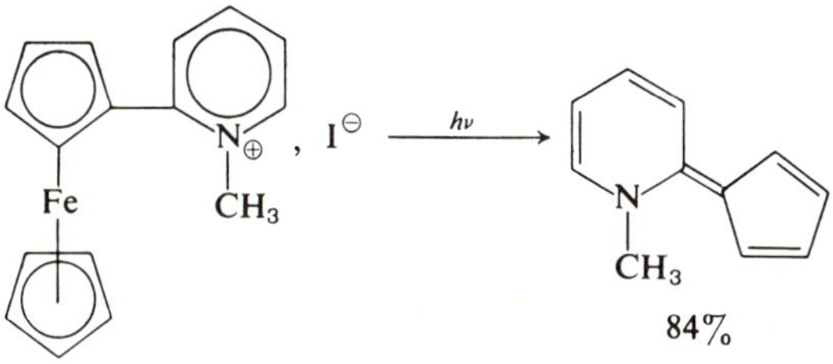

$$\text{(2)} \quad 84\%$$

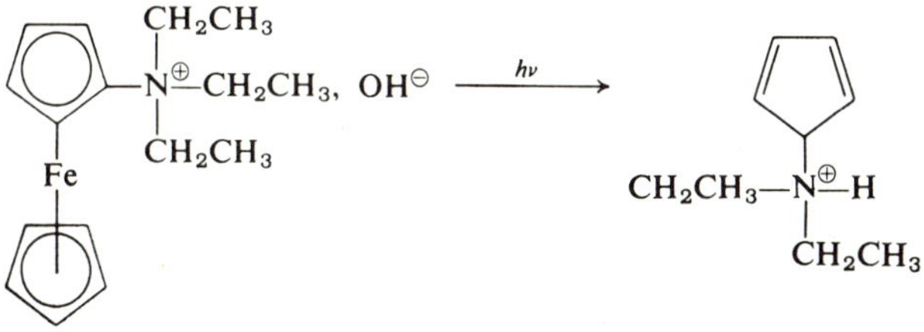

$$\text{(3)}$$

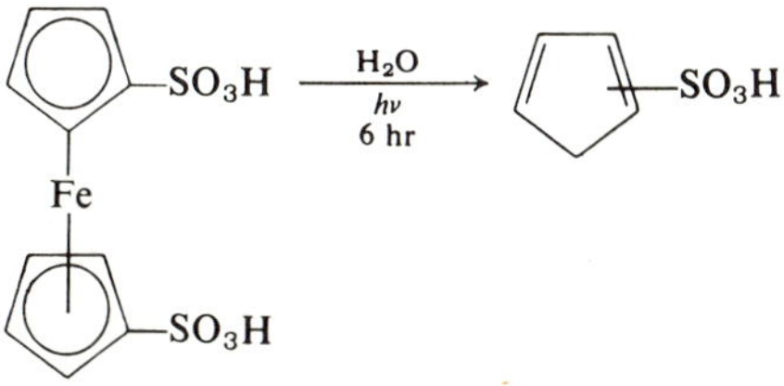

$$\text{(4)}$$

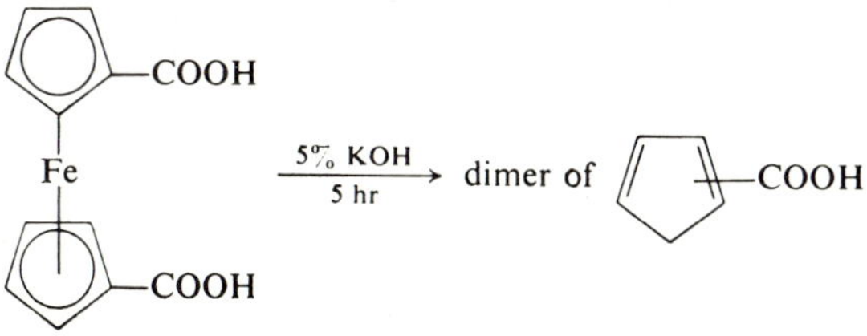

$$\text{(5)} \quad \text{dimer of } \text{Cp–COOH}$$

Conditions:
- (2) hν
- (3) hν
- (4) H₂O, hν, 6 hr
- (5) 5% KOH, 5 hr

Equation (2) proceeds as written when irradiation is carried out either with incandescent light (from a 300-W bulb for 1 hr) or sunlight.[41] Besides the N-methyl compound some cyclopentadiene is also reported. This photolysis is carried out in dilute aqueous sodium hydroxide.

Equations (3) and (4) represent work in two later papers.[42,43] Low energy radiation is likewise used and the solvent is predominantly water.

Equation (5) is incomplete only with respect to the structure of the dimer as the authors note:[44] "we did not set ourselves the task of establishing the structure" of the final acid.

In any case, it is clear that these photochemical reactions effect the rupture of the metal–ring bond and electromagnetic radiation of relatively low energy promotes the transformation. A common feature of all eight reactions is an electrophilic or potentially electrophilic grouping on or very near the ferrocene ring.†

The most recent published report of a photochemical transformation of a substituted ferrocene is by this reviewer.[45] Compound **12** is irradiated in an aliphatic hydrocarbon and one of the products is benzoylferrocene. This trans-

$$\Phi CHCH_2 CO\phi \xrightarrow{h\nu} \Phi\overset{O}{\underset{\|}{C}}\phi$$
$$\underset{\phi}{|}$$

12

formation most likely proceeds, as does Finnegan's, through an excited state of the carbonyl, in contrast to Nesmeyanov's work. (It was noted earlier that the triplet quenching ability of ferrocene is good, but the ferrocenyl moeity apparently has little or no effect in this reaction. One possibility is that alkyl substituted ferrocenes do not quench with high efficiency and/or the excited benzoyl carbonyl is a singlet, or still another possibility is that the ferrocene's "unsaturated groups" cannot align geometrically to quench.) A tentative mechanism involving α-cleavage has been postulated but remains unproven.[46]

Several workers have attempted to induce benzoylferrocene (in many ways an organometallic analog of benzophenone) to undergo or demonstrate some "clean" photochemistry. Unfortunately, up to the present, there are no published cases of any useful (or even tractable) photoproducts. Rausch, Vogel, and Rosenberg[47] reported that in an attempted photochemical reduction of benzoylferrocene in isopropyl alcohol, "nearly all the ketone was recovered after ten days exposure to sunlight, and only a very small amount of insoluble material resulted. This appeared to contain mostly iron oxide together with a trace of carbonaceous material." Likewise, Weliky and Gould[48] were unable to effect a reaction when they irradiated a solution of the ketone in isopropyl alcohol in sunlight for a month.

Ferrocene-C(O)Φ $\xrightarrow[\text{isopropyl alcohol}]{h\nu}$ no reaction

† These reactions are described in *Organometallic Chemistry*, I.U.P.A.C., 3rd International Symposium, Plenum, New York, 1968, pp. 217–220.

Tarr and Wiles,[15] attempting to explain the fact that acylferrocenes are mildly photolytically unstable in all hydroxylic solvents, postulate Eq. (10) leading to the unstable, substituted ferricenium ion.

$$\text{Fc-C(=O\cdots HOCH_3)-}\phi \xrightarrow{h\nu} \text{Fc}^{\oplus}\text{-C(=O)-}\phi + \text{HO:}^{\ominus} + \text{CH}_3\cdot \qquad (10)$$

The apparent low stability (or nonexistence) of an $n \to \pi^*$ excited state leading to photoproducts of the carbonyl in benzoylferrocene may be due to internal quenching or "masking" by the charge transfer bands which lie in the customary $n \to \pi^*$ region (see Table I). This latter postulate might be tested in benzoylosmocene where the charge transfer bands are significantly lower.[49] Irradiation of benzoylferrocene in acid has been tentatively reported[46] to lead to stable, isolable noniron containing compounds and this may be consistent with the Nesmeyanov work cited above and involve the species **13**. Formal removal of an electron pair from the iron with resultant transfer to the cyclopentadienyl moiety is a logical consequence of the "MC" bands and provides the charged resonance form of a stable fulvenol. (Reference 43 includes the note that acidic solutions of acetylferrocene and its oxime are decomposed under the action of light, but no further information is given.)

13

Richards and Pisker-Trifunac[50] have irradiated *cis*-styrylferrocene (**14**) under several different sets of conditions and in all cases high conversion to the *trans*-isomer was the only reaction observed. It is noteworthy that no cyclobutane or oxetane derivatives are formed. The same seems to apply to ferrocenyl analogs of stilbene. That is, no ring formation is observed and the photostationary state consists almost exclusively of the *trans*-isomer.

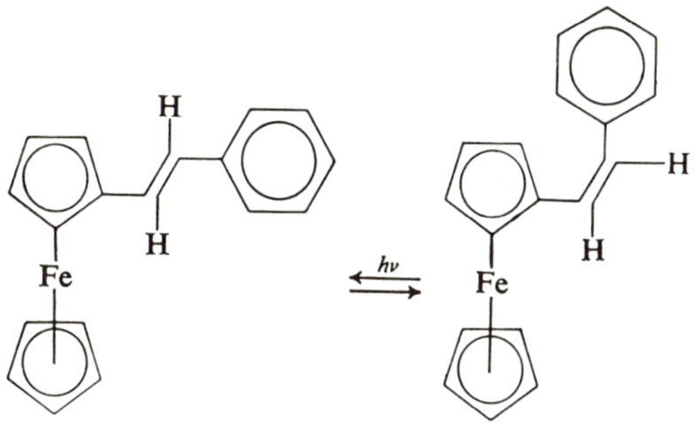

14

B. Ferrocene Derivatives as Protective Ultraviolet Absorbers

Tarr and Wiles, in a continuing study on retarding the photochemical depolymerization of fiber-forming macromolecules, report that ferrocenes are effective in protecting cotton cellulose.[15] Beinoravichute[51] has investigated the addition of ferrocene to polyethylene film. Wiles and Suprunchuk[52] have prepared some vividly colored ferrocenes as ultraviolet absorbers. The most extensive studies of the latter type have been reported by Schmitt and Hirt of American Cyanamid.[53] A variety of polymer substrates were used by them with several ferrocene derivatives, the most effective against embrittlement being *o*-hydroxybenzoylferrocene. Schmitt and Hirt were unable to pinpoint the reason for the effectiveness, but they did consider several possibilities, including an energy transfer mechanism. Kwei has reported[54] incorporation of ferrocene chemically into PVC with deleterious effect on polymer stability.

C. Other Metallocenes

Cymantrene (**6**) demonstrates a considerable amount of organic chemistry just as does ferrocene. No organic photochemistry has been found in cymantrenyl systems, however, due to the customary complication of photochemically induced decarbonylation. E. O. Fischer[55] has investigated this in some detail and three representative reactions are listed here.

Few substituted cymantrene derivatives have been investigated for their photochemical behavior, and whether decarbonylation will always be the preferred path is unknown. Cuingnet and Adalberon† attributed, without

---

† This work is reported in E. Cuingnet and M. Adalberon, *C. R. Acad. Sci., Paris*, **257**, 461 (1963).

proof, the rapid photochemical decomposition of acetylcymantrene in isopropanol to decarbonylation with no pinacolization.

When Balzani[56] photolyzed the *bis*-benzenechromium(I) hydroxide two characterizable products were chromium(III) and benzene (Eq. 11).

$$[(\text{C}_6\text{H}_6)_2\text{Cr}]^+ \text{OH}^- \xrightarrow{h\nu} \text{C}_6\text{H}_6 + \text{Cr(III)} \qquad (11)$$

Razuvaev found that *bis*-benzenechromium itself could be photolyzed[57] in the presence of an alkyl halide to obtain the halide of *bis*-benzenechromium(I) plus hydrocarbons derived most likely from the free radical corresponding to the original alkyl group. Most likely the decomposition of ferrocene[19] also affords a variety of hydrocarbons, but this has apparently not been investigated.

No reports have been made on photochemical transformations in other metallocenes such as uranocene and ruthenocene. These are doubtless forthcoming for the metallocenes and their derivatives offer fertile grounds for photochemical study.

Acknowledgments

This review was written during the 1969 Summer Quarter at the Department of Chemistry, University of California (Berkeley), whose entire staff was most helpful. The author also wishes to express his gratitude to Mrs. Margaret Cope for typing the manuscript.

REFERENCES

1. T. J. Kealy and P. L. Pauson, *Nature*, **168**, 1039 (1951). It has been claimed that ferrocene was first obtained (but not announced) in 1948. For more details about this, see G. Wilkinson and F. A. Cotton in *Progress in Inorganic Chemistry*, Vol. 1, Wiley-Interscience, New York, 1959, p. 3.
2. For a complete definition see M. D. Rausch, *Can. J. Chem.*, **41**, 1291 (1963).
3. See, for example, P. E. Eaton, *Accounts Chem. Res.*, **1**, 50 (1968).
4. Two representative books are N. J. Turro *Molecular Photochemistry*, Benjamin, New York, 1965; and A. Schönberg, *Präparative Organische Photochemie*, Springer, Berlin, 1968.
5. A. W. Adamson, *Rec. Chem. Progr.*, **29**, 191 (1968).
6. D. Valentine, Jr., *Advan. Photochem.*, **6**, 123 (1967).
7. M. Rosenblum, *Chemistry of the Iron Group Metallocenes* Part 1, Wiley-Interscience, New York, 1965.
8. J. T. Baker Co., *Photosensitizer and Quencher Kit*, Comm. No. Z901, Phillipsburg, New Jersey, 1967, p. 6.
9. R. T. Lundquist and M. Cais, *J. Org. Chem.*, **27**, 1167 (1962).
10. A. N. Nesmeyanov, B. M. Yavorskii, G. B. Zaslavskaya, and N. S. Kochetkova, *Dokl. Akad. Nauk. SSSR*, **160**, 837 (1965).
11. D. R. Scott and R. S. Becker, *J. Chem. Phys.*, **35**, 516 (1962).
12. A. T. Armstrong, F. Smith, E. Elder, and S. P. McGlynn, *J. Chem. Phys.*, **46**, 4321 (1967.)
13. S. Yamada, A. Nakahara, and R. Tsuchida, *J. Chem. Phys.*, **22**, 1620 (1954).
14. S. Yamada, A. Nakahara, and R. Tsuchida, *Bull. Chem. Soc. Jap.*, **28**, 465 (1955).
15. A. M. Tarr and D. M. Wiles, *Can. J. Chem.*, **46**, 2725 (1968).
16. J. P. Guillory, C. F. Cook, and D. R. Scott, *J. Amer. Chem. Soc.*, **89**, 6776 (1967).
17. M. Rosenblum, J. O. Santer, and W. G. Howells, *J. Amer. Chem. Soc.*, **85**, 1453 (1963).
18. G. Wilkinson, *J. Amer. Chem. Soc.*, **74**, 6146 (1952).
19. J. C. D. Brand and W. Snedden, *Trans. Faraday Soc.*, **53**, 894 (1957).
20. For various references, see E. Adman, M. Rosenblum, S. Sullivan, and T. N. Margulis, *J. Amer. Chem. Soc.*, **89**, 4540 (1967).
21. J. J. Dannenberg and J. H. Richards, *J. Amer. Chem. Soc.*, **87**, 1626 (1965).
22. V. Weinmayr, *J. Amer. Chem. Soc.*, **77**, 3009 (1955).
23. D. W. Hall and J. H. Richards, *J. Org. Chem.*, **28**, 1549 (1963).
24. K. L. Rinehart and R. E. Bozak, Unpublished results.
25. S. Yamada, H. Nakamura, and R. Tsuchida, *Bull. Chem. Soc. Jap.*, **30**, 647 (1957).
26. S. Yamada, H. Yamazaki, H. Nishikawa, and R. Tsuchida, *Bull. Chem. Soc. Jap.*, **33**, 481 (1960).
27. F. Scandola, V. Balzani, and V. Carassiti, *J. Inorg. Chem.*, **5**, 700 (1966).
28. A. J. Fry, R. S. H. Liu, and G. S. Hammond, *J. Amer. Chem. Soc.*, **88**, 4781 (1966).

29. W. G. Herkstroeter and G. S. Hammond, *J. Amer. Chem. Soc.*, **88**, 4769 (1966).
30. J. A. Bell and H. Linschitz, *J. Amer. Chem. Soc.*, **85**, 528 (1963).
31. Ref. 7, p. 148.
32. J. H. Richards, *Proc. Paint Res. Inst.*, **31**, 1433 (1964); **39**, 569 (1967).
33. M. Rosenblum, R. W. Fish, and C. Bennett, *J. Amer. Chem. Soc.*, **86**, 5166 (1964).
34. E. A. K. von Gustorf, 153rd National Meeting, American Chemical Society, Miami Beach, Florida, April 1967.
35. J. Brauman, Personal communication.
36. B. K. Krotoszynski, *J. Chem. Phys.*, **41**, 2220 (1964).
37. R. A. Finnegan and A. W. Hagen, *Tetrahedron Lett.*, **1963**, 365.
38. R. A. Finnegan and J. J. Mattice, *Tetrahedron*, **21**, 1015 (1965).
39. Footnote 16 in Ref. 38.
40. H. Kobsa, *J. Org. Chem.*, **27**, 2293 (1962).
41. A. N. Nesmeyanov, V. A. Sazonova, A. V. Gerasimenko, and N. S. Sazonova, *Dokl. Akad. Nauk SSSR*, **149**, 1354 (1963).
42. A. N. Nesmeyanov, V. A. Sazonova, V. I. Romanenko, N. A. Rodionova, and G. P. Zolnikova, *Dokl. Akad. Nauk SSSR*, **155**, 1130 (1964).
43. A. N. Nesmeyanov, V. A. Sazonova, and V. I. Romanenko, *Dokl. Akad. Nauk SSSR*, **152**, 1358 (1963).
44. A. N. Nesmeyanov, V. A. Sazonova, V. I. Romanenko, and G. P. Zolnikova, *Izv. Akad. Nauk SSSR, Ser Khim*, **1965**, 1694.
45. R. E. Bozak, *Chem. Ind.* (London), **1969**, 24.
46. R. E. Bozak, 5th Western Regional Meeting, American Chemical Society, Anaheim, California, October 1969.
47. M. Rausch, M. Vogel, and H. Rosenberg, *J. Org. Chem.*, **22**, 903 (1957).
48. N. Weliky and E. S. Gould, *J. Amer. Chem. Soc.*, **79**, 2742 (1957).
49. M. D. Rausch, E. O. Fischer, and H. Grubert, *J. Amer. Chem. Soc.*, **82**, 76 (1960).
50. J. H. Richards and N. Pisker-Trifunac, *Proc. Paint Res. Inst.*, **41**(533), 363 (1969).
51. *Chem. Abstr.* citation: Beinoravichute, *Mater Resp. Tekh. Konf. Vop. Issled. Primen. Polim. Mater.*, 6th Vilnius, 91 (1965).
52. D. M. Wiles and T. Suprunchuk, *Can. J. Chem.*, **46**, 1865 (1968).
53. R. C. Schmitt and R. C. Hirt, *J. Appl. Polym. Sci.*, **7**, 1565 (1963).
54. K. P. S. Kwei, Abstracts, 153rd National Meeting, American Chemical Society, Miami Beach, Florida, April, 1967, p. 5094.
55. E. O. Fischer and M. Herberhold, *Experientia*, *Suppl.*, **9**, 259 (1964).
56. V. Balzani, Unpublished results.
57. G. A. Razuvaev and G. A. Domrachev, *Tetrahedron*, **19**, 341 (1963).

Complications in Photosensitized Reactions

PAUL S. ENGEL,† *Department of Chemistry, Rice University, Houston, Texas 77001*

and

BRUCE M. MONROE, *Explosives Department, Experimental Station, E. I. du Pont de Nemours and Company, Wilmington, Delaware 19898*

CONTENTS

I. Introduction	246
II. Mechanism of a Sensitized Reaction	247
III. Reactions of Sensitizers	248
A. Unimolecular Sensitizer Reactions	248
1. α-Cleavage of Ketones (Type I)	248
2. β-Cleavage of Ketones (Type II)	249
3. Photoenolization	250
4. Dissociation of Halogen-Containing Compounds	252
B. Sensitizer–Sensitizer Interactions	253
1. Self-Quenching and Excimer Formation	253
2. Dimerization	255
C. Sensitizer–Substrate Interactions	256
1. Oxetane Formation—The Paterno-Buchi Reaction	256
2. Additions to Aromatic Hydrocarbons	257
3. Hydrogen Abstraction Reactions	258
4. Reversible Energy Transfer	259
D. Sensitizer Induced Photoreduction—Chemical Sensitization	262
E. Complications in Olefin Isomerization	268
1. *Cis-Trans* Isomerization via Radicals	268
2. The Schenck Mechanism	270
3. Nonvertical Energy Transfer	272
IV. Energy Transfer from Other than the Lowest Triplet State	273
A. Involvement of Sensitizer Singlets	273
1. Kinetic Considerations	274
2. Förster Transfer	275
3. Heavy Atom Effects	275
4. Quenching of Aromatic Hydrocarbon Singlets	277
a. Dienes	277
b. Strained Hydrocarbons	278

† Most of this chapter was written while this author was at the National Institutes of Health, Bethesda, Maryland.

    c. Sulfoxides 279
    d. Amines 280
    e. Azo Compounds 281
    f. Azides 285
    g. Peroxides 287
    h. Maleic Anhydride–Benzene 287
    i. Ketones. 287
   5. Quenching of Singlets of Carbonyl-Containing Compounds . . . 288
    a. Acetone. 288
    b. Fluorenone. 289
    c. Biacetyl 291
    d. Other Carbonyl Compounds. 291
  B. Involvement of Upper Triplet States 292
   1. Anthracene and Its Derivatives 292
   2. Bimolecular Reactions from Upper Triplet States 294
V. General Remarks on Selecting a Photosensitizer 295
VI. Tables of Sensitizer Properties 296
References 307

      Things are seldom what they seem,
      Skim milk masquerades as cream;
      Highlows pass as patent leathers;
      Jackdaws strut in peacock's feathers.
        W. S. Gilbert, "H.M.S. Pinafore," Act 2

I. INTRODUCTION

Triplet sensitization, the direct formation of triplet states by energy transfer, is a powerful tool for the study of photochemical reactions.[1–3] By a proper choice of sensitizer the triplet state of a molecule can often be generated selectively so that processes arising from the singlet state are eliminated. It has been possible to investigate the triplet reactions of molecules which have low quantum yields of intersystem crossing; for example, the study of the photochemistry of olefins[4–8] has been greatly facilitated by this technique. Product formation in the presence of light and a triplet sensitizer is often assumed to be unassailable evidence that the reaction proceeds via the triplet state.

As sensitization has come into widespread use, it has become apparent that energy transfer is by no means the only process which can occur when a sensitizer is irradiated in the presence of an acceptor. Other unwanted and often totally unexpected interactions between the sensitizer and itself, the substrate, or the solvent can produce confusing or misleading results. These interactions, which include chemical reaction of sensitizers, involvement of singlet excited states, and numerous other complications, will constitute the subject matter of this review. As is often true in a rapidly expanding field, as many complications have been discovered by reinvestigation of old photochemical systems as by the study of new ones.

Since the photochemistry of many compounds that have been used as triplet sensitizers has been well studied, we will not attempt to cover these reactions in detail. Unless the investigator is unaware of them, common photochemical processes such as the Norrish Type II cleavage are not ordinarily a complication and as will be mentioned later, they can actually serve as mechanistic probes. A discussion of the mechanisms of triplet energy transfer[1,3,9] is beyond the scope of this review as are other specific reactions which have been recently covered elsewhere.

II. MECHANISM OF A SENSITIZED REACTION

The following mechanism, in which S is the sensitizer, A the acceptor, and superscripts indicate the multiplicity of the excited states, describes a general photosensitized reaction:

$$S \xrightarrow{h\nu} S^{*1} \quad \text{excitation} \tag{1}$$

$$S^{*1} \xrightarrow{k_f} S + h\nu' \quad \text{fluorescence} \tag{2}$$

$$S^{*1} \xrightarrow{k_{ds}} S \quad \text{radiationless decay} \tag{3}$$

$$S^{*1} \xrightarrow{k_{ic}} S^{*3} \quad \text{intersystem crossing} \tag{4}$$

$$S^{*3} \xrightarrow{k_d} S \quad \text{radiationless decay} \tag{5}$$

$$S^{*3} \xrightarrow{k_p} S + h\nu'' \quad \text{phosphorescence} \tag{6}$$

$$S^{*3} + A \xrightarrow{k_e} S + A^{*3} \quad \text{energy transfer} \tag{7}$$

$$A^{*3} \xrightarrow{k_r} \text{products} \quad \text{reaction} \tag{8}$$

In the ideal case, the only function of the sensitizer is to absorb light, undergo intersystem crossing to its lowest triplet state, and transfer the energy to the acceptor; that is, only reactions (1), (4), (7), and (8) are significant and if the others occur, they are easily accounted for. Inherent in this scheme are two important assumptions: (*1*) the only interaction of sensitizer and substrate is energy transfer and (*2*) energy is transferred from the lowest triplet state. Unfortunately, these assumptions often break down and the resultant complications will be discussed in the following pages. Section III will cover other interactions of the sensitizer with itself, with the substrate, or with the solvent and will include such diverse reactions as decomposition of the sensitizer, self-quenching, hydrogen abstraction, and transfer of energy from substrate to sensitizer. Section IV will encompass energy transfer from other than the lowest triplet state and so include transfer from singlet states and upper triplets. Finally, tables of useful sensitizer parameters will indicate what complications might be expected from particular compounds.

III. REACTIONS OF SENSITIZERS

Although it may seem obvious that an ideal photosensitizer should be photostable, it is doubtful that a compound exists which will not undergo some photochemical reaction if irradiated under the right set of conditions. Thus, when choosing a sensitizer it is necessary to consider not only its triplet energy, intersystem crossing yield, and extinction coefficient but also its possible photochemical reactions. In favorable cases a reaction of the sensitizer may be suppressed or eliminated completely by making the concentration of acceptor sufficiently high to quench all of the sensitizer triplet before this reaction takes place. In this section we will examine some of the photoreactions of commonly used sensitizers, point out under what conditions they might lead to complications, and show how some of these reactions might be quenched.

A. Unimolecular Sensitizer Reactions

1. α-Cleavage of Ketones (Type I). Although acetone undergoes efficient decarbonylation when irradiated in the vapor phase, the quantum yield for CO production from irradiation in the liquid phase is only $\sim 10^{-4}$.[10,11] At least twelve products, five of which have been identified, have been detected from the liquid phase irradiation of acetone.[12] Since some of these products

$$CH_3-\overset{O}{\underset{\|}{C}}-CH_3 \xrightarrow[\text{quartz, } N_2, \text{ 4 hr}]{\text{450 W Hg lamp}} CH_3-\overset{O}{\underset{\|}{C}}-H + \underset{O}{\overset{\displaystyle\diagup}{\biggl|}}OH +$$

$$\underset{\underset{CH_3}{|}}{\overset{\overset{OH}{|}}{CH_3-C-CH_2}}-\overset{O}{\underset{\|}{C}}-CH_3 + \underset{\underset{H_3C}{|}\ \underset{CH_3}{|}}{\overset{\overset{OH}{|}\ \overset{OH}{|}}{CH_3-C-C-CH_3}} +$$

$$CH_3-\overset{O}{\underset{\|}{C}}-CH_2-\overset{O}{\underset{\|}{C}}-CH_3 + 7 \text{ other products}$$

total over-all yield = 0.7%

must be formed by radical pathways proceeding from an initial hydrogen abstraction rather than an α-cleavage, complications could arise in the study of reactions, such as olefin isomerizations, which can be catalyzed by radicals. Although no quantum yields have been reported for formation of these products, decomposition of acetone appears to be an inefficient process which should not lead to complications in most reactions, particularly if the acetone triplet is efficiently quenched by the acceptor.

Most other ketones are similar to acetone in that they show no significant amount of α-cleavage in solution. Generally cleavage arises from an upper excited state, but in solution the excited molecule is collisionally deactivated to its lowest triplet state which does not have sufficient energy to undergo bond breaking. For example, it has been estimated that 65 kcal/mole is barely enough energy to cleave biacetyl[13] but the lowest triplet of biacetyl lies in the 55–57 kcal/mole range.[14] Hence it is not surprising that the quantum yield for CO production from the irradiation of biacetyl in perfluorinated solvents is $\sim 10^{-2}$ to 10^{-4}.[15] For most ketones generally used as sensitizers, α-cleavage is not a problem unless the initial products are particularly stable free radicals; for example, t-butyl alkyl ketones show Type I cleavage, but these compounds would not ordinarily be used as sensitizers.[16]

A ketone that can cleave to give relatively stable radicals is dibenzylketone, whose inclusion in a table of potentially useful photosensitizers[17] seemed inconsistent with the report[18] that it gave a 100% yield of CO upon irradiation. Investigation by two groups[19,20] has shown that the quantum yield for CO formation is greater than 0.7 and that at least 80% of the decarbonylation occurs from a triplet state with a lifetime of 2×10^{-10} sec. This would indicate that the cleavage is too fast to be efficiently quenched and, hence, dibenzyl ketone cannot be used as a photosensitizer. Attempts to sensitize decarbonylation using compounds of higher triplet energy than the 72.2 kcal reported[14] for dibenzyl ketone failed, and further investigation revealed that energy was not transferred. This anomalous behavior was accounted for when it was found that a facile reaction in the presence of air and light formed benzaldehyde, which was entirely responsible for the phosphorescence.[19] A better estimate of the triplet energy of dibenzylketone was given as 79 kcal.

Benzylmethylketone undergoes similar α-cleavage[21] but the acyl radical does not lose carbon monoxide.

2. β-Cleavage of Ketones (Type II). Ketones which possess γ-hydrogen atoms can undergo cleavage to olefins and smaller carbonyl compounds. This reaction, which has been extensively studied by several groups,[22–26] is known as the β-cleavage or Type II split. Since most of the commonly used ketone photosensitizers do not contain γ-hydrogens, this reaction is seldom a problem; in fact, it can be used as a test for energy transfer. If a sensitizer such as butyrophenone transfers energy to an acceptor, the Type II process will be quenched; lack of quenching indicates that energy is not being transferred. In this regard it should be noted that the Type II cleavage of n-butyrophenone and n-valerophenone proceeds entirely from the triplet state while the photoelimination of 2-pentanone, 2-hexanone, and 2-octanone has been shown to occur from both the singlet and triplet states and is, therefore, only partially suppressed by triplet quenchers.[27-29]

3. Photoenolization. o-Substituted aldehydes and ketones with the general structure **1** undergo photoenolization by triplet intramolecular hydrogen abstraction, a process formally analogous to the first step of the Type II cleavage of aryl alkyl ketones.[30-36] Apparently the presence of a γ-hydrogen, which permits a cyclic six-membered transition state, is necessary since 2-t-butylbenzophenone does not undergo this process. Use of a sensitizer with general structure **1** may lead to such complications as reduction in the measured quantum yield because of the energy wasted by enolization, competitive absorption of light by the enol, sensitization of the reaction by the enol, quenching of reactive intermediates by the enol, or reaction of the enol with the substrate. Most of these have not been observed primarily because few attempts have been made to use photoenolizable compounds as sensitizers.

Very early in the study of photosensitization it was discovered that salicylaldehyde, 2,4-dihydroxybenzophenone, and 2-hydroxy-4-methoxy-benzophenone do not sensitize piperylene isomerization, indicating that enolization is much faster than quenching by the diene.[37] Later it was shown that 2-hydroxybenzophenone would sensitize the dimerization of isoprene in concentrated solution, but that the reaction was much less efficient than when benzophenone was used as a sensitizer (Table I).[36]

TABLE I
2-Hydroxybenzophenone-Sensitized Dimerization of Isoprene[a]

Sensitizer	Solvent	Relative yield of dimers
Benzophenone	Benzene	1.00
2-Hydroxybenzophenone	Cyclohexane	~0.03
2-Hydroxybenzophenone	Ethanol	~0.15

[a] Samples consisted of sensitizer plus equal volumes of solvent and isoprene.

The enhanced reaction in ethanol is attributed to weakening of the tight, intramolecular hydrogen bond which is believed to facilitate photoenolization of 2-hydroxybenzophenone. *o*-Hydroxybutyrophenone undergoes photoenolization in preference to Type II cleavage.[25] The enol of *o*-benzylbenzophenone (2) forms a Diels-Alder adduct with dimethyl acetylenedicarboxylate, indicating that reaction with a reactive substrate is possible.[30]

In a study of steric hindrance to energy transfer, Herkstroeter, Jones, and Hammond observed photoenolization of the hindered benzophenones 3 and 4.[38]

3 R = CH$_3$
4 R = CH(CH$_3$)$_2$

To determine whether the differences in stilbene photostationary state they observed with these sensitizers were due to sensitization or to quenching by the enol, its formation was studied by flash kinetic spectroscopy. The observation that both piperylene and stilbene strongly quenched enol formation suggests that enolization of a ketone in which an abstractable hydrogen is bonded to carbon is much more easily quenched than enolization of one in which it is bonded to oxygen, an effect which is no doubt due to the strong intramolecular hydrogen bond in the latter case. Intramolecular hydrogen abstraction is also shown by o-nitrotoluene.[39]

Biacetyl (5) is reported to undergo enolization when irradiated in water, methanol, or hexane solution, presumably via a highly strained four-membered transition state instead of the usual six-membered one.[40-43] From the wavelength dependence of enol formation, it was concluded that enolization occurred from the second triplet state.

$$CH_3-\underset{\underset{O}{\|}}{C}-\underset{\underset{O}{\|}}{C}-CH_3 \longrightarrow CH_3-\underset{\underset{O}{\|}}{C}-\overset{\overset{OH}{|}}{C}=CH_2$$

$$5 \qquad\qquad 6$$

4. Dissociation of Halogen-Containing Compounds. Halogenated compounds, particularly those containing bromine or iodine, are likely to undergo photoelimination of the halogen atom. p-Bromobutyrophenone (7) yields butyrophenone, HBr, and butyrophenones with unsaturated alkyl side chains when irradiated in toluene.[25,44] The quantum yield for butyrophenone

$$Br-\underset{}{\bigcirc}-\underset{\underset{}{\|}}{\overset{O}{C}}-C_3H_7 \xrightarrow[\phi CH_3]{h\nu,\ 313\ nm} \underset{}{\bigcirc}-\underset{\underset{}{\|}}{\overset{O}{C}}-C_3H_7 + HBr +\ \text{unsaturated butyrophenones}$$

$$7 \qquad\qquad \Phi = 0.25$$

production is 0.25 and no Type II cleavage is observed. Photolysis of 7 in pentene and piperylene leads to polymerization, but since the disappearance of 7 is not inhibited, these compounds do not quench the reaction.

Similar decomposition is observed in p-bromoacetophenone, o-bromo-, p-bromo-, and p,p'-dibromobenzophenone, and p-iodobenzophenone[44] but not in the fluoro- and chloro-substituted compounds. This order of reactivity follows the bond dissociation energies for aromatic halides which are about 90 kcal/mole for chlorobenzene, 70 kcal/mole for bromobenzene, and 60 kcal/mole for iodobenzene. The lowest-lying triplet of p-bromoacetophenone is 71.2 kcal[45] while that of the substituted benzophenones is slightly lower since benzophenone itself has a lower triplet energy than acetophenone. p,p'-Dibromobenzophenone was the least reactive of the compounds that photoeliminated halogen atoms.

Because of the ease with which they undergo photodissociation, aromatic iodides have been used extensively to generate free radicals.[46] The rate constant for the dissociation of p-diiodobenzene triplet has been measured to be about 4×10^{12} to 4×10^{13} sec^{-1} which would give a lifetime of 2.5×10^{-13} to 2.5×10^{-14} sec.[47] Obviously a sensitizer with this short a dissociative lifetime could not be quenched by an acceptor.

B. Sensitizer–Sensitizer Interactions

1. Self-Quenching and Excimer Formation. It is often desirable to use a high concentration of sensitizer to ensure that it will absorb all of the light, particularly if the substrate has any significant absorption at the wavelength of irradiation. Recent studies have shown that quantum yields determined under these conditions can be anomalously low if the sensitizer acts as a bimolecular quencher of its own triplets.

In a study of the photoisomerization of 4,4-dimethyl-2-cyclohexenone (**8**) to 6,6-dimethylbicyclo[3.1.0]-hexane-2-one (**9**) and 3-isopropyl-2-cyclopentenone (**10**), Chapman and Wampfler[48] accounted for the pronounced effect of sensitizer concentration by such a self-quenching mechanism. The phenomenon was exhibited by four ketones with (π^*,π) lowest triplet states but not by benzophenone or acetophenone which have (π^*,n) triplet states.[49]

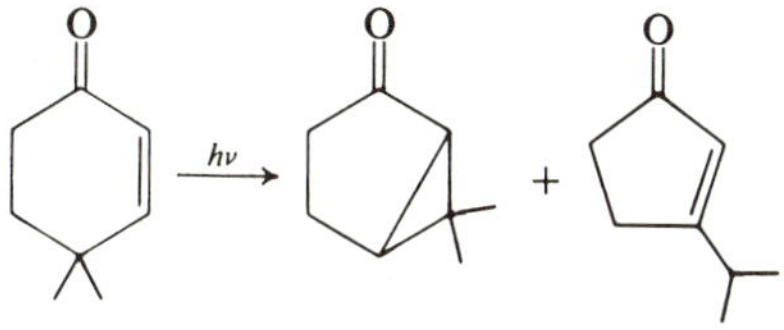

In cases of self-quenching, the acceptor and ground state sensitizer will compete for the sensitizer triplet as illustrated in the following scheme.

$$S \xrightarrow{h\nu} S^{*1} \qquad \text{excitation} \qquad (1)$$

$$S^{*1} \xrightarrow{k_{ic}} S^{*3} \qquad \text{intersystem crossing} \qquad (4)$$

$$S^{*3} + S \xrightarrow{k_{sq}} 2S \qquad \text{self quenching} \qquad (9)$$

$$S^{*3} + A \xrightarrow{k_e} S + A^{*3} \qquad \text{energy transfer} \qquad (7)$$

$$A^{*3} \xrightarrow{k_r} \text{products} \qquad \text{reaction} \qquad (8)$$

Notice that this is the same as the mechanism written in Section II except that step (9), self-quenching, has been added. If we assume that the quantum yield of intersystem crossing for the sensitizer is unity and that all the sensitizer triplet is quenched either by the acceptor or by ground state sensitizer, then the following expression for Φ_a, the actual quantum yield for reaction of the substrate triplet, can be derived:

$$\frac{\Phi_a}{\Phi_m} = 1 + \left(\frac{k_{sq}}{k_e[A]}\right)[S] \qquad (10)$$

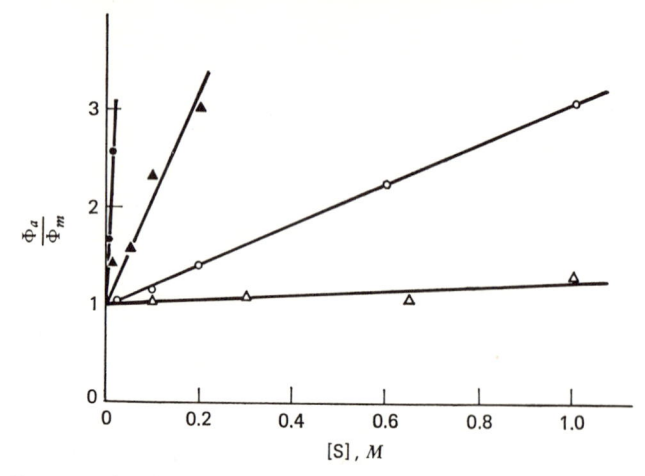

Fig. 1. Concentration dependence of the efficiency of energy transfer: △, p-methoxyacetophenone; ○, m-methoxyacetophenone; ▲, 3,4-methylenedioxyacetophenone; ●, thioxanthone. (From Chapman and Wampfler[48] with permission of the American Chemical Society.)

Φ_m is the measured quantum yield of product formation. Ratios of k_{sq} to k_e can be calculated from the slopes in Figure 1, and if we assume that energy transfer is diffusion controlled, the self-quenching rate constants may be determined.

Equation (10) allows one to calculate Φ_a by determining the quantum yield of reaction as a function of sensitizer concentration. However, it should also be remembered that if the substrate absorbs part of the light, Φ_m may need to be corrected for changes in the amount of light absorbed by the sensitizer.

TABLE II
Sensitizer Self-Quenching Parameters

Sensitizer	k_{sq}/k_e	k_{sq} M^{-1} sec^{-1} [a]	Φ_r [b]	τ_r, sec [c]
Benzophenone	0	0	0.67	7.1×10^{-3}
Acetophenone	0	0	0.41	5×10^{-3}
p-Methoxyacetophenone	3×10^{-2}	9×10^7	0.04	0.38
m-Methoxyacetophenone	2×10^{-1}	8×10^8	6×10^{-3}	0.71
3,4-Methylenedioxyacetophenone	1	4×10^9	2×10^{-3}	1.2
Thioxanthone	16	6×10^{10}		

[a] $[A] = 0.10M$; assumes $k_{diff} = 3.7 \times 10^9 M^{-1}$ sec^{-1} for t-butyl alcohol.
[b] Quantum yield for photoreduction from Ref. 49.
[c] Radiative lifetime in isopropyl alcohol from Ref. 49.

Self-quenching can be avoided by using a sensitizer which is a poor competitor for its own triplet state or, in mechanistic terms, has a small k_{sq}. Then the term in parentheses in Eq. (10) will approach zero and Φ_m will approach Φ_a. Alternatively, the ratio [S]/[A] can be made small but this may allow excessive light absorption by the substrate.

One further complication should be noted as well; namely, a sensitizer that contains a quenching impurity will display identical kinetics, since the impurity concentration will be proportional to the sensitizer concentration.

It is apparent from Table II that the rate of self-quenching increases with decreasing reactivity in photoreduction and with increasing triplet lifetime. The mechanism of self-quenching is not known but formation of an excimer, a dimer which is stable only in the excited state, has been suggested.[48] This was supported by a shift to longer wavelength of the phosphorescence maximum of m-methoxyacetophenone at high concentration. Thioxanthone may have some additional ground state interaction, since it quenches at sixteen times the diffusion controlled rate.

Excimers have been suggested in several other photochemical processes. The triplet lifetime of o-xylene decreases from ~900 nsec at 0.01M in methylcyclohexane to 13 nsec in 8M solution.[50] The inefficiency of the triplet sensitized dimerization of indene has been attributed to formation of an indene excimer.[51] The emission from acetone originally[52] believed to arise from an excimer has now been shown[53] to be the true monomer fluorescence while that attributed to the monomer appears to have been due to an impurity.

A large number of aromatic hydrocarbons, including benzene, naphthalene, anthracene, pyrene, 1,2-benzanthracene, 1,2-benzpyrene, 3,4-benzpyrene, perylene, and many of their derivatives, are known to form singlet excimers.[9,54-57] Fluorescence of solutions of these compounds is often characterized by a broad, structureless band at longer wavelength than that of the structured monomer band. The amount of excimer fluorescence increases with the concentration of hydrocarbon as the monomer fluorescence decreases. Since binding energies are 10 kcal/mole or less, excimer formation is reversible in fluid solution at room temperature. Further discussion with leading references is given in a chapter by Lamola[9] and in reviews by Birks[54] and by Förster.[55]

2. Dimerization. Many polycyclic aromatic hydrocarbons form photodimers via 4 + 4 cycloadditions.[58,59] Anthracene (**11**), for example, dimerizes with a limiting quantum yield of 0.3 when irradiated in benzene.[59,60] The reaction takes place from the singlet state and competes with fluorescence, which drops to zero at high anthracene concentrations. More detailed coverage of these reactions is found in the reviews by Bowen[59] and by Trecker.[58]

[Scheme: anthracene 11 + hν/φH → dibenzobarrelene-type adduct; Φ ≤ 0.3]

C. Sensitizer–Substrate Interactions

1. Oxetane Formation—The Paterno-Büchi Reaction. A large number of carbonyl compounds, primarily aldehydes, ketones, and quinones, form oxetanes by photocycloadditions to olefins.[61-63] In general, it is observed that (1) carbonyl compounds which have low-lying (π^*,n) triplet states and which are photoreduced in isopropyl alcohol form oxetanes most readily, and (2) oxetane formation takes place when energy transfer from the carbonyl compound to the olefin is unfavorable because of the relative location of their triplet levels.[64,65] Hence, oxetanes are most readily formed from simple olefins and allenes[63,66] but are seldom formed from dienes.[67] An extensive review by Arnold[63] covers the mechanism and scope of this reaction.

Recent work by several groups has shown that formation of oxetanes from alkyl aldehydes and ketones can take place from the singlet as well as from the triplet state.[68]

Turro[69] has reported that several alkyl ketones react with 1,2-dicyanoethylene (12) by nucleophilic attack of the ketone (π^*,n) singlet state to give oxetanes stereospecifically. Addition of diene quenchers which are known to deactivate ketone triplets do not affect the rate of oxetane formation. This

[Scheme: R₁R₂C=O + (CN)CH=CH(CN) (12) → oxetane with R₁, R₂, CN, H, CN, H substituents]

and the fact that the fluorescence of cyclopentanone was strongly quenched by **12** argue in favor of a reactive singlet state.

Further work by Turro and coworkers[70] led to the discovery of a singlet complex in this reaction. A plot of the reciprocal of the quantum yield for oxetane formation versus the reciprocal of the concentration of **12** according to Eq. (34) p. 275 gave a slope of 2.6 but an intercept of 13.2. This suggests that oxetane formation proceeds via an intermediate (complex) which is frequently deactivated. Since the value of K_S obtained from this plot agreed with the K_Q from quenching of acetone fluorescence, it follows that most fluorescence quenching leads to complex formation.

Photochemical addition of acetone to *cis* and *trans* 1-methoxy-1-butene involves both acetone singlets and triplets.[71] Since spin inversion in the 1,4-biradical is relatively slow, predominant loss of stereochemistry is observed in the oxetane derived from triplet acetone. On the other hand, stereochemistry is partially retained when acetone singlets attack the substituted olefin.

In 1966, it was noted[72] that irradiation of cyclopentadiene and acetone gave an oxetane but apparently no special significance was attached to this observation. In fact, it constitutes a surprising exception to generalization (2) above since the triplet energy of ketones is expected to be substantially higher than that of dienes. From recent work of two groups, it now appears that as in the cases above, the singlet state of the ketone is the reactive species. Dowd[73] found that 3-methylene cyclobutanone added stereospecifically to piperylene at the more highly substituted double bond, features which distinguish the reaction from what might be expected of a triplet diradical intermediate. Kubota[73a] reported that in the photoreaction of propionaldehyde with cyclohexadiene the rate of oxetane formation increased with increasing cyclohexadiene concentration while the diene dimer yield decreased. Since the dimerization is known to occur from the triplet state, the aldehyde singlet is a likely precursor of the oxetane. These results should be taken into account when attempting to sensitize a diene photoreaction with an aliphatic ketone.

2. Additions to Aromatic Hydrocarbons. A variety of photochemical additions to aromatic hydrocarbons have been reported. Benzene and its derivatives add to maleic anhydride[74-76] as well as to simple olefins,[77-80] isoprene,[81] acetylene derivatives,[79,82] and alcohols.[83] The mechanism of the maleic anhydride-benzene reaction is discussed in Section IV.A.4. Naphthalene forms a photoadduct with dimethyl acetylenedicarboxylate[82] and with acrylonitrile[82a] while anthracene behaves similarly with maleic anhydride[84] and with 1,2-benzanthracene.[85] The photoaddition of several aromatic amines to anthracene has been reported to proceed via a charge transfer complex[86,87]; in fact, the majority of these addition reactions may proceed in this manner.

Photoaddition reactions can destroy the original sensitizer, form a new sensitizer, or form a substance that quenches the desired reaction completely. Since substrate would also be consumed, its measured rate of disappearance will be anomalously high and that of product appearance too low. The importance of ensuring that the sensitizer acts as a true catalyst in a photoreaction is demonstrated by a study[88] in which 1,2-benzanthracene (**13**) was used to sensitize the dimerization of cyclohexadiene. The quantum yield of dimer formation was found to decrease at higher sensitizer concentration;

however, further investigation revealed that this was due in part to formation of a sensitizer-substrate photoadduct.

3. Hydrogen Abstraction Reactions. Hydrogen abstraction by the (π^*,n) triplet state of carbonyl compounds is one of the best known and most thoroughly studied photochemical reactions.[1,49,89-91] When it occurs during an attempted sensitization experiment, however, the resulting free radicals can cause a host of undesirable effects. These include formation of radical derived products from solvent, sensitizer, and substrate; induced decomposition, chemical sensitization (Section III.D), and olefin isomerization (Section III.E.1). Although abstraction is normally quenched by energy transfer to the substrate, it can still occur when the triplet states of sensitizer and substrate are unfavorably disposed for efficient energy transfer.[92,93] For example, benzophenone ($E_T = 68.5$ kcal) sensitization of cyclopentene dimerization gave only products derived from hydrogen abstraction while acetone ($E_T = 80$ kcal) led to a 56% yield of dimers in addition to the hydrogen abstraction products.[92] This demonstrates the importance of knowing the relative triplet energies of sensitizer and acceptor before a photosensitization experiment is attempted.

Hydrogen abstraction by ketone "sensitizers" forms the basis of an elegant alkylation procedure for ethers, amides, lactones, amino acids, and peptides developed by Elad and his group.[94] Terminal olefins, for example, can be added to γ-butyrolactone (15) in good yield.[95]

$$\text{15} + n\text{-C}_6\text{H}_{13}\text{—CH}=\text{CH}_2 \xrightarrow[\text{acetone}]{\text{sunlight}} n\text{-C}_8\text{H}_{17}\text{—(lactone)} \quad 66\%$$

There is still considerable interest in the products of irradiation of hydrogen abstracting ketones; for example, Plank[96] has reported that benzophenone in ether gives benzophenone pinacol in higher yield when naphthalene is added. It was suggested that the pinacol is formed from some intermediate which is photochemically destroyed in the absence of a quencher. Challand[97] has isolated products from addition of ether to acetophenone.

It has been shown that the benzophenone sensitized decomposition of benzoyl peroxide is due in part to formation of the benzophenone ketyl radical, which induces decomposition.[98,99] Hydrocarbon sensitized peroxide decomposition is discussed in Section IV.A.4. The formation of benzonitrile from the benzophenone sensitized irradiation of benzalazine, which was originally attributed to hydrogen abstraction by benzophenone,[100] actually results from a photooxidation.[101]

4. Reversible Energy Transfer. It is almost always assumed that the transfer of triplet energy from the sensitizer to substrate is irreversible. Back transfer, transfer of triplet energy from the acceptor back to the sensitizer, can, however, take place under the right set of circumstances. The following scheme illustrates the mechanism for back transfer.

$$S \xrightarrow{h\nu} S^{*1} \quad \text{excitation} \quad (1)$$

$$S^{*1} \xrightarrow{k_{ic}} S^{*3} \quad \text{intersystem crossing} \quad (4)$$

$$S^{*3} \xrightarrow{k_d} S \quad \text{triplet decay} \quad (5)$$

$$S^{*3} + A \xrightarrow{k_e} A^{*3} + S \quad \text{energy transfer} \quad (7)$$

$$A^{*3} + S \xrightarrow{k_{-e}} A + S^{*3} \quad \text{back transfer} \quad (11)$$

$$A^{*3} \xrightarrow{k_a} A \quad \text{acceptor decay} \quad (12)$$

$$A^{*3} \xrightarrow{k_r} \text{products} \quad \text{reaction} \quad (8)$$

Decay of triplet sensitizer and acceptor are included in steps (5) and (12), respectively, but for simplicity, singlet decay is neglected. Making the usual steady state assumption and solving for the observed quantum yield of product formation, Φ_m, one obtains the following expression:

$$\frac{1}{\Phi_m} = \left(\frac{k_r + k_a}{k_r}\right)\left(1 + \frac{k_d}{k_e[A]}\right) + \frac{k_d k_{-e}[S]}{k_r k_e[A]} \tag{13}$$

The quantum yield for product formation from the sensitizer triplet, Φ_a, is given by the expression:

$$\Phi_a = \frac{k_r}{k_r + k_a} \tag{14}$$

Substituting into Eq. (13), we obtain:

$$\frac{\Phi_a}{\Phi_m} = 1 + \frac{k_d}{k_e[A]} + \frac{k_d k_{-e}[S]}{k_r k_e[A]} \tag{15}$$

Equation (15) permits us to examine the circumstances under which back transfer will become important. If we assume that energy transfer to the acceptor is the only mode by which the sensitizer triplet decays to the ground state (i.e., $k_d \sim 0$), Eq. (15) reduces to:

$$\Phi_m = \Phi_a \tag{16}$$

This demonstrates that unless the sensitizer has a route of decay to the ground state other than energy transfer to the acceptor, all the energy which goes into the sensitizer will eventually be measured as a reaction of the acceptor, and the multiple energy transfers will be kinetically undetectable. However, it is unlikely that a sensitizer with $k_d = 0$ will be found, for it would have an infinite triplet lifetime.

When the rate constants are such that the last term in Eq. (15) is not negligible, determination of the quantum yield of reaction as a function of sensitizer concentration will reveal back transfer; this experiment also tests for sensitizer self-quenching (Section III.B). In the most frequently encountered case, forward transfer is much faster than back transfer (i.e., $k_e \gg k_{-e}$) so that this term is small. The dependence of Φ_m on sensitizer concentration disappears and Eq. (15) reduces to the usual expression for quantum yield as a function of [A].

$$\frac{\Phi_a}{\Phi_m} = 1 + \frac{k_d}{k_e[A]} \tag{17}$$

The sensitizer concentration dependence is lost and back transfer becomes unimportant when product formation from the excited acceptor (k_r) is much faster than decay of the triplet sensitizer to its ground state (k_d). As an extreme example, an acceptor which dissociates on the first vibration after

TABLE III
Rate Constants for Energy Transfer to and back from Biacetyl[a,b]

Acceptor	E_T, cm^{-1}	$\log k_e$	$\log k_{-e}$	$\left(\log k_e - \dfrac{\Delta E_T}{2.3RT}\right)$[c]	$\left(\log k_{-e} - \dfrac{\Delta E_T}{2.3RT}\right)$[c]
Pyrene	16,930	9.88	4.23	3.98	
1,2-Benzpyrene	18,510	9.77	7.69	7.23	
Fluoranthene	18,510	9.72	7.33	7.17	
2,2′-Dinaphthyl	19,560	9.47	9.13	9.17	
1,5-Dinitronaphthalene	19,900	9.37			
1,8-Dinitronaphthalene	20,000	9.41			
1-Iodonaphthalene	20,500	7.54			
1-Chloronaphthalene	20,645	7.46	9.59		7.58
1-Bromonaphthalene	20,650	7.47	9.53		7.51
2-Iodonaphthalene	21,040	6.77			
Naphthalene	21,180	6.42	9.95		6.80

[a] $E_T = 19,700$ cm^{-1}.
[b] Benzene solvent (log $k_{\text{diff}} = \log[8RT/3000\eta] = 10$).
[c] $\Delta E_T \equiv E_T(\text{biacetyl}) - E_T(\text{acceptor})$.

energy transfer would not be quenched, even though back transfer was extremely facile.

In a classic study of back transfer, Sandros determined the forward and reverse energy transfer rate constants for biacetyl with a number of compounds of varying triplet energies.[102,103] These values are given in Table III.[102]

The key conclusion from his study was that the rate of back transfer can be expressed by the equation:

$$\log k_{-e} = \log k_e - \frac{\Delta E_T}{2.303RT} \qquad (18)$$

where ΔE_T is the difference in triplet energies between donor and acceptor, R the gas constant, and T the absolute temperature.

If ΔE_T is about 1000 cm^{-1} (2.86 kcal), then k_e is faster than k_{-e} by a factor of 100. Thus the axiom that triplet energy transfer is efficient only when it is exothermic by at least 3 kcal arises because an energy gap of this size is needed to eliminate back transfer. In fact, forward energy transfer is moderately efficient even when it is slightly endothermic. (Note the rate values for 1,5- and 1,8-dinitronaphthalene where transfer is about 0.3 kcal endothermic.)

In a study of the benzophenone photoreduction, Wagner[104] found that with biphenyl, fluorene, or triphenylene as quenchers, the quantum yield of photoreduction depended upon the benzophenone concentration. This indicates that reversible quenching of the reaction, a phenomenon similar to the reversible sensitization observed by Sandros, is appreciable in this system.

D. Sensitizer Induced Photoreduction—Chemical Sensitization

Ketones with (π^*,n) lowest triplet states, particularly benzophenone, are the most frequently used sensitizers. Although these compounds have intersystem crossing yields near unity and relatively high triplet energies, they are, as mentioned earlier, also efficient hydrogen abstractors. Severe complications can arise when these compounds are used to sensitize photoreductions, since the sensitizer ketyl radicals formed by hydrogen abstraction are themselves good reducing agents. Reduction by ketyl radicals or "chemical sensitization"† may lead to such effects as enhancement of the quantum yield of photoreduction, change in products, or photoreduction of compounds which are stable on direct irradiation. Moreover, in several studies to be mentioned

---

† Fischer[105] appears to have been the first to use the term "chemical sensitization" for a sensitizer induced photoreduction. Unfortunately the term has recently been applied to entirely unrelated phenomena, namely, Schenck type olefin isomerization[2] and reactions sensitized by chemically produced excited states.[106] To avoid confusion, it should be restricted to reduction by sensitizer-derived radicals.

later, these effects were mistakenly attributed to production of a reactive triplet state of the substrate.

When a ketone is excited in hydrogen containing solvent, the solvent competes with acceptor for the sensitizer triplet. A higher acceptor concentration favors energy transfer (k_e) over hydrogen abstraction (k_H) as expressed by the following equation:

$$\frac{\text{Rate of energy transfer}}{\text{Rate of abstraction}} = \frac{k_e[A]}{k_H[CH_2]} \qquad (19)$$

[A] is the concentration of acceptor and [CH_2] is the solvent concentration. The k_e to k_H ratio has been measured for benzophenone in isopropyl alcohol, the most commonly used system for sensitized photoreductions. From a study of naphthalene quenching of benzophenone photoreduction, Beckett and Porter[107] found that $k_e = 192 k_H[CH_2]$.

Substituting this result into equation 19 we obtain:

$$\frac{\text{Rate of energy transfer}}{\text{Rate of abstraction}} = 192[A] \qquad (20)$$

At $0.1 M$ acceptor, for example, the rate of energy transfer is 19.2 times the rate of abstraction. However, this means that about 5% of the benzophenone triplet will still abstract from the solvent to form ketyl radicals. A compound which is itself a poor hydrogen abstractor may show a greatly enhanced quantum yield of photoreduction under these conditions.

In a study of the benzophenone sensitized photoreduction of camphorquinone (16), Monroe and Weiner[108–110] found that the quantum yield of photoreduction decreased from 0.386 to 0.087 as [16] increased from 0.003 to $0.20 M$.

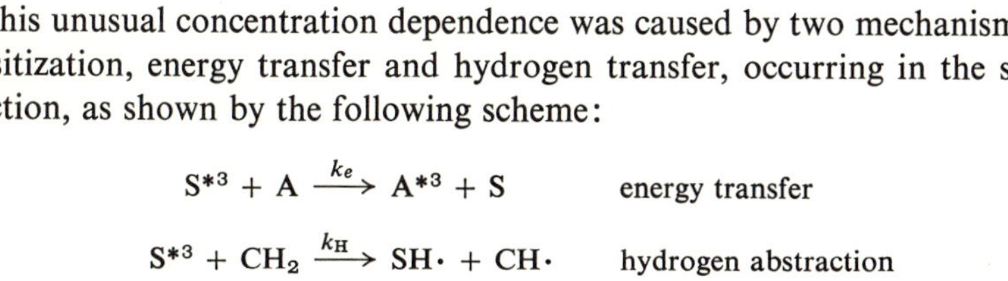

This unusual concentration dependence was caused by two mechanisms of sensitization, energy transfer and hydrogen transfer, occurring in the same reaction, as shown by the following scheme:

$$S^{*3} + A \xrightarrow{k_e} A^{*3} + S \qquad \text{energy transfer} \qquad (7)$$

$$S^{*3} + CH_2 \xrightarrow{k_H} SH\cdot + CH\cdot \qquad \text{hydrogen abstraction} \qquad (21)$$

$$A^{*3} + CH_2 \xrightarrow{k_c} CH\cdot + AH\cdot \qquad \text{hydrogen abstraction} \qquad (22)$$

$$A^{*3} \xrightarrow{k_d} A \qquad \text{triplet decay} \qquad (5)$$

$$SH\cdot + A \xrightarrow{k_1} S + AH\cdot \qquad \text{hydrogen transfer} \qquad (23)$$

$$CH\cdot + A \xrightarrow{k_2} C + AH\cdot \qquad \text{hydrogen transfer} \qquad (24)$$

$$2AH\cdot \xrightarrow{k_3} A + AH_2 \qquad \text{disproportionation} \qquad (25)$$

CH_2 is isopropyl alcohol, $CH\cdot$ the acetone ketyl radical, $SH\cdot$ the benzophenone ketyl radical, and $AH\cdot$ the camphorquinone ketyl radical. Solution of the rather complex steady state kinetics gives

$$\Phi_m \left(\frac{[A]}{1 - \Phi_m} \right) = \Phi_a \left(\frac{[A]}{1 - \Phi_m} \right) + \frac{k_H[CH_2]}{k_e} \qquad (26)$$

where Φ_m is the observed quantum yield of photoreduction and Φ_a is the actual efficiency of photoreduction of the camphorquinone triplet.

It is apparent from Eq. (26) that both the actual quantum yield and the ratio of k_H to k_e may be obtained from a plot of $(\Phi_m[A])/(1 - \Phi_m)$ versus $[A]/(1 - \Phi_m)$. This is shown in Figure 2. The values thus obtained were in excellent agreement with those determined by other methods[107,108] (cf. Table IV).

The deviation from the straight line seen in Figure 2 at low concentrations of **16** was attributed to quenching of the benzophenone triplet by the inter-

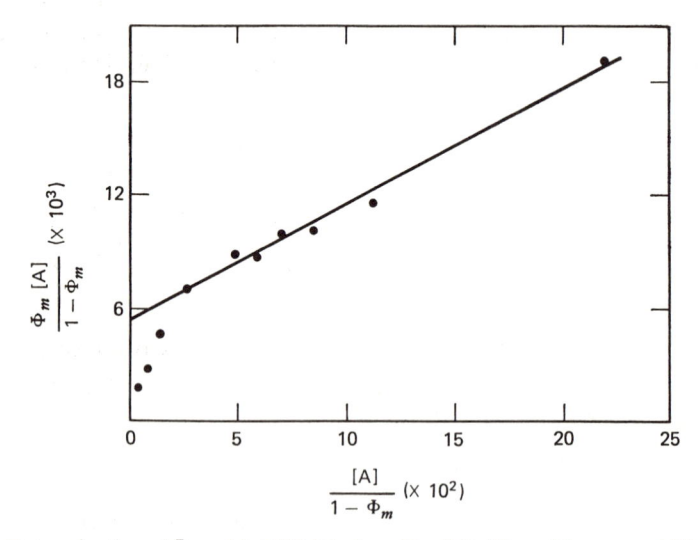

Fig. 2. Determination of Φ_a and $k_H[CH_2]/k_e$ from Eq. (26). (From Monroe and Weiner[108] with permission of the American Chemical Society.)

TABLE IV
Comparison of Parameters for the Sensitized Reduction of 16

	Φ_a	$k_H[CH_2]/k_e$
From Fig. 2	0.061	$5.21 \times 10^{-3}M$
Independent measurement[107,108]	0.059	$5.37 \times 10^{-3}M$

mediate radicals. This explanation has precedent in the work of Yang and Murov on the photoreduction of benzophenone in isopropanol.[111]

In a closely related study, Rubin also concluded that photoreduction was sensitized by both energy transfer and hydrogen transfer.[112] He found that the quantum yield for the benzophenone sensitized reduction of camphorquinone in p-xylene decreased with increasing acceptor concentration, reaching the quantum yield for the unsensitized reaction as a lower limit. Significantly, the product formed by cross-coupling of the p-methylbenzyl and the benzophenone ketyl radicals was not observed, indicating that when benzophenone abstracts hydrogen from xylene, the resultant radicals do not undergo cage recombination.

An interesting example where erroneous mechanistic conclusions were drawn from sensitization experiments is the photoreduction of acridine (17). Although irradiation of acridine in a variety of hydrogen containing solvents gave measurable reduction to acridan and diacridan, addition of benzophenone or acetophenone greatly enhanced the rate of this reaction.[113]

Photoreduction was quenched by high concentrations of biacetyl, slightly retarded by iodonaphthalene, but not affected by azulene or anthracene.[113] These observations led to the unsatisfying conclusion that reduction proceeded via a triplet state which could be only selectively quenched. However, later work[114] using flash photolysis showed that the benzophenone ketyl radical was generated upon irradiation of solutions of benzophenone and acridine, and that its predominant mode of disappearance was by reaction with acridine. Hence, acetophenone and benzophenone sensitize the reaction by hydrogen transfer. The anomalous effect of quenchers could be readily explained when it was demonstrated that the acridine singlet, rather than the triplet, is the reactive state in hydrogen abstraction.[115] Neither azulene nor anthracene quenches the acridine singlet, but iodonaphthalene enhances intersystem crossing by a heavy atom effect (cf. Section IV.A.3). The results with biacetyl may be due to "chemical quenching," that is, accepting the hydrogen atom from the intermediate acridine radical (18).

In a study of the quenching of benzophenone photoreduction with biacetyl, it was observed that the biacetyl was consumed.[116] This phenomenon also is readily explained by "chemical quenching," transfer of a hydrogen atom from the benzophenone ketyl radical to biacetyl to form the biacetyl ketyl radical and ultimately the pinacol derived from biacetyl.

The observation that azoxybenzene (19) rearranges to 2-hydroxyazobenzene (20) on direct irradiation but is reduced to azobenzene (21) in the presence of benzophenone was originally explained by a difference in reactivity between the singlet and triplet states.[117] However, three new lines of evidence argue in favor of chemical sensitization in this system: observation of the benzophenone ketyl radical upon flash photolysis of mixtures of 19 and benzophenone, reduction of 19 to 21 by nonphotochemically generated ketyl radicals, and the failure of nonhydrogen abstracting sensitizers to convert 19 to 21.[118]

Recent work on the photochemistry of ketimines has shown that they do not undergo reduction unless ketones are present. Thus chemical sensitization is entirely responsible for the photoreduction of benzophenone methylimine (22)[105] while intramolecular chemical sensitization has been suggested as the mechanism for reduction of the acylketimine (23).[119] In related work,

Padwa, Bergmark, and Pashayan were able to demonstrate that 22 and other ketimines are photochemically unreactive and that all of the reduction observed upon direct irradiation is due to hydrogen abstraction by the corresponding ketone formed either by hydrolysis or by oxidation of the imine.[120] The results with 23 may have a similar explanation.

Ketyl radicals have also been implicated in the photoreduction of azo dyes,[121] pyrazolone azomethene dye (24),[122] 4-(diethylamino)-4'-nitro-azobenzene (25),[123] and quinizarin (26).[124] The reduction of azobenzene by

ethyl chlorophyllide[125] and by disodium anthraquinone-2,6-disulfonate,[126] as well as the pyrochlorophyll sensitized reduction of nitro compounds,[127] probably proceed by similar mechanisms.

As a final example we would like to note the reactions of quinone methide **27**, which remains unchanged upon irradiation in methanol solution through a Pyrex or Vycor filter.[128] Reduction is observed when **27** is irradiated in isopropyl alcohol with benzophenone or acetophenone,[129] in methanol with acetophenone, or in methanol with no sensitizer through quartz.[128] While the effect of ketone sensitizers is due to chemical sensitization,[129] the "wavelength effect" is much more surprising. Further investigation showed that irradiation of methanol alone in a quartz vessel yielded ethylene glycol. The 185 nm line of the mercury arc, which is absorbed by Vycor and Pyrex, decomposes

methanol and other alcohols[130]; in this case, **27** scavenges the solvent

$$CH_3OH \xrightarrow[185\ nm]{h\nu} [H\cdot + \cdot CH_2OH] \longrightarrow \begin{array}{c} CH_2OH \\ | \\ CH_2OH \end{array} + H_2$$

radicals. Thus photoreductions observed on prolonged irradiation of dilute solutions with unfiltered light may, in fact, be caused by solvent decomposition and unusual "wavelength effects" may have a similar origin.

E. Complications in Olefin Isomerization

1. Cis-Trans Isomerization via Radicals. The reversible addition of free radicals to olefins can induce *cis-trans* isomerization of the double bonds.[131–134] This reaction can be quite clean, converting the olefin to a thermodynamic *cis-trans* equilibrium mixture with very little double bond migration.[131,134] *cis,cis*-Cyclotetradecane-1,8-diene (**28**), for example, is readily isomerized upon irradiation in the presence of diphenyl disulfide, with or without added sensitizer.[131] When iodine was used, isomerization was accompanied by extensive double bond migration.

cis,cis **28** → trans,trans 92% + cis,trans 6% + **28** 2%

In a photosensitized reaction, radical induced isomerization can occur if the sensitizer undergoes either homolytic decomposition or hydrogen abstraction, or if the system contains impurities which give radicals on irradiation. The result may be to shift the measured photostationary state in the direction of thermodynamic equilibrium and to give anomalously high values of the quantum yields for *cis-trans* isomerization.

In the initial work[135,136] on the sensitized isomerization of stilbene, it was found that quinone sensitizers gave photostationary states which were *trans*-rich relative to those predicted by energy transfer (Fig. 3). The original plot of sensitizer triplet energy versus isomer ratio of the photostationary state therefore showed several maxima and minima. Later work on the rates of energy transfer from various sensitizers to stilbene proved that the plot was in fact a smooth curve and that the photostationary states observed with quinones were not the true ones.[137] Irradiation of benzene solutions of

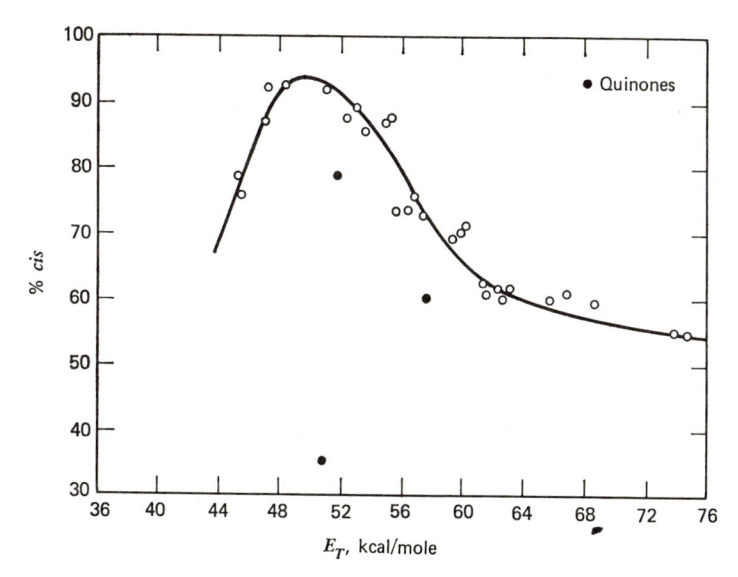

Fig. 3. Photostationary states obtained in the photosensitized isomerization of stilbene as a function of triplet energy of the sensitizer. (From Wagner and Hammond.[1])

quinones and stilbene or substituted stilbenes in the cavity of an esr spectrometer showed signals characteristic of semiquinone radicals.[138,139] Moreover, 2,3-diphenylpropene was converted to a mixture of *cis* and *trans*-1,2-diphenylpropene when irradiated with chloranil, indicating that hydrogen abstraction had taken place.[138] These results strongly suggest that the originally observed photostationary states were *trans*-rich because of some radical induced isomerization.

The benzophenone sensitized isomerization of the three isomers of 2,4-hexadiene (**29**) was originally reported to have quantum yields as high as 56 and this led the authors to postulate a quantum chain mechanism for isomerization.[140] The highest quantum yields were observed for the isomerization of *cis,cis*-2,4-hexadiene to its *cis,trans* isomer, a conversion that would tend to bring the system to thermal equilibrium. Further, the largest quantum yields were observed for one-bond isomerizations (i.e., *cis,cis* to *cis,trans*), while those of two-bond isomerizations (i.e., *cis,cis* to *trans,trans* and *trans,trans* to *cis,cis*) were never greater than one. Reinvestigation of this work using carefully purified diene failed to reproduce this chain effect.[141] All the quantum yields were well below one and their sum was two as predicted by theory.[142] It appears, therefore, that the originally observed chain isomerization was due to radicals formed from impurities in the starting dienes.

trans,trans   *cis,trans*   *cis,cis*

29

2. The Schenck Mechanism. Two mechanisms were originally proposed for photosensitized olefin reactions. One suggested that sensitization occurred by exchange energy transfer from an excited donor (S) to yield ground state donor and excited state acceptor.[24,136]

$$S^{*3} + A \longrightarrow A^{*3} + S \searrow \text{products}$$

The second, due to Schenck, proposed that there was bond formation between the donor and acceptor to give a partially or completely bonded complex which either underwent reaction or came apart to give excited acceptor and ground state donor.[143]

$$S^* + A \longrightarrow \cdot A\text{---}S\cdot \longrightarrow S + A^* \longrightarrow \text{products}$$
$$\downarrow B$$
$$AB + S$$

The success of the energy transfer theory in correlating data for a number of systems with sensitizers of widely varying structure, particularly aromatic hydrocarbons for which bond formation between donor and acceptor does not appear to be a reasonable process, led to its acceptance as the general mechanism for photosensitized olefin isomerization. However, there are special cases in which another mechanism of sensitization is operative.

Yang[144] has reported a study of 3-methyl-2-pentene (30), which undergoes isomerization and oxetane formation when irradiated with benzophenone or benzaldehyde. The quantum yields for these processes are shown in Table V.

$$\phi\diagdown \atop R\diagup C{=}O \ + \ {CH_3 \diagdown \atop H \diagup}C{=}C{\diagup C_2H_5 \atop \diagdown CH_3} \ \xrightarrow{h\nu} \ {CH_3 \diagdown \atop H\diagup}C{=}C{\diagup CH_3 \atop \diagdown C_2H_5} \ + \ \text{oxetanes}$$

R = φ, H   30

TABLE V
Quantum Yields for 3-Methyl-2-pentene Isomerization

Sensitizer	E_T, kcal	Φ_{oxetanes}	$\Phi_{t \to c}$	$\Phi_{c \to t}$
Benzaldehyde	71.5	0.45	0.12	0.18
Benzophenone	68.0	0.16	0.24	0.36
Triphenylene	66.5	—	0.004	—

Benzophenone was more efficient in sensitizing isomerization than was benzaldehyde, although the energy transfer mechanism would predict a higher efficiency for the sensitizer with the greater triplet energy. Triphenylene, which cannot react with the olefin to form an oxetane, did not sensitize isomerization. Quenching studies showed that the triplet state of the carbonyl compound was the common intermediate for both isomerization and oxetane formation and that the bimolecular rate constants for energy transfer were substantially less than diffusion controlled.

Similar observations were made by Saltiel in a study of the isomerization of 2-pentene.[145] In contrast to the stilbene system, a single decay ratio could not account for the results with different sensitizers (Table VI). This implies that different intermediates were produced with different sensitizers.

Other systems for which the results cannot be explained by energy transfer include the ketone sensitized isomerization of β-methylstyrene, 1,2-dichloroethylene and 2-butene studied by Caldwell,[147,148] deuterium exchange between acetone-d_6 and tetramethylethylene reported by Japar, Pomerantz, and Abrahamson,[149] and the acetone sensitized isomerization of 1-methoxy-1-butene examined by Turro and Wriede.[71]

TABLE VI
Sensitized Isomerization of 2-Pentene

Sensitizer	E_T, kcal	Decay ratio[a]
Benzene	84	1.00[b]
Acetone	80	1.17
Acetophenone	74	1.90

[a] Amount of decay of intermediate to *trans*-olefin/amount of decay to *cis*-olefin.
[b] Assumed from values for butenes, Ref. 146.

All the olefins involved in these studies are simple olefins to which energy transfer from ketones should be endothermic. Although the details of the mechanism of isomerization are still a matter of some debate,[5,71] it is generally agreed that isomerization takes place by addition of the sensitizer to the olefin to form a new intermediate which may be of a biradical nature. Thus if energy transfer is not favored, there is another mechanism by which ketone sensitizers can induce olefin isomerization, and the observed quantum yields and photostationary states may differ sharply from those predicted by the energy transfer mechanism.

A related phenomenon has been observed in the benzophenone sensitized isomerization of *cis*-piperylene.[150] The measured quantum yield of *cis* to *trans* isomerization increased from 0.55 to 0.90 as the concentration of piperylene increased from 0.08 to 10M. This observation can be rationalized as arising from addition of the piperylene triplet to a ground state diene molecule to give a biradical intermediate which can either cyclize to the dimer[151] or dissociate to give two molecules of the more thermodynamically stable *trans*-isomer. This mechanism predicts that the quantum yield for the isomerization of *trans*-piperylene to *cis*-piperylene should decrease with increasing diene concentration, an experiment that has not yet been reported.

3. Nonvertical Energy Transfer. The concept of nonvertical energy transfer was introduced by Hammond and Saltiel to explain the observation that low energy sensitizers would photoisomerize stilbene, despite the fact that energy transfer was expected to be highly unfavorable.[135-137] The experimental evidence and reasoning that led to this conclusion has been discussed in detail elsewhere.[1,9,152] In essence, nonvertical transfer provides a mechanism for energy transfer which is expected to be strongly forbidden on an energetic basis. Unlike Schenck type transfer, it correlates the results from a large number of sensitizers including aromatic hydrocarbons, on the basis of a single intermediate decaying to *cis*- and *trans*-isomers. This intermediate is believed to be a twisted, non-Franck-Condon stilbene triplet which is of lower energy than the spectroscopic state. If nonvertical sensitization takes place, the *cis-trans* ratio observed at the photostationary state may differ sharply from that observed with a high energy sensitizer since it is now dependent upon the relative rates of energy transfer to the two isomers as well as to the decay ratio of the intermediate.

It should be noted that this interpretation was recently questioned by Bylina.[153] On the basis of an assumed Gaussian shape for both sensitizer and acceptor spectra, he has shown that the observed rates of energy transfer can be accounted for in terms of the overlap of their $T_1 \leftarrow S_0$ transition bands.

IV. ENERGY TRANSFER FROM OTHER THAN THE LOWEST TRIPLET STATE

A. Involvement of Sensitizer Singlets

In the most appealing case, triplet sensitizers undergo extremely rapid intersystem crossing from the excited singlet to the triplet manifold.

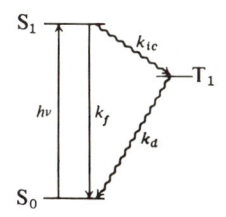

Indeed it has been shown[154] that k_{ic}, the rate of intersystem crossing, is greater than 10^{10} sec^{-1} for benzophenone and presumably for other aromatic ketones[155]; however, in many compounds it is only about 10^7 sec^{-1}.[156,157] Thus if their reaction rates with S_1 are close to diffusion controlled, certain substrates can interfere with intersystem crossing. If the sensitizer fluoresces, singlet interactions can be detected by a decrease in the intensity and lifetime of the emission. The interaction may or may not lead to chemical reaction in the substrate, but kinetic complications will arise in either case. In several of the examples to be discussed, a charge transfer complex between acceptor and singlet excited donor has been implicated.

The manifestations of singlet interactions need not be subtle; for example, attempts to prepare 1,3-cyclohexadiene photodimers using naphthalene as a sensitizer will give an infinitesimal yield.[158] Moreover, the reported[159] use of triphenylene to form triplet radical pairs from azo compounds is incorrect since only singlet radical pairs are formed.[160]

1. Kinetic Considerations. The important processes involved in singlet state interactions are shown below:

$$S \xrightarrow{h\nu} S^{*1} \quad \text{excitation} \quad (1)$$

$$S^{*1} \xrightarrow{k_f} S^{*1} + h\nu \quad \text{fluorescence} \quad (2)$$

$$S^{*1} \xrightarrow{k_{ds}} S \quad \text{radiationless decay} \quad (3)$$

$$S^{*1} \xrightarrow{k_{ic}} S^{*3} \quad \text{intersystem crossing} \quad (4)$$

$$S^{*1} + A \xrightarrow{k_{es}} S + A^{*1} \quad \text{energy transfer} \quad (7')$$

The effect of singlet quenchers on the fluorescence intensity of the sensitizer is derived as follows:

$$\Phi_f = \frac{k_f}{k_{ds} + k_f + k_{es}[A] + k_{ic}} \equiv \frac{k_f}{k_\Sigma + k_{es}[A]} \quad (27)$$

k_Σ is the total rate of sensitizer singlet decay in the absence of quencher. The singlet lifetime is defined as $\tau_S = k_\Sigma^{-1}$. In the absence of quencher, $\Phi_f^0 = k_f/k_\Sigma$. Then

$$\frac{\Phi_f^0}{\Phi_f} = 1 + \frac{k_{es}[A]}{k_\Sigma} = 1 + k_{es}\tau_S[A] = 1 + K_Q[A] \quad (28)$$

where K_Q is called the quenching constant. Equation (28) is the usual Stern-Volmer form which predicts a linear relationship between the quencher concentration and the reciprocal of the fluorescence quantum yield (i.e., intensity).

The effect of singlet quenchers on the sensitizer triplet yield (Φ_t) is derived similarly.

$$\Phi_t = \frac{k_{ic}}{k_\Sigma + k_{es}[A]} \quad (29)$$

$$\Phi_t^0 = k_{ic}/k_\Sigma \quad (30)$$

$$\frac{\Phi_t^0}{\Phi_t} = 1 + \frac{k_{es}[A]}{k_\Sigma} = 1 + K_Q[A] \quad (31)$$

It will be noted that

$$\Phi_t^0/\Phi_t = \Phi_f^0/\Phi_f \quad (32)$$

When energy transfer occurs from a short lived state such as a singlet, the efficiency of transfer often depends on the acceptor concentration, even when the latter is above $10^{-2}M$. If the acceptor singlet leads to product with a constant efficiency Φ_a, the quantum yield of product formation Φ_m will be related to the acceptor concentration as shown below.

$$\Phi_{A^{*1}} = \frac{k_{es}[A]}{k_{es}[A] + k_\Sigma} \qquad \Phi_m = \Phi_{A^{*1}}\Phi_a \qquad (33)$$

$$\frac{1}{\Phi_m} = \frac{1}{\Phi_a}\left(1 + \frac{k_\Sigma}{k_{es}[A]}\right) = \frac{1}{\Phi_a}\left(1 + \frac{1}{K_S[A]}\right) \qquad (34)$$

K_S is the sensitization constant and in the absence of other complications, it should be equal to K_Q.

We shall frequently refer to Eq. (27)–(34) in the following pages.

2. Förster Transfer. Singlet energy transfer has actually been known for many years, as evidenced by numerous studies of the sensitized fluorescence of dyes.[161,162] In these cases strong overlap of donor emission with acceptor absorption results in Förster type transfer but the acceptor undergoes no photochemistry.

The only recent example of Förster transfer of photochemical importance is the demonstration by Saltiel[163] that the ability of azulene to increase the photostationary *trans/cis* ratio in direct photoisomerization of the stilbenes is due entirely to radiationless transfer of excitation from *trans*-stilbene singlets to azulene. As expected for Förster transfer, this "azulene effect" did not depend upon solvent viscosity. The experimental value of R_0, the critical radius of transfer in Förster's formula,[161] was 18 Å, in good agreement with the value calculated from the overlap of stilbene emission and azulene absorption.

In the balance of the cases of singlet sensitization to be discussed, the acceptor generally undergoes some photochemical reaction; however, we shall also consider examples where quenching occurs without formation of the acceptor excited state. Sensitizers whose singlet state undergoes photochemistry have been treated in Section III.

3. Heavy Atom Effects. By virtue of their ability to enhance spin-orbit coupling, heavy atoms promote both radiative and nonradiative spin forbidden processes.[164] Thus heavy atom solvents have been used to increase the extinction coefficient of ground state to triplet absorption and thereby render these transitions visible.[165]

An interesting example of an external heavy atom effect was reported recently by Fischer.[166] Stilbene was irradiated in methyl iodide at the wavelength (436 nm) corresponding to its triplet energy. Although the extinction coefficient of $S_0 - T_1$ absorption in *trans*-stilbene would normally be infinitesimal, use of the heavy atom solvent raised it to 0.011, making it possible to determine that isomerization to *cis* occurred with a quantum yield of 0.75 ± 0.25.

Several examples of heavy atom quenching of aromatic hydrocarbon states are known; for example, carbon tetrabromide is an efficient quencher of the fluorescence of anthracene[167] and carbon tetrachloride behaves similarly with *p*-terphenyl.[168] Since quenching results in formation of the triplet state, it has been possible to use the heavy atom effect to measure intersystem crossing efficiencies (Φ_t^0). Because of the elegance of this technique[169] and the importance of the results in photochemistry, we shall cover it in some detail.

Experimentally the key to the method is determining the relative population of triplet molecules by the optical density of triplet-triplet absorption following flash excitation. To derive the kinetic expressions needed to determine Φ_t^0, the scheme is the same as that written previously with the additional proviso that the singlet interaction lead to intersystem crossing by the sensitizer.

$$S^{*1} + Q \xrightarrow{k_q} S^{*3} + Q \qquad \text{heavy atom effect} \qquad (35)$$

$$S^{*3} \xrightarrow{k_d} S \qquad \text{radiationless decay} \qquad (5)$$

While Eq. (30) is still true, Eqs. (28) and (29) will be replaced respectively by

$$\frac{\Phi_f^0}{\Phi_f} = 1 + k_q \tau_S [Q] \qquad (36)$$

$$\Phi_t = \frac{k_{ic} + k_q [Q]}{k_\Sigma + k_q [Q]} \qquad (37)$$

Using the relationship $\tau_S = k_\Sigma^{-1}$, we divide Eq. (37) by Eq. (30) and introduce the fact that the T—T absorbance (D) is proportional to the yield of triplets.

$$\frac{\Phi_t}{\Phi_t^0} = \frac{1 + \dfrac{k_q}{k_{ic}}[Q]}{1 + k_q \tau_S [Q]} = \frac{D}{D_0} \qquad (38)$$

The derivation is completed by eliminating $k_q[Q]$ with Eq. (36) and reintroducing Eq. (30).

$$\frac{\Phi_f^0}{\Phi_f} = \left(\frac{D}{D_0} \frac{\Phi_f^0}{\Phi_f} - 1 \right) \Phi_t^0 + 1 \qquad (39)$$

Plots of Φ_f^0/Φ_f versus the quantity in parentheses were linear for all eleven aromatic hydrocarbons studied and the slope yielded Φ_t^0 values ranging from 0.32 to 0.85. Since the sum of Φ_f^0 and Φ_t^0 was unity, the important conclusion was drawn that internal conversion from the first excited singlet to the ground state is negligible in these molecules.

Heavy atom enhancement of intersystem crossing has been used to determine the mechanism of acridine photoreduction in ethanol.[115] It was found that addition of sodium iodide decreased the fluorescence intensity and the rate of disappearance of acridine to the same extent, confirming that the singlet state is responsible for photoreduction. From the increase in triplet state absorption upon addition of iodide it was found that Φ_t^0 for acridine was 0.76. Thus the short singlet lifetime (0.8 nsec) of acridine is due to rapid intersystem crossing to unreactive triplet states.

Attempts to observe heavy atom effects on the Type II elimination of aliphatic ketones[170] and on the photodimerization of coumarin[171] were unsuccessful, probably because spin-orbit perturbation due to the carbonyl group is already large.[172] Cowan and Drisko,[173] however, were able to enhance intersystem crossing in acenaphthylene and thereby change the ratio of photodimers by using heavy atom solvents. Since the *cis* dimer is formed predominantly via a singlet excimer while the *trans* arises from the triplet state, addition of heavy atoms should result in more *trans* dimer. In fact the *cis*:*trans* ratio was 5:1 in cyclohexane while in *n*-propyl bromide it was 2:5.

4. Quenching of Aromatic Hydrocarbon Singlets. *a. Dienes.* The Pandora's box of photochemically significant singlet complications was opened by Hammond and coworkers[158] who noted that in several cases Φ_t values measured by the triplet counting technique of Lamola and Hammond[174] were less than those obtained by other methods.[156,175] Thus Lamola[174] obtained 0.39 for naphthalene while Wilkinson,[156] using the heavy atom technique, obtained 0.82.

In the triplet counting technique the substance whose intersystem crossing yield is to be determined is used as a sensitizer for the photoisomerization of piperylene (1,3-pentadiene). Under conditions where essentially all of the sensitizer triplets are intercepted by diene, the quantum yield of isomerization is directly proportional to the intersystem crossing yield of the sensitizer. If the diene interferes with sensitizer singlets however, the yield of triplets will fall, resulting in an anomalously low value for Φ_t.

Stephenson monitored the quenching of aromatic hydrocarbon singlet states by observing a decrease in the sensitizer fluorescence intensity with added diene.[158] A marked effect of the structure of both the sensitizer and the quencher was noted, with values of k_q ranging from 4×10^9 to 8.7×10^6; however, no quantitative correlation between k_q and several factors that might influence the stability of the proposed excited complex could be obtained (see below). The important observation that singlet quenching led to no observable diene photochemistry ruled out both singlet energy transfer and enhancement of intersystem crossing as the quenching mechanism. Thus when 1-methylnaphthalene was used to sensitize dimerization of 1,3-cyclohexadiene, the dimer yield decreased at high diene concentrations. This was attributed to the diene interfering with production of 1-methylnaphthalene triplets needed to sensitize dimerization. In another experiment, naphthalene was irradiated in the presence of myrcene (31) which gives different products from its singlet and triplet excited states. No singlet product could be detected and the triplet products were produced in only small yields which again decreased with increasing diene concentration. In view of the efficiency with which cyclohexadiene quenches naphthalene singlets, it is surprising that Baldwin was able to sensitize dimerization of α-phellandrene (32) with naphthalene.[176]

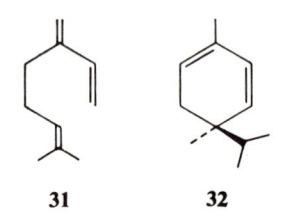

Recently two groups of workers[177,178] found that dienes quench the fluorescence of rigid unstrained azo compounds. However, since azo compounds show little promise as sensitizers, this interesting observation falls outside the scope of this review.

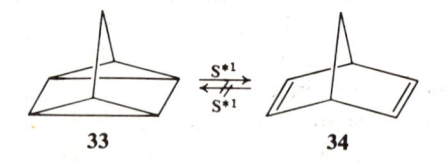

b. *Strained Hydrocarbons.* The quenching of aromatic hydrocarbon fluorescence is not restricted to dienes; a remarkable example is the report of Hammond and Murov[179,180] that quadricyclene (33) has this property. An important difference from the dienes is that 33 undergoes a photochemical reaction, conversion to norbornadiene (34). The mechanism of this transformation is a subject of considerable interest and debate though it is generally agreed that it must involve some kind of complex of acceptor and excited singlet sensitizer. Although no satisfactory explanation has yet been advanced as to how the complex, which possesses some 90 kcal of excess energy, goes to products, the predominant fate of a free vibrationally excited species in solution should be collisional deactivation.[181] Solomon, Steel, and Weller[181] have demonstrated the importance of charge transfer from quadricyclene to naphthalene in the complex with the report that the rate of fluorescence quenching increases in the order 1-methoxynaphthalene < naphthalene < 1-cyanonaphthalene. Any general explanation for singlet quenching must consider that exciton interactions and reverse charge transfer affect the rate of complex formation and that the observed quenching efficiency may depend upon the rate at which the complex decays.[180]

Paralleling the behavior of quadricyclene is that of 1,2-diphenylcyclopropane (35) which undergoes photochemical isomerization as shown below. This reaction, which proceeds via nonvertical triplet energy transfer with such

$$\underset{35}{\overset{\phi\quad\phi}{\underset{H\quad H}{\triangle}}} \;\overset{S^*}{\rightleftarrows}\; \overset{\phi\quad H}{\underset{H\quad\phi}{\triangle}}$$

sensitizers as benzophenone, has now been shown to involve singlets when aromatic hydrocarbon sensitizers are used.[179] Piperylene has only a weak quenching effect on the naphthalene sensitized reaction, which would not be expected if sensitizer triplets were involved. When naphthalene substituted with an optically active group was used as sensitizer, considerable asymmetric induction was found; this was originally[182] taken as evidence that nonvertical triplet energy transfer requires intimate interaction between donor and acceptor.[183] This result no longer bears on the mechanism of triplet energy transfer but it is certainly consistent with an excited complex mechanism. It would perhaps be of interest to see whether asymmetric induction occurs with an optically active benzophenone or other appropriate triplet sensitizer.

c. *Sulfoxides.* In 1965 Mislow and coworkers[184] reported the sensitized intermolecular and intramolecular photoracemization of optically active sulfoxides. Because little was known about singlet complications at that time, several early conclusions[185] about the mechanism of the photoracemization require modification.

Since the quantum yield for conversion of optically pure 36 to its enantiomer was 0.55–0.60 while the value taken for intersystem crossing in naphthalene was less than 0.5, it was originally concluded that some singlet excitation must be transferred from the naphthalene nucleus to the *p*-toluenesulfinyl group. However, as mentioned earlier, the intersystem crossing efficiency of naphthalene is actually[156] 0.82 so that this conclusion is not justified. The fact that piperylene could quench reaction (40) was taken to mean that triplet naphthalene was the sensitizing species; but since it is now known[158] that piperylene quenches naphthalene singlets, there is no longer support for the triplet mechanism.

The confusion generated in the initial report[185] on photoracemization of sulfoxides has recently been removed with the postulate that naphthalene singlet forms an excited complex with sulfoxides.[186] Thus, despite the fact that the singlet state of **36** lies at 113 kcal, some 23 kcal above that of naphthalene, **36** quenches the fluorescence of this hydrocarbon with $k_{es} = 3.2 \times 10^7 M^{-1} sec^{-1}$. From the dependence of the quantum yield of racemization on sulfoxide concentration (Eq. 34), a value of $k_{es} = 2.3 \times 10^7 M^{-1} sec^{-1}$ was deduced. Since these values are the same within experimental error, it follows that the singlet state of naphthalene is responsible for photoracemization.

d. Amines. Overwhelming evidence for the importance of charge transfer in the quenching of perylene singlets by amines has been obtained by three separate groups.[187-190] Because this case is particularly suited to detailed study, it serves as a model for photochemically important charge transfer interactions, several examples of which have already been mentioned. Weller[188] has shown that the efficiency of fluorescence quenching increases when amines of lower ionization potential are used and when solvent polarity is high. In a series of experiments in which a solution of perylene and dimethylaniline was subjected to an intense flash while observing the absorption spectrum, transient bands were observed at 580 and 490 nm. The former was assigned to the perylene radical anion by comparison with the absorption spectrum of perylene reduced by metallic sodium while the 490 nm band was ascribed to the triplet-triplet absorption of perylene. The effect of amine ionization potential and solvent polarity on the band intensities were in accord with the mechanism shown in Figure 4. Excited singlet perylene (P*[1]) and amine (A) form a very short lived charge transfer complex ($P^- \cdot A^+$) which can disappear in three ways: by emission of light[188] ($h\nu_{ct}$), by back

COMPLICATIONS IN PHOTOSENSITIZED REACTIONS 281

$$P^{*1} + A \rightleftarrows P^- \cdot A^+ \longrightarrow P^-_{solv} + A^+_{solv}$$

$$\downarrow \qquad\qquad \downarrow$$

$$P + h\nu_f \qquad P + A + h\nu_{ct} \quad P + A \quad P^{*3} + A$$

Fig. 4.

transfer of an electron to give ground state or triplet perylene, or by simultaneous separation and solvation of the radical ions. In solvents of low dielectric constant, the latter process is not favored so that charge transfer emission and perylene triplet-triplet absorption rather than the absorption of the radical anion are seen.

This mechanism has been confirmed by Ware,[189] who investigated the decay kinetics of P^{*1} and of the charge transfer complex. It was shown that $(P^- \cdot A^+)$ can indeed dissociate to give back P^{*1} and that this process has an activation energy of 10.2 kcal.

Charge transfer is also important in the quenching of acridine[187] and pyrene[191] singlets by amines.[192]

e. Azo Compounds. Although singlet sensitization in azo compound decomposition was only recently discovered, several cases are already known and it has provided an explanation for a number of poorly understood aspects of azo photochemistry.

Bartlett and Engel[160] have presented five lines of evidence that sensitized decomposition of azo-2-methyl-2-propane (**38**) proceeds via singlet energy transfer: (*1*) **38** quenches the fluorescence of a variety of aromatic hydrocarbons at a diffusion controlled rate; (*2*) the efficiency of triphenylene sensitized decomposition of **38** falls at lower azo concentration ($10^{-3}M$), a type of behavior inconsistent with energy transfer from a long lived triplet; (*3*) producing triphenylene triplets by benzophenone sensitization results in virtually no azo decomposition; (*4*) addition of piperylene, an effective quencher of triphenylene triplets but not singlets, does not alter the quantum yield of decomposition of **38**; and (*5*) sensitization with 9,10-diphenylanthracene, whose fluorescence quantum yield is unity, leads to decomposition with a quantum yield of 0.26. Perdeuteroacetone was the only ketone sensitizer which decomposed azomethane and **38** with reasonable efficiency.[193] Since acetone is unlike aromatic ketones in that its singlet lives long enough to

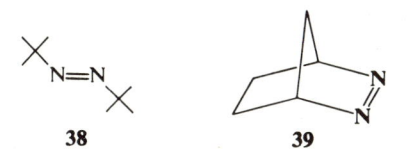

38   **39**

fluoresce,[52] and since the lowest triplet state of acyclic azo compounds appears to be stable, it is likely that singlet transfer occurs with acetone.

Steel and coworkers[194] have also noted that 2,3-diazabicyclo[2.2.1]-heptene-2 (**39**) quenches the fluorescence of naphthalene. A more extensive study of this compound was carried out by Engel[195] who reported that there was a sharp drop in the efficiency of sensitized decomposition and in the rate of sensitizer fluorescence quenching as the sensitizer singlet energy fell below that of **39**. This points out a key difference between azo compounds and the cases discussed previously; namely, there is a spectroscopic azo singlet state lying below that of the sensitizer.

The discovery that azo compounds undergo singlet sensitized decomposition is particularly relevant to the problem of spin correlation effects in free radical reactions. Any radical pair precursor that gives a difference in products depending upon whether it is produced as a singlet or triplet excited state is said to show a spin correlation effect.

In one of the earliest spin correlation studies, Hammond and Fox[159a] produced cyanocyclohexyl radicals by photolysis of azo-1-cyanocyclohexane (**40**) and its related ketenimine (**41**) (cf. Fig. 5). The radical pairs that do not

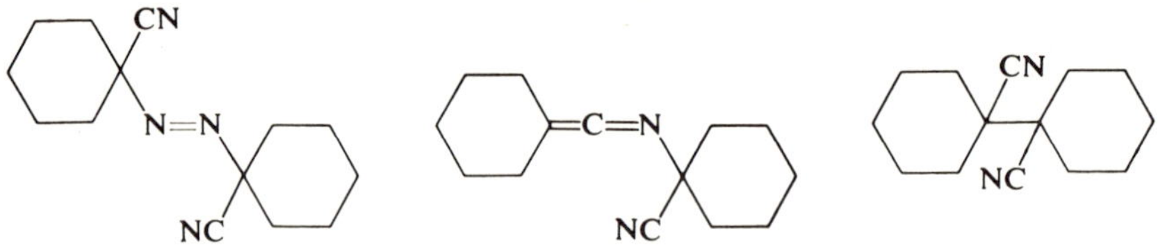

RN$_2$R
40

RR′
41

RR

Fig. 5.

	Yield of RR
RR′ $\xrightarrow{hv}$ R↓ ↑R ⟶ RR	24.1%
RR′ $\xrightarrow{S^{*3}}$ R↑ ↑R ⟶ RR	8.3%
RN$_2$R $\xrightarrow{hv}$ R↑N$_2$↓R ⟶ RR	19.9%
RN$_2$R $\xrightarrow{S^{*3}}$ R↑N$_2$↑R ⟶ RR	17.7%

recombine diffuse out of the solvent cage and are scavenged by the cumene solvent. The fact that a spin correlation effect was observed with ketenimine but not with the azo compound was explained with the notion that the intervening nitrogen molecule reduces interaction between the radical centers and causes spin inversion to be rapid. The question then arises, however, that if interposing a nitrogen molecule in the singlet radical pair reduces the cage effect from 24.1 to 19.9%, why does not the same hold for the triplet pair and lower the last figure in Figure 5 to something less than 8.3%?

The above results are easily explained when it is realized that triphenylene, the sensitizer used in these studies, is one of the best singlet sensitizers. Thus no triplet radical pairs were ever produced from the azo compound. Furthermore, the quantum yields of sensitized ethylazoisobutyrate photolysis, which at first were puzzling,[159a] are now readily understood. Aromatic hydrocarbons, which have long singlet lifetimes, give the highest quantum yields while ketones, which transfer only triplet energy to azo compounds, do not result in decomposition because the azo triplet produced is stable.

No spin correlation effect was seen in photolysis of azocumene,[159b] perfluoroazomethane,[196] or in azomethane itself.[160,193] These results are again explained by the lack of any real triplet sensitized decomposition.

The behavior of cyclic azo compounds differs considerably from that of the acyclic ones in that spin correlation effects have been seen in several cases. Since singlet decomposition gives a different product distribution than triplet sensitization, these cases provide an elegant test of the sensitization mechanism.

Bartlett and Porter[197] studied the stereospecificity of the photochemical decomposition of **42**. Direct irradiation led to nearly complete retention of stereochemistry in the cyclobutane products but thioxanthone, xanthone, or

p-methoxyacetophenone sensitization resulted in considerable loss of stereospecificity. This is explained with the postulate that the triplet biradical lives longer than the singlet and consequently has time to undergo bond rotation before ring closure. Triphenylene sensitized photolysis of **42** gave relatively high stereospecificity[198] which suggests singlet sensitized decomposition. Although the overlap of triphenylene fluorescence and the absorption of **42** prevented fluorescence intensity quenching experiments, it was shown that **42** quenches the fluorescence of 9,10-diphenylanthracene at a diffusion controlled rate.

The photochemical decomposition of bicyclic azo compound **43** gave the results shown in Table VII.[199,200] The dependence of product composition on the stereochemistry of the starting azo compound, which was observed in direct photolysis, is removed by benzophenone or triphenylene sensitization since the intermediate triplet biradical undergoes randomization of stereochemistry before closure. Addition of piperylene to the benzophenone reaction lowered the rate of product formation but did not alter the product composition. The same experiment in the case of triphenylene changed the product composition back to that characteristic of direct (singlet) decomposition. Thus piperylene quenched triphenylene triplets and left only singlet sensitization. It is somewhat surprising that triphenylene gave a triplet type product distribution in the absence of piperylene in view of the high efficiency with which compound **39** quenches triphenylene fluorescence. The quenching constant K_Q was greater than 185 so that at an azo concentration of $0.005M$ we have

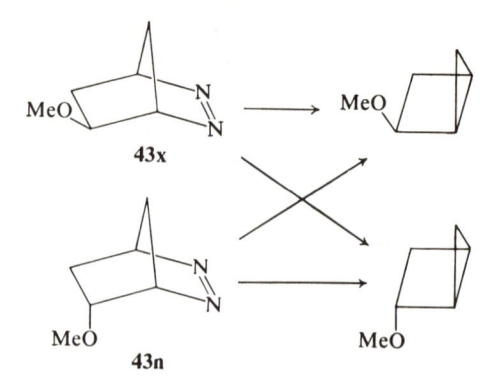

$$\frac{\Phi_f^0}{\Phi_f} = \frac{\Phi_t^0}{\Phi_t} = 1 + K_Q[A] = 1 + 185(0.005) = 1.93$$

TABLE VII
Photolysis of **43x** and **43n** in Cyclohexane

Compound, concn, M	Conditions	% cis	% trans
43x, 0.01	Direct irradiation	47	53
43n, 0.01	Direct irradiation	82	18
43x, 0.01	Benzophenone, $0.05M$	76	24
43n, 0.01	Benzophenone, $0.05M$	76	24
43x, 0.005	Triphenylene, $0.009M$	73	27
43n, 0.005	Triphenylene, $0.009M$	78	22
43x, 0.01	Benzophenone, $0.05M$ Piperylene, $0.1M$	77	23
43n, 0.01	Benzophenone, $0.009M$ Piperylene, $0.1M$	78	22
43x, 0.005	Triphenylene, $0.009M$ Piperylene, $0.1M$	44	56
43n, 0.005	Triphenylene, $0.009M$ Piperylene, $0.1M$	80	20

From the values for triphenylene of $\Phi_t^0 = 0.92$ and $\Phi_f^0 = 0.08$, it is calculated that $\Phi_t = 0.48$ and $\Phi_f = 0.04$. If the rest of the sensitizer singlets transfer energy, singlet sensitization should be about as important as triplet sensitization; however, the product composition from **43x** appears heavily weighted in favor of the latter. Perhaps the methoxyl group renders **43** a much less effective singlet quencher than **39**, or more likely, not all singlet quenching leads to azo decomposition as has been noted for compound **38**.[193]

Another cyclic azo compound which shows a spin correlation effect is the pyrazoline **44**[201] (Fig. 6). The product distribution under various conditions

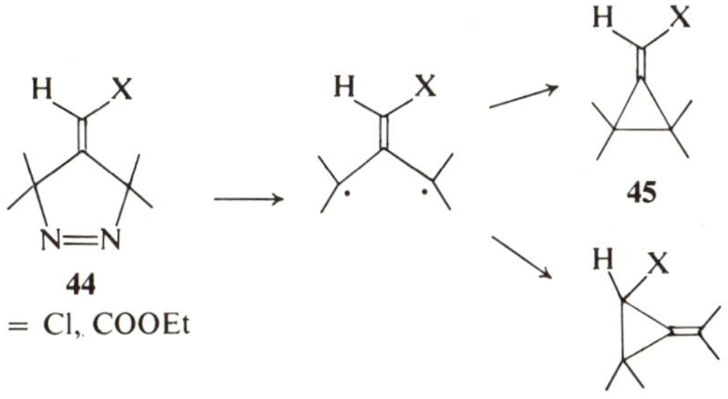

Fig. 6.

is shown in Table VIII. The fact that triphenylene gave a product distribution closer to that obtained in direct photolysis than did benzophenone was ascribed to singlet energy transfer. As supporting evidence, the authors have noted that **44** quenches the fluorescence of triphenylene.

TABLE VIII
Products from the Photolysis of **44** (X = Cl) in Pentane

Conditions	% **45**	% **46**
Predicted statistically	33	67
Direct photolysis	77	23
Benzophenone sensitized	25	75
Triphenylene sensitized	32	68

f. Azides. In a study of the photosensitized decomposition of alkyl azides,[202] an apparently linear relationship between sensitizer triplet energy and the log of the rate constant for energy transfer was obtained. Phenanthrene, however, gave a somewhat larger value than expected for this rate constant, which was determined from the variation in quantum yield of

TABLE IX
Quantum Yield for Sensitized Hexylazide Disappearance

Piperylene concn, M	Phenanthrene, Φ	Benzophenone, Φ
0.0	0.24	0.054
0.01	0.20	0.008
0.10	0.20	0.001

azide decomposition with azide concentration. Further investigation[203] revealed that azides quench the fluorescence of aromatic hydrocarbons even though singlet transfer is strongly endothermic. A plot of the log of the quenching rate constant versus sensitizer singlet energy was linear and could be extrapolated to a value of 91–92 kcal for the singlet of hexylazide. Since the rate of endothermic transfer was much faster than predicted for classical energy transfer, a mechanism involving vertical excitation of a vibrationally excited bent azide ground state to form a low energy bent excited state was favored.

Further evidence for singlet sensitization was the effect of added piperylene on the sensitized photolysis of hexylazide, as shown in Table IX. Although piperylene quenches phenanthrene triplets, azide decomposition still occurs by singlet sensitization. The benzophenone singlet is very short lived so triplet sensitized azide decomposition is strongly inhibited by addition of piperylene. This reasoning is analogous to that used to demonstrate singlet sensitization in compound **39**.

Recently Swenton and coworkers[204] have discovered singlet energy transfer in the decomposition of 2-azidobiphenyl (**47**). Aromatic ketone sensitizers gave **49** in about 40% yield and almost no carbazole **48**, while

aromatic hydrocarbons such as triphenylene and naphthalene gave predominantly **48**. Addition of 0.2M piperylene did not diminish the rate of triphenylene sensitized azide disappearance. The photochemistry of this system is probably complicated by the fact that **49** should be a good acceptor of both singlet[160,205] and triplet energy.[206]

g. *Peroxides.* As early as 1955 Szwarc[207] showed that benzene, anthracene, and acenaphthene would sensitize the photodecomposition of acetyl peroxide, $(CH_3CO)_2O_2$. No further investigation in this area was reported until 1963 when Vasil'ev[208,209] published a study of the sensitized photolysis and radiolysis of benzoyl peroxide. At high peroxide concentration, the quantum yield of both toluene and 2,5-diphenyloxazole (DPO) sensitized photolysis approached the value of 0.4 found in direct irradiation. From the effect of peroxide concentration on quantum yield (cf. Eq. 34), K_S was calculated as 450 ± 80 for toluene and 40 ± 10 for DPO. Fluorescence quenching experiments gave a value of $K_Q = 45 \pm 4$ for DPO and the good agreement with K_S is strong evidence that singlet DPO is the effective sensitizer. Since the peroxide singlet energy is much higher than that of the sensitizer, the authors postulated an excited complex where redistribution of energy breaks the weakest bond.

The most recent work[210] on sensitized decomposition of peroxides dealt primarily with ketones (cf. Section III.C.3) although it was also demonstrated that anthracene could sensitize decomposition of benzoyl peroxide. In view of the earlier work[208,209] and the fact that benzoyl peroxide quenches anthracene fluorescence at a rate of $1.0 \times 10^{10} M^{-1} \sec^{-1}$ in benzene,[211] a singlet exciplex mechanism is strongly implicated.

h. *Maleic Anhydride–Benzene.* Hardham and Hammond[212] have reported that the sensitized photoaddition of maleic anhydride to benzene proceeds via a complex of the two reactants. Sensitizers with a triplet energy greater than 66 kcal led to adduct formation while those below 66 kcal gave no adduct. Triphenylene behaved anomalously since no adduct was formed even though its triplet energy is 66.6 kcal. Since maleic anhydride was reported to quench triphenylene fluorescence, this anomaly may be attributed to quenching of triphenylene singlets, thus removing the triplets needed to sensitize addition. In view of the fact that the complex always absorbs some light, it is not clear why the unsensitized reaction did not proceed with triphenylene present.

i. *Ketones.* Zimmerman and Swenton[213] have noted that transfer of singlet energy from naphthalene to dienone **50** results in the same rearrangement as direct irradiation. The quantum yield of product formation increased at higher dienone concentration; however, because **50** still absorbed a considerable amount of light even at high naphthalene concentration, it is not clear how much of this effect was due to increased efficiency of energy transfer. Addition of 0.063M piperylene to the naphthalene sensitized reaction only lowered the quantum yield from 0.28 to 0.26; since triplet naphthalene transfers energy to piperylene,[174] this argues against triplet sensitization. Energy transfer from singlet naphthalene to **50** must be much faster than the $10^8 M^{-1} \sec^{-1}$ reported by Stephenson[158] for transfer to piperylene since the diminution in quantum yield is small despite the large ratio of piperylene to dienone. This view is also supported by the high efficiency with which **50** quenched naphthalene fluorescence.[213]

The steroidal ketone **51** quenches acenaphthene fluorescence at a diffusion controlled rate[211] and the rearrangement shown[214] is sensitized by acenaph-

thene. Since triplet transfer would very likely be endothermic, this is best interpreted as singlet sensitization.

In a careful study of cyclopentenone photocycloaddition reaction, DeMayo and coworkers[215] have noted that ketone sensitizers of triplet energy less than 71.2 kcal did not sensitize cycloaddition to cyclohexene. Triphenylene (E_T = 66.6 kcal) and acenaphthene (E_T = 59.3 kcal) were exceptional since they resulted in quantum yields of 0.10 and 0.21, respectively. This behavior, as well as the fact that 0.1M cyclopentenone quenched acenaphthene fluorescence by 90% in an EPA (ether-isopentane-alcohol) glass at 77°K strongly implicate singlet energy transfer.

5. Quenching of Singlets of Carbonyl-Containing Compounds. *a. Acetone.* Recent reports have shown that singlet complications are not restricted to hydrocarbons; for example, the photodecomposition of 1,4-dichlorobutane (**52**) to free radicals is sensitized by the (π^*,n) singlet state of acetone.[216] Besides the observations that **52** quenches acetone fluorescence and that addition of triplet quenchers does not suppress the photoreaction, kinetic arguments support the singlet sensitization hypothesis. Even at high concentrations of **52**, the quantum yield of HCl formation is low, which was rationalized on the basis of energy transfer from a short-lived excited state. Thus in Eq. (33) k_Σ is much greater than $k_{es}[A]$ so that a plot of Φ_m versus [A] is expected to be linear, as was in fact observed. At lower acetone concentration, the quantum yield for HCl formation rose sharply, pointing to the monomer [52] form as the effective photosensitizer. Since the existence of acetone excimers has recently been questioned,[53] this point may require reinterpretation.

It has recently been shown that 1,3-pentadiene quenches the fluorescence of acetone and other aliphatic ketones,[217] a phenomenon similar to that discussed in Section IV.A.4.a. In related work,[19] cyclohexa-1,3-diene has been found to quench dibenzylketone fluorescence eight times as efficiently as 1,3-pentadiene but still at only one twelfth of the diffusion controlled rate. Besides introducing difficulties into intersystem crossing yield determinations, the discovery that dienes quench aliphatic ketone singlets may account for cases previously explained on the basis of a short-lived triplet, in which only slight inhibition of a ketone photoreaction was observed. With dibenzylketone, singlet quenching manifests itself by producing a lower yield of cyclohexadiene photodimers[218] than would be predicted from the short triplet lifetime of the sensitizer.[19] Further implications of the above observations are given in the paper by Wettack, Turro, and coworkers.[217]

b. Fluorenone. Elegant kinetic studies[219] of the photocycloaddition of ketenimines to fluorenone have led to the discovery of an exciplex similar to that from dienes and aromatic hydrocarbons. Equation (34) was applied to the competition between cycloaddition and decay of the short lived fluorenone triplet. As shown in Figure 7, the expected linear relationship between $1/\Phi_{\text{adduct}}$ and $1/[A]$ was seen except at high ketenimine concentration. The sharp upward curvature in this region and the observation that ketenimine quenched fluorenone fluorescence were at first[220] explained by deactivation of fluorenone singlets by **53**. Since singlet energy transfer would be highly endothermic, a complex of fluorenone singlet and ketenimine was postulated.

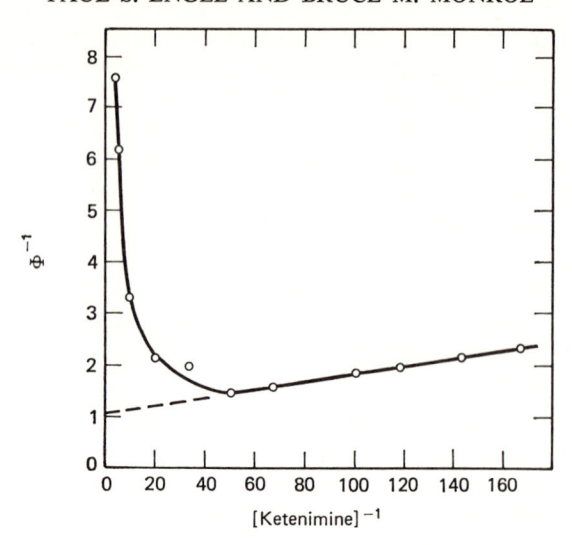

Fig. 7. Dependency of the reciprocal of the quantum yield for adduct formation on ketenimine concentration in the fluorenone–dimethyl-N-(cyclohexyl)ketenimine reaction. Slope = $0.006M$; intercept = 1.08. (From Singer and Davis[220] with permission of the American Chemical Society.)

More recent experiments[219] with added di-t-butylnitroxide, a quencher of both singlets and triplets, and detailed kinetic analysis of the results led to the conclusion that the complex proceeds to adduct with about 60% efficiency but that complex formation still could not account for the sharp rise in Figure 7. Deactivation of the complex by ketenimine would explain this curvature, but at present this idea is only speculative.[220]

In a recent study, Singer[221] showed that triethylamine quenches fluorenone fluorescence with a greater efficiency in more polar solvents. According to Eq. (28), this could be caused by an increase in k_{es} or by a decrease in k_Σ. Since the fluorescence yield of fluorenone was also found to increase dramatically in more polar solvents, it is clear from Eq. (27) that k_Σ must decrease. This was rationalized with the idea that increasing solvent polarity reorders the electronic states so that intersystem crossing changes from $(\pi^{*1},\pi) \to (\pi^{*3},n)$ to $(\pi^{*1},\pi) \to (\pi^{*3},\pi)$. According to El-Sayed,[222] the former can be as much as 10^3 faster than the latter which would raise the lifetime and fluorescence quantum yield in polar solvents.

The above ideas fit in nicely with the work of Cohen and Guttenplan[223] who reported the photoreduction of fluorenone by triethylamine. Dilution of neat triethylamine with cyclohexane increased the quantum yield of reduction by decreasing the amount of singlet quenching.[221]

c. *Biacetyl.* Like all α-diketones,[224] biacetyl (CH$_3$COCOCH$_3$) has the unique property of emitting both fluorescence and phosphorescence in solution. It has therefore found extensive use both as a sensitizer and a quencher in testing the multiplicity of photochemical reactions.[102,113,225-227] Dubois[228] used biacetyl as a quencher of fluorescence of many substances ranging from benzene to acetone. That singlet energy transfer was occurring was shown unambiguously by observing concomitant sensitization of biacetyl fluorescence. By this method, it was possible to obtain donor singlet lifetimes in good agreement with values from direct measurement. Turro[229] has recently observed the quenching of biacetyl fluorescence and phospherescence by a variety of substances. Fluorescence was quenched by phenols and amines but not by hydrogen donors such as benzhydrol. It was concluded that quenching may involve either electron or hydrogen abstraction and that either of these processes may be reversible.

d. *Other Carbonyl Compounds.* In the course of an investigation[230] of the photochemistry of pyruvonitrile (CH$_3$COCN), it was found that the apparent intersystem crossing yield measured by the technique of Lamola and Hammond[174] varied with the concentration of *cis*-piperylene. At high diene concentration, the value of Φ_t decreased, showing that piperylene captures the precursor to the lowest pyruvonitrile triplet state. Since no measurable loss of starting materials was observed and since electronic singlet energy transfer was ruled out for energetic reasons, an excited singlet complex mechanism was postulated. The absence of detectable fluorescence from pyruvonitrile obviated the use of fluorescence quenching as a test for involvement of singlets. From the effect of solvent viscosity on the intersystem crossing yield, the rate of quenching was judged to be diffusion controlled. However, as in the work of Singer[221] on fluorenone, it was suggested that the rate of intersystem crossing may also vary with solvent.

Little evidence is available to determine whether aromatic ketones can act as singlet sensitizers. If they can do so, it will be only at high acceptor concentrations because the lifetime of benzophenone singlets,[154] for example, is less than 2 × 10^{-10} sec and that of several other aromatic ketones is estimated to be less than 10^{-10} sec.[155] Yang[231] has noticed that the photochemical addition of benzophenone to 2,3-dimethyl-2-butene is less efficient in neat olefin than in 4M olefin, which may indicate deactivation of benzophenone singlets. Golub[216] has mentioned that benzophenone is a singlet sensitizer in the photolysis of 1,4-dichlorobutane.

The room temperature phosphorescence spectra of benzophenone and its substituted analogues show a weak band about 1600 cm^{-1} higher in frequency than the first phosphorescence band. Because the temperature dependence of its intensity corresponds to an activation energy approximately equal to the spectroscopic singlet-triplet splitting, Saltiel and coworkers[232] have assigned this emission to activation controlled delayed fluorescence.[233a] The report of Jones and Calloway[233b] on the emission of benzophenone in polymer matrices supports this conclusion. Assuming that there is a finite population of aromatic ketone singlets in equilibrium with triplets, one can imagine that if reaction from S_1 competes effectively with fluorescence, some aromatic ketone photoreactions might actually proceed through the singlet state. Moreover, singlet sensitization by an aromatic ketone is possible when the acceptor does not possess an unreactive low-lying triplet state which would act as an energy wasting sink.

B. Involvement of Upper Triplet States

1. Anthracene and Its Derivatives. Liu[234] has convincingly demonstrated that several reactions sensitized by anthracene and its derivatives proceed through the second excited triplet state, T_2. The first case studied was the photosensitized rearrangement of **54** to yield the products shown below. A break in the efficiency of triplet sensitized reaction was found between

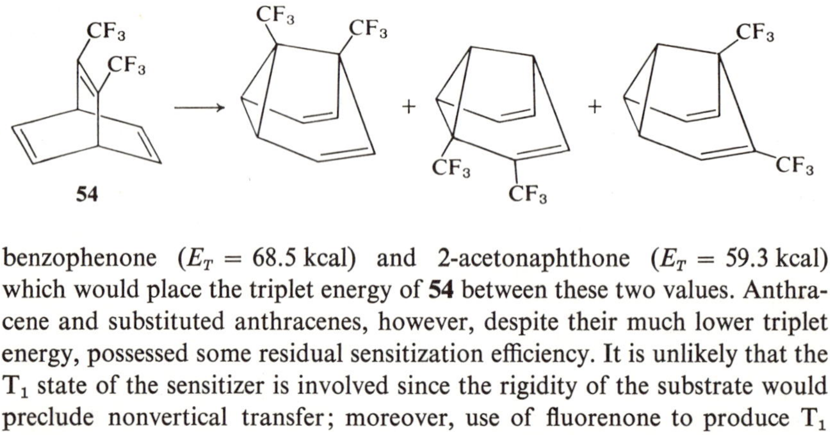

benzophenone ($E_T = 68.5$ kcal) and 2-acetonaphthone ($E_T = 59.3$ kcal) which would place the triplet energy of **54** between these two values. Anthracene and substituted anthracenes, however, despite their much lower triplet energy, possessed some residual sensitization efficiency. It is unlikely that the T_1 state of the sensitizer is involved since the rigidity of the substrate would preclude nonvertical transfer; moreover, use of fluorenone to produce T_1 selectively did not result in rearrangement of **54**. Involvement of S_1 was ruled out with the demonstration that addition of **54** did not affect the lifetime or quantum yield of anthracene fluorescence.[235] Because of the short lifetime of T_2, internal conversion should compete with energy transfer to **54** and Eq. (34) should be applicable. In fact, Liu[234] found that for 9,10-dibromoanthracene, the plot of $(\Phi_1)^{-1}$ versus $[54]^{-1}$ was linear and that K_S was 2.4. This implies that the lifetime of T_2 is 0.1 nsec. Three other rigid systems related to **54** were studied and gave similar results.

Transfer from T_2 of anthracene has also been invoked in several nonrigid systems[235,236]; for example, the product composition from the well-studied[237] photodimerization of butadiene agreed with that of high energy sensitizers. In the photosensitized isomerization of 1,3-pentadiene, anthracene failed to follow a monotonic decrease in quantum yield with sensitizer energy. A plot of Eq. (34) behaved as with the rigid system **54** and gave the more accurate value of τ_{T_2} as 0.22 nsec. The efficiency of the photodimerization of acrylonitrile varies in a manner that suggests involvement of the T_2 state of anthracene. The fact that the anomalous behavior of butadiene and 1,3-pentadiene under sensitization by anthracenes is explained with involvement of T_2 makes it unnecessary to postulate nonvertical energy transfer[237] (cf. Section III.E.3) in these cases.

A particularly interesting case of transfer from T_2 has been noted[238] in the 9,10-dichloroanthracene (DCA) sensitized isomerization of stilbene. As shown below, the DCA ground state which results from quenching of T_2 by stilbene (St) can act as a quencher of stilbene triplets and thereby alter the stilbene *cis*:*trans* photostationary state.

$$DCA(T_2) + St(S_0) \longrightarrow DCA(S_0) + St(T_1) \longrightarrow DCA(T_1) + St(S_0)$$

A kinetic analysis of the effect of DCA concentration on the photostationary state showed that in benzene, 28% of the intimately associated pairs [DCA + *trans*-stilbene*] underwent energy transfer before separating.

Spectroscopic evidence has also been adduced for the ability of the anthracene T_2 to transfer energy to other substances.[239] Selective excitation of guest anthracene in a host dibenzofuran crystal also containing napthalene-d_8 as a guest resulted in naphthalene phosphorescence. As shown in Figure 8, the path of the energy is $S_0 \to S_1$ excitation of anthracene, intersystem crossing

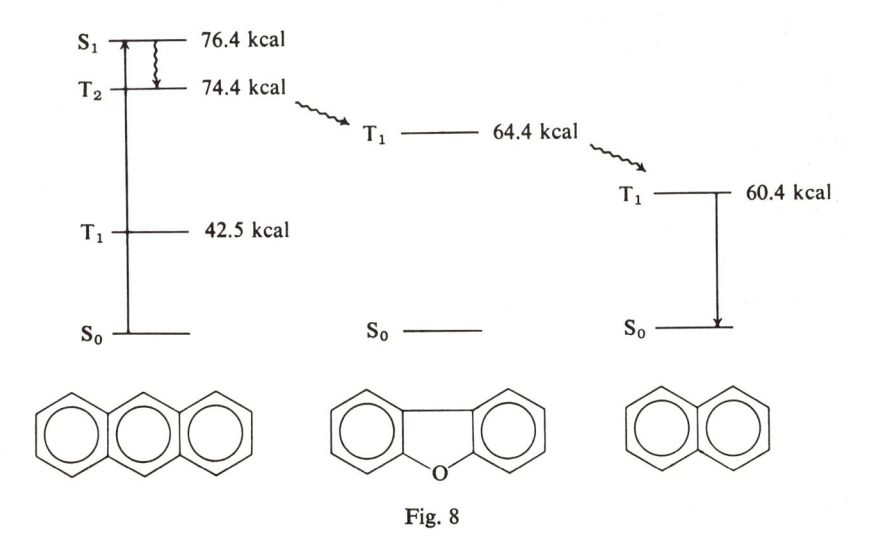

Fig. 8

to T_2, energy transfer to host T_1 and exciton migration away, trapping by naphthalene-d_8, and finally phosphorescence.

2. Bimolecular Reactions from Upper Triplet States. Other cases of sensitization by second excited triplet states have not yet come to light; however, several bimolecular reactions of this sort have been reported. Since an upper excited state that lives long enough to undergo a bimolecular reaction should also be capable of transferring energy, these reactions will be discussed briefly.

DeMayo and coworkers[215] have reported that although benzophenone did not sensitize the cycloaddition of cyclopentenone to cyclohexene, the enone quenched the photoreduction of benzophenone in isopropanol. His conclusion that cycloaddition proceeds via an upper triplet state of the cyclopentenone has elicited objections and alternative interpretations. In a recent review,[240] the slope of the Stern-Volmer plot for quenching of the benzophenone-isopropanol photoreduction was mistaken for k_q/k_r (rate of quenching/rate of hydrogen abstraction by benzophenone), leading to the suggestion that the inefficiency of sensitized addition is low because sensitizer decay competes with energy transfer to the reactive cyclopentenone triplet. Wagner[240a] has produced more substantial evidence against the two triplet mechanism by demonstrating that the lifetime of the cyclopentenone triplet which dimerizes is the same as the one which transfers energy to 1,3-pentadiene. Furthermore, the diene *cis-trans* photostationary state, though it is anomalously rich in *trans*, did not depend on the diene concentration. Other mechanisms besides energy transfer which could explain the observed quenching of the benzophenone-benzhydrol photoreduction are hydrogen atom transfer (cf Section III.D) or complex formation between benzophenone and cyclopentenone.[240b] The latter can now be eliminated, however, for Loutfry and deMayo[240c] have shown that cyclopentenone quenches the phosphorescence of benzophenone in solution[232] with very low efficiency. This deals the final blow to the upper triplet mechanism for cyclopentenone cycloaddition.

From the fact that formation of the two cycloadducts from 4,4-dimethyl-2-cyclohexenone (**55**) and 1,1-dimethoxyethylene was quenched with different efficiencies by di-*t*-butylnitroxide, Chapman[241] concluded that two triplet states were involved. This paramagnetic quencher is known, however, to deactivate both singlets and triplets.

Work of Yang[242] and of Warwick and Wells[243] has shown that two excited states are involved in the photochemical addition of 9-anthraldehyde (56) to 2,3-dimethyl-2-butene. Quenching by di-*t*-butylnitroxide gave a curved Stern-Volmer plot that could be analyzed to yield lifetimes of 1.0 and

0.3 nsec for the two states. Neither of these was the lowest triplet (π^*,π) or the lowest singlet but were most likely the T_2 and the (π^*,n) triplet states.

A recent report of Schuster and Sussman[244] indicates that the lowest triplet of eucarvone sensitizes cyclohexadiene dimerization while an upper triplet apparently undergoes cycloaddition to this diene.

V. GENERAL REMARKS ON SELECTING A PHOTOSENSITIZER

Nearly all the classes of compounds generally used as sensitizers can lead to complications. Aromatic hydrocarbons may interact with substrate via singlet and upper triplet states or they may self-quench, dimerize, and form adducts. Ketones with (π^*,π) triplet states can self-quench while quinones and those with (π^*,n) triplet states may show complications due to oxetane formation, Schenck type isomerization, hydrogen abstraction, and chemical sensitization.

With the possibility of so many side effects in photosensitized reactions, the question may well be asked, "How can one make an intelligent choice of sensitizer?" Obviously the approach is more sophisticated than "dump in some benzophenone." Although the answer will depend a great deal on the system under study, we shall attempt to provide a few general rules for the selection of a sensitizer.

The necessity of considering sensitizer light absorption characteristics, intersystem crossing yield, triplet lifetime, and triplet energy has been discussed elsewhere[3] but the latter is worthy of further comment. Many of the complications, including oxetane formation, cycloaddition, hydrogen abstraction, back transfer, Schenck type isomerization, and nonvertical energy transfer, become much more likely when the triplet levels of sensitizer and acceptor are not properly disposed for efficient energy transfer. Subject to the exceptions noted in Section III.C.4, the general rule of thumb that a 3 kcal energy difference is needed for irreversible energy transfer is not a bad one to follow.

The use of ketones with (π^*,n) triplet states to sensitize photoreductions will be attended by chemical sensitization; for example, with benzophenone in isopropyl alcohol and $0.5M$ acceptor, 1% of the sensitizer triplet will still abstract from the solvent even if the acceptor quenches at the diffusion controlled rate. Failure to determine that the quantum yield of reduction was greater than 0.01 might lead to the conclusion that triplet sensitization was occurring.

Particular attention should be paid to the possibility of singlet interactions when aromatic hydrocarbons and other compounds which fluoresce are employed as sensitizers. Unambiguous evidence for this phenomenon can be obtained from the effect of acceptor on sensitizer fluorescence and it can be minimized by the use of low acceptor concentrations. Interestingly, not all aromatic hydrocarbons engage in the type of quenching[158] that seems to involve charge transfer; for example, triphenylene fluorescence is not affected by piperylene.[160]

In general, fewer complications are observed with (π^*,π) triplet state ketones than with either (π^*,n) ketones, quinones, or aromatic hydrocarbons. The only serious problem observed to date with these compounds is self-quenching, which, however, is readily revealed by the effect of sensitizer concentration of quantum yield.

VI. TABLES OF SENSITIZER PROPERTIES

In Tables X–XII we have collected useful constants for selected potential sensitizers and have indicated what difficulties have been reported from the use of a particular compound. The definitions of the abbreviations used in the last column are

a = addition reactions
cs = chemical sensitization
d = dimerization
dec = decomposition
e = excimers
en = photoenolization
ha = hydrogen abstraction

ox = oxetanes
s = singlet interactions
sch = Schenck mechanism
sq = self-quenching
ut = upper triplets
II = Type II cleavage

Explanation of Table: E_T = triplet energy in kcal/mole. E_S = singlet energy in kcal/mole from lowest energy absorption band; * in the E_S column indicates highest energy fluorescence emission. Φ_{ic} = efficiency of intersystem crossing in nonpolar solvents; * in the Φ_{ic} column indicates ethanol as solvent. Φ_f = fluorescence quantum yield in degassed nonpolar solvent; * in the Φ_f column indicates ethanol. τ_f = fluorescence lifetime in nanoseconds in degassed nonpolar solvent.

Compound	E_T	E_S	Φ_{ic}	Φ_f	τ_f	Complications
Benzene	84.4,[r] 84.1,[e] 85.2[bb]	105*,[d] 108,[e] 108,[r] 108.7[rr]	0.25[vv]	0.07,[d] 0.20,[b] 0.04,[e] 0.11,[cc] 0.05[ee]	30.3,[c] 32.0,[h] 29,[d] 26[dd]	e, a
Cumene	83.0[a]	105*[d]		0.12,[d] 0.09[ee]	22[d]	e
Ethylbenzene	82.8[a]	105*[d]		0.18,[d] 0.12[ee]	31[d]	e
Toluene	82.3[a]	105*[d]	0.53[vv]	0.12,[ee] 0.17,[d] 0.23[cc]	41.2,[c] 38.4,[h] 34,[d] 26[ff]	s, e
o-Xylene	81.9[a]	105*[d]		0.3,[e] 0.19,[d] 0.15[ee]	32.2[d]	e
Triphenylmethane	81.3[a]			0.23[e]		
m-Xylene	80.1[a]	104*[d]		0.3,[e] 0.17,[d] 0.13[ee]	30.8[d]	e
p-Xylene	80.1[a]	104*[d]		0.3,[e] 0.40,[d] 0.415[cc]	34.5,[h] 30,[d] 28.0[dd]	e
Mesitylene	80.1[a]	104*[d]	0.63[vv]	0.17[d]	44.2,[c] 40.4,[h] 36.5[d]	e
Durene	79.8[a]			0.5[e]		
Hexamethylbenzene				0.037[cc]	6.0[dd]	
Fluorene	67.6,[g] 67.9[r]	95.7,[tt] 95.0[d]	0.32*,[1] 0.31,[j] 0.37,[b] 0.10[u]	0.68*,[1] 0.54,[k] 0.80,[d] 0.52[u]	10[d]	
Triphenylene	66.6,[1] 67.4[r]	83.2,[d] 83.4*,[r] 83.5[rr]	0.89,[m] 0.95,[j] 0.85,[uu] 0.86[vv]	0.06,[b] 0.08,[d] 0.09,[m] 0.15[uu]	31,[n] 36.6[d]	s
Biphenyl	65.5,[f] 65.5,[ss] 69.5,[oo] 65.2[bb]	~99,[d] 95.9[o]	0.51,[u] 0.81[vv]	0.18,[d] 0.23,[e] 0.12[u]	16.0[d]	

continued

297

TABLE X continued

Compound	E_T	E_S	Φ_{ic}	Φ_f	τ_f	Complications
1,3,5-Triphenylbenzene	65.0,[ss] 64.8[f]	~89[d]		0.27[d]	42.6[d]	
Phenanthrene	62.2,[g] 61.8,[f] 61.8,[e] 62.2[r]	81.0,[r] 83.1,[e] 82.9,[tt] 82.5[r]	0.80*,[m] 0.76,[j] 0.88,[uu] 0.70,[p] 0.85,*[l] 0.82[vv]	0.15,[e] 0.1*,[k] 0.12,[uu] 0.2*,[e] 0.13*[m]	56.0,[h] 63[q]	s
Naphthalene	60.9,[bb] 60.5,[z] 60.9,[r] 60.9,[g,f] 60.8,[e] 60.6[ss]	90.8,[e] 90.7,[r] 90.5,[rr] 90.2[d]	0.80*,[l] 0.71*,[m] 0.40,[j] 0.82[l] 0.68[vv]	0.19*,[e] 0.28,[l] 0.21*,[m] 0.40,[b] 0.23,[d] 0.38,[cc] 0.10,[k] 0.12*[k]	103,[dd] 100,[n] 96,[d] 110[gg]	e, a, s
2-Methylnaphthalene	60.9[ss]	89.4[d]	0.51[j]	0.32[d]	59[d]	s
1-Methylnaphthalene	59.7,[a] 60.1[o]	90.6,[d] 90.0[o]	0.48[j]	0.25[d]	67,[d] 77[gg]	s
2,3-Dimethylnaphthalene		89.4[d]		0.38[d]	78[d]	s
2,6-Dimethylnaphthalene		88.0[d]		0.45[d]	38[d]	s
Acenaphthene	59.5,[a] 59.3[ss]	89.1,[d] 85.2[rr]	0.45*,[m] 0.47,[j] 0.58*,[l] 0.46[vv]	0.60,[d] 0.31,[k] 0.39*[m]	46[d]	
p-Terphenyl	58.7,[aa] 58.3,[f] 58.4[ss]	~91.1[d]	0.11,[u] 0.07[uu]	0.93,[d] 1.0,[e] 0.93[uu]	0.95,[d] 2.2,[jj] 1.3[h]	
p-Quaterphenyl	43[f] (theory)	~86.1[d]	0.13[u]	0.89[d]	0.8[d]	
9,10-Dihydrophenanthrene				0.48[u]		
1,2,3,4,5,6,7,8-Tetra-benzanthracene	58.7[r]	74.8[tt]				
1,2,6,7-Dibenzpyrene	58.4[tt]	76.9[tt]				
Picene	57.4[r]	76.1[tt]				
3,4-Benzphenanthrene	57.2[r]	76.4,[rr] 76.9[r]				
Chrysene	57.2,[r,bb] 56.7[f]	79.2,[d] 78.6,[rr] 79.5*[r]	0.81,[u] 0.85*,[l] 0.82*,[m] 0.67[j]	0.14,[d] 0.17*[m]	40.2,[n] 44.7[d]	s
2,2'-Binaphthyl	56.1,[l] 56.0[aa]	85.6[d]		0.41[d]		
1,1'-Binaphthyl		87.2[d]		0.77[d]	3.0[d]	
Coronene	55.5,[r] 55.6,[e] 54.5[tt]	69.8,[e] 67.2*,[r] 66.8[tt]	0.56*[l]	0.3[e] (CHCl$_3$) 0.23*[l]		

299

1,2-Benzpyrene	53.2,[r] 53.0,[aa] 53.0,[aa] 52.8[z]	76.0*,[r] 78.2[tt] ~80[tt]			e
Fluoranthene	52.9[r]	72.5[r]		80[n]	
1,2,7,8-Dibenzanthracene	52.2,[1] 52.4[r,f]	72.8[r]	0.89,[J] 1.03[p]	27.1[n]	
1,2,5,6-Dibenzanthracene	50.8[l,z]	76.5[tt]		43.0[n]	s, e
1,2,3,4-Dibenzanthracene	48.2,[1] 48.7,[g] 47.0,[f] 48.5[z]	76.9,[d] 77.1[t]	0.08,[u] 0.38*,[1] 0.27*,[m] 0.10[ll]	435,[ee] 480[ee]	s, e
Pyrene	47.2,[1] 47.3,[f] 47.3[r]	74.5,[r] 73.8[rr]	0.79,[u] 0.82*,[1] 0.77,[1] 0.55,[x] 0.87[ll]	0.32,[d] 0.65*,[v] 0.72*,[m] 0.68[ee]	s, e
1,2-Benzanthracene				0.18,[u] 0.22*,[1] 0.20,[x] 0.19[ee]	
	46.5[r]	70.5,[r] 70.3[tt]		44.1[ee]	e
1,12-Benzperylene			0.03*[pp]		
9,10-Dimethylanthracene	44.4[a]			11.0*,[kk] 16.2[nn] 0.89*[pp]	
9,10-Dimenthyl-1,2-benzanthracene				26.5[ee]	
Anthracene	42.5,[mm] 42.7,[e,r] 42.0,[f] 42.1[bb]	75.4,[d] 75.7,[e] 75.2*,[r] 75.5,[rr] 75.2[mm]	0.58,[p] 0.75,[b] 0.72*,[1] 0.6[w]	0.25,[b] 0.28,[e] 0.36,[d] 0.35,[e] 0.30*,[k] 0.24,[hh] 0.30*,[m] 0.33,[p] 0.24[cc]	11.1,[q] 4.9,[d] 4.0[ee]	e, d, a, s, ut
9-Phenylanthracene		72.2*[d]	0.505*,[v] 0.47[pp]	0.49,[pp] 0.49,[d] 0.45,[v] 0.45[kk]	5.1,[kk] 6.5[d]	
9-Methylanthracene	40.6,[nn] 41.4[z]	73.4[d]	0.67[pp]	0.35,[d] 0.29,[kk] 0.33[pp]	4.6,[d] 5.2[kk]	d, ut, e
9,10-Diphenylanthracene		70.1[d]	0.03[pp]	0.89,[pp] 1.00,[d] 1.0,[e] 0.85[ee]	9.35,[d] 7.3,[ee] 8.9,[ll] 10.0[h]	s
3,4,9,10-Dibenzpyrene	40.2[r]	66.1[tt]		0.84[kk]	6.8,[kk]	s, e
3,4-Benzpyrene	41.9,[1] 41.9[r]	71.1*,[r] 71.0[tt]			36[n]	d, ut
2-Methylanthracene	40.6[nn]			0.32[e]		e
Perylene	36.0[r]	64.9,[d] 65.1,[tt] 64.7[r]	0.06[s]	0.89,[ee] 0.89,[s] 0.94,[d] 0.98[e]	6.4,[h] 7.8,[n] 6.4,[d] 4.9[ee]	
1,6-Diphenylhexatriene		73.9[d]		0.78,[d] 0.80[ee]	12.4,[d] 6.9,[ee] 15.2[h]	

continued

TABLE X continued

3,4,8,9-Dibenzpyrene	34.5[r]	64.2*,[r] 63.4[tt]		6.4[n]
Anthanthrene	33.9[r]	66.5,[r] 66.0[tt]	0.23[s]	8.1[n] e, s
Azulene	33–37[b]	80.6*[d]	0.24[d]	1.4[d]
Tetracene (naphthacene)	29.3,[f] 29.2,[e]	60.6*,[d] 59.6,[e]	0.63[t]	0.16,[t] 0.21[d] 6.5,[n] 6.4,[d] d
	30.0[r]	60.5*,[r] 60.4[rr]		5.5,[ee] 9.2[h]
Pentacene	22.0,[e] 22.9[r]	49.5,[e] 48.9[rr]		d

[a] W. G. Herkstroeter, A. A. Lamola, G. S. Hammond, and S. L. Murov, Unpublished.
[b] A. A. Lamola, Private communication to G. S. Hammond.
[c] O. E. Wagner, L. G. Christophorou, and J. G. Carter, *Chem. Phys. Lett.*, **4**, 224 (1969).
[d] I. Berlman, *Handbook of Fluorescence Spectra of Aromatic Molecules*, Academic Press, London, 1965.
[e] E. J. Bowen, *Advances in Photochemistry*, Vol. 1, W. A. Noyes, Jr, G. S. Hammond, and J. N. Pitts, Jr., Eds., Wiley-Interscience, New York, 1963, p. 23.
[f] J. S. Brinen and M. K. Orloff, *Chem. Phys. Lett.*, **1**, 276 (1967).
[g] W. Herkstroeter, A. A. Lamola, and G. S. Hammond, *J. Amer. Chem. Soc.*, **86**, 4537 (1964).
[h] C. D. Amata, M. Burton, W. P. Helman, P. K. Ludwig, and S. A. Rodemeyer, *J. Chem. Phys.*, **48**, 2374 (1968).
[i] A. R. Horrocks and F. Wilkinson, *Proc. Roy. Soc., Ser. A*, **306**, 257 (1968).
[j] G. S. Hammond and A. A. Lamola, *J. Chem. Phys.*, **43**, 2129 (1965).
[k] G. Weber and F. W. J. Teale, *Trans. Faraday Soc.*, **53**, 646 (1957).
[l] W. G. Herkstroeter and G. S. Hammond, *J. Amer. Chem. Soc.*, **88**, 4769 (1966).
[m] C. A. Parker and T. A. Joyce, *Trans. Faraday Soc.*, **62**, 2785 (1966).
[n] S. L. Murov, Unpublished.
[o] V. L. Ermolaev, *Sov. Phys. Usp.*, **80**, 333 (1963).
[p] G. Porter and P. Bowers, *Proc. Roy. Soc., Ser. A*, **299**, 348 (1967).
[q] G. L. Powell, *J. Chem. Phys.*, **47**, 95 (1967).
[r] R. Nurmukhametov, *Russ. Chem. Rev.*, **35**, 473 (1966).
[s] B. Stevens and B. E. Algar, *Chem. Phys. Lett.*, **1**, 219 (1967).
[t] B. Stevens and B. E. Algar, *Ibid*, **1**, 58 (1967).
[u] W. Heinzelmann and H. Labhart, *Ibid*, **4**, 20 (1969).

[x] H. Labhart, *Helv. Chem. Acta*, **47**, 2279 (1964).
[y] A. R. Horrocks, A. Kearvell, K. Tickle, and F. Wilkinson, *Trans. Faraday Soc.*, **62**, 3393 (1966).
[z] D. F. Evans, *J. Chem. Soc.*, **1957**, 1351.
[aa] E. Clar and M. Zander, *Ber.*, **89**, 749 (1956).
[bb] G. N. Lewis and M. Kasha, *J. Amer. Chem. Soc.*, **66**, 2100 (1944).
[cc] E. J. Bowen and A. H. Williams, *Trans. Faraday Soc.*, **35**, 44 (1939).
[dd] T. V. Ivanova, P. I. Kudryashov, and B. Sveshnikov, *Sov. Phys. Doklady*, 6, 407 (1961).
[ee] J. B. Birks and I. H. Munro, in *Progress in Reaction Kinetics*, Vol. 4, G. Porter, Ed., p. 281. Pergamon Press, Oxford, 1967.
[ff] T. V. Ivanova, G. A. Mokeeva and B. Sveshnikov, *Opt. Spectry.*, **11**, 325 (1961).
[v] C. A. Parker and C. G. Hatchard, *Trans. Faraday Soc.*, **59**, 284 (1963).
[w] P. G. Bowers and G. Porter, *Proc. Roy. Soc., Ser. A*, **296**, 435 (1967).
[gg] N. Mataga, N. Tomura, and H. Nishimura, *Mol. Phys.*, **9**, 367 (1965).
[hh] B. L. Van Duuren, *Chem. Rev.*, **63**, 625 (1962).
[ii] W. R. Ware, *J. Phys. Chem.*, **66**, 455 (1962).
[jj] M. Burton, P. Ludwig, and J. T. D'Allessio, *Acta Phys. Polon.*, **26**, 517 (1964).
[kk] A. S. Cherkasov, V. A. Molchanov, T. M. Vember, and K. G. Voldaiking, *Sov. Phys. Doklady*, **1**, 427 (1956), values in ethanol.
[ll] By the method of Hammond and Lamola,[j] F. D. Lewis and W. H. Saunders, Jr., *J. Amer. Chem. Soc.*, **90**, 7033 (1968).
[mm] S. P. McGlynn, T. Azumi, and M. Kasha, *J. Chem. Phys.*, **40**, 507 (1964).
[nn] R. S. H. Liu and J. R. Edman, *J. Amer. Chem. Soc.*, **91**, 1492 (1969).
[oo] P. J. Wagner, *Ibid.*, **89**, 2820 (1967).
[pp] C. A. Parker and T. A. Joyce, *Chem. Commun.*, **1967**, 744.
[qq] R. Cundall, *Ibid.*, **1969**, 116.
[rr] H. B. Klevens and J. R. Platt, *J. Chem. Phys.*, **17**, 470 (1949).
[ss] A. P. Marchetti and D. R. Kearns, *J. Amer. Chem. Soc.*, **89**, 768 (1967).
[tt] E. Clar, *Polycyclic Hydrocarbons*, Academic Press, New York, 1964.
[uu] R. E. Kellogg and R. G. Bennett, *J. Chem. Phys.*, **41**, 3042 (1964).
[vv] K. Sandros, *Acta Chem. Scand.*, **23**, 2815 (1969).

TABLE XI
Aromatic Carbonyl Compounds

Explanation of Table: E_T = triplet energy in kcal/mole from O—O phosphorescence band. Φ_{ic} = intersystem crossing yield from sensitized olefin isomerization method. τ_p = phosphorescence lifetime in a rigid glass at $-196°$ (sec). Repetition of the data compiled by Arnold[63] has in general been avoided.

Compound	E_T	Φ_{ic}	τ_p	Complications
Propiophenone	74.6,[c] 74.8[d]		0.0038[d]	ha, cs, ox, sch
Xanthone	74.2,[c] 70.9[d]		0.02[d]	ox, (sq)
Acetophenone	73.6,[c] 74.1,[a] 74.3,[k] 73.7,[d] 74.3[e]	1.0[b]	0.0023,[d] 0.004[a] 0.008[h]	ha, cs, sch, ox
p-Methylacetophenone	72.8[a]		0.084[a]	
m-Methylacetophenone	72.5[a]		0.074[a]	
m-Methoxyacetophenone	72.4[a]		0.25[a]	sq
p-Methylbutyrophenone	72.2,[e] 73.7[m]		0.009[l,m]	II
p-Chlorobutyrophenone	72.0,[e] 72.4[m]		0.045[m]	II
Anthrone	72.0[d]		0.0015[d]	
Benzaldehyde	71.9,[c] 71.5,[d] 71.9,[e] 72.1[k]		0.0015[d]	ox, sch, ha, cs
Butyrophenone	71.9,[e] 74.7[m]		0.002,[l] 0.005[m]	II
3,4-Dimethylacetophenone	71.5[a]		0.17[a]	sq
p-Methoxyacetophenone	71.5,[a] 70.7[e]		0.26[a]	sq
3,5-Dimethylacetophenone	71.3[a]		0.11[a]	
p-Bromoacetophenone	71.2[e]			dec
3,4,5-Trimethylacetophenone	70.8[a]		0.20[a]	
p-Chlorobenzaldehyde	70.8[d]			
m-Iodobenzaldehyde	70.8[d]		0.00065[d]	(dec)
4,4'-Dimethoxybenzophenone	70.3[d]		0.065[l]	ox
p-Hydroxybutyrophenone	69.8[e]			
o-Chlorobenzaldehyde	69.6[d]			
p-Acetamidobutyrophenone	69.4[e]			
2,4-Dibromoacetophenone	69.2[e]			(dec)
Benzophenone	68.5,[c] 69.3,[d]	1.00[b]	0.0047,[d] 0.05–0.08[l]	ha, cs, ox, sch

3,4-methylenedioxyacetophenone	65.8[a]		sq
Thioxanthone	65.5[c]		sq
Benzylideneanthrone	62.6[f]		
Anthraquinone	62.4,[c] 62.8[d]	1.00[g]	ha, ox, sch
α-Naphthoflavone	62.2[c]	0.90[b]	
Flavone	62.0[c]		
Michler's ketone	61.0,[c] 62.0[k]	1.00[b]	
4-Acetylbiphenyl	60.6[f]		(sq)
m-Nitroacetophenone	60.1,[f] 58.1[k]		
2-Naphthylphenyl ketone	59.6[c]		ox, (sq)
2-Naphthaldehyde	59.5,[c] 58.4[e]		ox, (sq)
2-Acetonaphthone	59.3[c]		(sq)
3-Acetophenanthrene	59.0[j]		
9-Acetophenanthrene	58.4[j]		
1-Naphthylphenyl ketone	57.5[c]		(sq)
1-Acetonaphthone	56.4,[c] 57.8[k]	0.84[b]	(sq)
1-Naphthaldehyde	56.3,[c] 56.9[k]		ox, (sq)
Fluorenone	53.3[c]	0.93[b]	s, ox
1-Phenyl-1,2-propanedione	53.2[f]		
		0.37[a]	
		0.27[l]	
		0.95[h]	

[a] N. C. Yang, D. S. McClure, S. L. Murov, J. J. Houser, and R. Dusenbury, *J. Amer. Chem. Soc.*, **89**, 5466 (1967).
[b] A. A. Lamola and G. S. Hammond, *J. Chem. Phys.*, **43**, 2129 (1965).
[c] W. G. Herkstroeter, A. A. Lamola, and G. S. Hammond, *J. Amer. Chem. Soc.*, **86**, 4537 (1964).
[d] V. L. Ermolaev, *Sov. Phys. Usp.*, **80**, 333 (1963).
[e] D. R. Kearns and W. A. Case, *J. Amer. Chem. Soc.*, **88**, 5087 (1966).
[f] W. G. Herkstroeter, A. A. Lamola, S. L. Murov, and G. S. Hammond, Unpublished data.
[g] F. D. Lewis and W. H. Saunders, Jr., *J. Amer. Chem. Soc.*, **90**, 7033 (1968).
[h] D. S. McClure, *J. Chem. Phys.*, **17**, 905 (1949).
[i] J. N. Pitts, H. W. Johnson, and T. Kuwana, *J. Phys. Chem.*, **66**, 2456 (1962).
[j] A. P. Marchetti and D. R. Kearns, *J. Amer. Chem. Soc.*, **89**, 768 (1967).
[k] G. N. Lewis and M. Kasha, *Ibid*, **66**, 2108 (1944).
[l] J. Calvert and J. Pitts, Jr., *Photochemistry*, Wiley, New York, 1966, p. 383.
[m] J. N. Pitts, D. R. Burley, J. C. Mani, and A. D. Broadbent, *J. Amer. Chem. Soc.*, **90**, 5902 (1968).

TABLE XII
Other Sensitizers

Explanation of Table: E_T = triplet energy in kcal/mole. E_S = singlet energy in kcal/mole from lowest energy absorption band; * in the E_S column indicates highest energy fluorescence emission. Φ_{ic} = efficiency of intersystem crossing in nonpolar solvents. Φ_f = fluorescence quantum yield in degassed nonpolar solvent; * in the Φ_f column indicates ethanol solvent. τ_f = fluorescence lifetime in nanoseconds in degassed nonpolar solvent; * in the τ_f column indicates ethanol solvent. τ_p = phosphorescence lifetime in seconds at $-196°$.

Compound	E_T	E_S	Φ_{ic}	Φ_f	τ_f	τ_p	Complications
Phenol	81.5[s]	102[c]		0.08,[c] 0.19*[o]	2.1,[c] 4.7*[o]	2.9[z]	
Diphenyl ether	80.7[a]	101[c]		0.03[c]	2.0[c]		
Anisole	80.7[a]	102[c]		0.29[c]	8.3[c]	3.0[z]	
Acetone	80[k]	88.8[l]	1.0[k]	0.001,[l] 0.01[k]	2.0[m]	0.0004[k]	ox, s, ha, sch
1,4-Dichlorobenzene	78.3,[a] 74.4[s], 80.2[n]						
Benzonitrile	77.0[s]						
1,4-Dibromobenzene	79.4[n]					0.00033[z]	
Aniline	76.6,[a] 76.6[s]	94.5[c]		0.08[c]	3.9[c]	4.7[z]	
Hydroquinone	74.8[a]				2.0*[o]		
Diphenylamine	71.0,[b] 72.1[s]	89[b]	0.38[l]				
Carbazole	70.1,[g] 70.4,[g] 70.1[s]	84,[b] 85.7*,[c] 84.4[g]	0.36[l]	0.38[c]	16.1[c]	7.6[g]	
Diphenyleneoxide	70.1[e]						
Triphenylamine	70.1,[e] 70.1[g]	84.4,[c] 83.0[g]	0.88[l]	0.03[c]		0.7[g]	
Dibenzothiophene	69.7[e]						
Quinoline	62.1,[s] 62.1[g]	91.5,[c] 91.3[g]	0.31[l]				
2-Chloronaphthalene	60.3[n]	89.0[c]			4.2[c]	0.47[z]	s
2-Bromonaphthalene	60.3[n]					0.021[z]	
2-Iodonaphthalene	60.3[n]					0.0025[z]	
2-Naphthol	60.3,[n] 60.4[s]	86.0[c]		0.32[c]	13.3[c]	1.30[z]	
1-Fluoronaphthalene	60.0[a]		0.63[l]				
1-Chloronaphthalene	59.2,[g] 58.6[n]	89.6[g]				0.29[g]	

Compound							
1-Bromonaphthalene	59.1,[g] 59.6[n]	89.5[g]					dec
1-Iodonaphthalene	58.6,[g] 59.2[n]					0.018[g]	
	59.9,[a] 60.3[s]					0.002[g]	
Nitrobenzene	58.3,[a] 58.6[s]	89.1[c]	0.27[1]		10.6[c]	1.9[z]	
1-Naphthol	54.3,[1] 57.5[s]	82.3[c]	0.15[1]		6.0[c]	1.5[z]	
1-Naphthylamine	55.8[e]						
5,12-Naphthacenequinone	54.9,[e] 56.4[s]	65[b]	0.98[b]	0.0027[w]	10[t]	0.0078[w,x]	s, en, ha, ox
Biacetyl						0.0046[ξ,x]	
Acetylpropionyl	54.7,[e] 53.2[s]		0.92[1]	0.0013[w]		0.0113[w,x]	ox, ha, sch
	53.7,[a] 53.0[a]						ha, ox, a
Benzil	50.0[a]						
Benzoquinone	49.2[q]	56.0[q]			5.5,[o,y] 2.0[r,p]		ha
Acridine orange	45.3[f]	73.5[c]	0.76[j]		0.9*,[c] 7.0[o,p]		ha
Acridine	43.8[s]						
Phenazine	43.0[q]	49.3[q]	0.71[u]		6.16*[o,y]		dec
Rhodamine B	43.0,[a] 42.6[s]			0.19,[h,u] 0.15[o,p] 4.7[o,p]			ut
Eosin	41.7,[a] 40.7[d]		0.48[d]	0.55,[c] 0.65,[o]	8.5,[c] 9.0,[d]		ut, s
1,5-Dichloroanthracene	40.2,[v] 40.5[a]	70.3*[c]		0.48*[o]	9.6,[c] 7.2*[o]		
9,10-Dichloroanthracene				0.10,[d] 0.095*[o]	1.8,[d] 1.9*[o]		dec, ut
9,10-Dibromoanthracene	40.2[v]	70.2[v]					
Thiobenzophenone	40.3[s]						
Crystal violet	38.9[s]						
Fluorescein	47.2[q]	55.0[q]	0.05[u]	0.66*[o]	5.7,*[o] 6.7,*[o]		
					4.3*[o]		

[a] W. Herkstroeter, A. A. Lamola, G. S. Hammond, and S. L. Murov, Unpublished values.
[b] A. A. Lamola, Unpublished data.
[c] I. Berlman, *Handbook of Fluorescence Spectra of Aromatic Molecules*, Academic Press, London, 1965.
[d] R. S. H. Liu and J. R. Edman, *J. Amer. Chem. Soc.*, **91**, 1492 (1969).
[e] W. G. Herkstroeter, A. A. Lamola, and G. S. Hammond, *Ibid.*, **86**, 4537 (1964).
[f] D. F. Evans, *J. Chem. Soc.*, **1957**, 1351.
[g] V. L. Ermolaev, *Sov. Phys. Usp.*, **80**, 333 (1963).

continued

TABLE XII continued

[h] G. Weber and F. W. J. Teale, *Trans. Faraday Soc.*, **53**, 646 (1957).
[i] A. A. Lamola and G. S. Hammond, *J. Chem. Phys.*, **43**, 2129 (1965).
[j] F. Wilkinson and J. T. Dubois, *Ibid.*, **48**, 2651 (1968).
[k] R. F. Borkman and D. R. Kearns, *Ibid.*, **44**, 945 (1966).
[l] M. O'Sullivan and A. C. Testa, *J. Amer. Chem. Soc.*, **90**, 6245 (1968).
[m] In aerated hexane: F. Wilkinson and J. T. Dubois, *J. Chem. Phys.*, **39**, 377 (1963).
[n] A. P. Marchetti and D. R. Kearns, *J. Amer. Chem. Soc.*, **89**, 768 (1967).
[o] J. B. Birks and I. H. Munro, in *Progress in Reaction Kinetics*, Vol. 4, G. Porter, Ed., p. 281, Pergamon Press, Oxford, 1967.
[p] In water; not degassed.
[q] R. W. Chambers and D. R. Kearns, *Photochem. Photobiol.*, **10**, 215 (1969), in MeOH–EtOH.
[r] R. F. Chen, G. G. Vorek, and N. Alexander, *Science*, **156**, 949 (1967).
[s] In EPA glass; O—O phosphorescence band: G. N. Lewis and M. Kasha, *J. Amer. Chem. Soc.*, **66**, 2108 (1944).
[t] N. J. Turro and R. Engel, *Ibid.*, **91**, 7113 (1969).
[u] G. Porter and P. Bowers, *Proc. Royal Soc., Ser. A*, **299**, 348 (1967), in aq. OH⁻.
[v] S. P. McGlynn, T. Azumi, and M. Kasha, *J. Chem. Phys.*, **40**, 507 (1964).
[w] M. Almgren, *Photochem. Photobiol.*, **6**, 829 (1967).
[x] At room temperature.
[y] Not degassed.
[z] D. S. McClure, *J. Chem. Phys.* **17**, 905 (1949).

REFERENCES

1. P. J. Wagner and G. S. Hammond, in *Advances in Photochemistry*, Vol. 5, W. A. Noyes, Jr., G. S. Hammond, and J. N. Pitts, Jr., Eds., Wiley-Interscience, New York, 1968, p. 21.
2. N. J. Turro, in *Energy Transfer and Organic Photochemistry (Technique of Organic Chemistry*, Vol. 14), P. A. Leermakers and A. Weissberger, Eds., Wiley-Interscience, New York, 1969, p. 133.
3. N. J. Turro, J. C. Dalton, and D. S. Weiss, in *Organic Photochemistry*, Vol. 2, O. L. Chapman, Ed., Dekker, New York, 1969, p. 1.
4. W. L. Dilling, *Chem. Rev.*, **69**, 845 (1969).
5. N. J. Turro, *Photochem, Photobiol.*, **9**, 555 (1969).
6. W. G. Dauben and W. T. Wipke, *Pure Appl. Chem.*, **9**, 539 (1964).
7. J. S. Swenton, *J. Chem. Educ.*, **46**, 7 (1969).
8. G. J. Fonken, in *Organic Photochemistry*, Vol. 1, O. L. Chapman, Ed., Dekker, New York, 1967, p. 197.
9. A. A. Lamola, in *Energy Transfer and Organic Photochemistry (Technique of Organic Chemistry*, Vol. 14), P. A. Leermakers and A. Weissberger, Eds., Wiley-Interscience, New York, 1969, p. 17.
10. W. A. Noyes, Jr., G. B. Porter, and J. E. Jolly, *Chem. Rev.*, **56**, 49 (1956).
11. J. G. Calvert and J. N. Pitts, Jr., *Photochemistry*, Wiley, New York, 1966, p. 393.
12. J. T. Przybytek, S. P. Singh, and J. Kagen, *Chem. Commun.*, **1969**, 1224.
13. W. A. Noyes, Jr., W. A. Mulac, and M. S. Matheson, *J. Chem. Phys.*, **36**, 880 (1962).
14. W. G. Herkstroeter, A. A. Lamola, and G. S. Hammond, *J. Amer. Chem. Soc.*, **86**, 4537 (1964).
15. S. A. Greenberg and L. S. Forster, *Ibid.*, **83**, 4339 (1961).
16. N. C. Yang and E. D. Feit, *Ibid.*, **90**, 504 (1968).
17. Reference 11, p. 298.
18. G. Quinkert, K. Opitz, W. Wiersdorff, and J. Weinlich, *Tetrahedron Lett.*, **1963**, 1863.
19. P. S. Engel, *J. Amer. Chem. Soc.*, **92**, 6074 (1970).
20. R. H. Eastman and W. Robbins, *Ibid.*, **92**, 6076 (1970).
21. Y. Ogata, K. Takagi, and Y. Izawa, *Tetrahedron*, **24**, 1617 (1968).
22. N. C. Yang, S. P. Elliott, and B. Kim, *J. Amer. Chem. Soc.*, **91**, 7551 (1969).
23. P. J. Wagner and P. A. Kelso, *Tetrahedron Lett.*, **1969**, 4151.
24. J. N. Pitts, Jr., D. R. Burley, J. C. Mani, and A. D. Broadbent, *J. Amer. Chem. Soc.*, **90**, 5900, 5902 (1968).
25. E. J. Baum, J. K. S. Wan, and J. N. Pitts, Jr., *Ibid.*, **88**, 2652 (1966).
26. F. D. Lewis and N. J. Turro, *Ibid.*, **92**, 311 (1970).
27. P. J. Wagner and G. S. Hammond, *Ibid.*, **88**, 1245 (1966).
28. T. J. Dougherty, *Ibid.*, **87**, 4011 (1965).
29. N. C. Yang and S. P. Elliott, *Ibid.*, **90**, 4194 (1968).
30. N. C. Yang and C. Rivas, *Ibid.*, **83**, 2213 (1961).
31. E. F. Zwicker, L. I. Grossweiner, and N. C. Yang, *Ibid.*, **85**, 2671 (1963).
32. A. Beckett and G. Porter, *Trans. Faraday Soc.*, **59**, 2051 (1963).
33. E. F. Ullman and K. R. Huffman, *Tetrahedron Lett.*, **1965**, 1863.
34. F. Neidel and W. Brodowski, *Chem. Ber.*, **101**, 1398 (1968).
35. W. A. Henderson, Jr., and E. F. Ullman, *J. Amer. Chem. Soc.*, **87**, 5424 (1965).

36. A. A. Lamola and L. J. Sharp, *J. Phys. Chem.*, **70**, 2634 (1966).
37. G. S. Hammond, N. J. Turro, and P. A. Leermakers, *Ibid.*, **66**, 1144 (1962).
38. W. G. Herkstroeter, L. B. Jones, and G. S. Hammond, *J. Amer. Chem. Soc.*, **88**, 4777 (1966).
39. H. A. Morrison and B. H. Migdalf, *J. Org. Chem.*, **30**, 3996 (1965).
40. J. Lemaire, *J. Phys. Chem.*, **71**, 2653 (1967).
41. J. Lemaire, M. Niclause, X. Deglise, J. Andre, G. Persson, and M. Bouchy, *C. R. Acad. Sci., Paris, Ser. C*, **267**, 33 (1968).
42. D. Phillips, J. Lemaire, C. S. Burton, and W. A. Noyes, Jr., in *Advances in Photochemistry*, Vol. 5, W. A. Noyes, Jr., G. S. Hammond, and J. N. Pitts, Jr., Eds., Wiley-Interscience, New York, 1968, p. 355.
43. B. M. Monroe, *Ibid.*, Vol. 8, p. 84.
44. E. J. Baum and J. N. Pitts, Jr., *J. Phys. Chem.*, **70**, 2066 (1966).
45. D. R. Kearns and W. A. Case, *J. Amer. Chem. Soc.*, **88**, 5087 (1966).
46. R. K. Sharma and N. Kharasch, *Angew. Chem.*, **80**, 69 (1968); *Ibid., Angew. Chem. Int. Ed. Engl.*, **7**, 36 (1968).
47. A. Marchetti and D. R. Kearns, *J. Amer. Chem. Soc.*, **89**, 5335 (1967).
48. O. L. Chapman and G. Wampfler, *Ibid.*, **91**, 5390 (1969).
49. N. C. Yang, D. S. McClure, S. L. Murov, J. J. Houser, and R. L. Dusenbery, *Ibid.*, **89**, 5466 (1967); S. Murov, Ph.D. Thesis, University of Chicago, 1967.
50. R. B. Cundall and A. J. R. Voss, *Chem. Commun.*, **1969**, 116.
51. C. DeBoer, *J. Amer. Chem. Soc.*, **91**, 1855 (1969).
52. M. O'Sullivan and A. C. Testa, *Ibid.*, **90**, 6245 (1968).
53. G. D. Renkes and F. S. Wettack, *Ibid.*, **91**, 7514 (1969).
54. J. B. Birks, *Nature*, **214**, 1187 (1967).
55. T. Förster, *Angew Chem.*, **81**, 364 (1969); *Ibid., Angew Chem. Int. Ed. Engl.*, **8**, 333 (1969).
56. J. B. Birks, D. J. Dyson, and T. A. King, *Proc. Royal Soc., Ser A*, **277**, 270 (1964).
57. J. B. Birks, D. J. Dyson, and I. M. Munro, *Ibid.*, **275**, 575 (1963).
58. D. J. Trecker, in *Organic Photochemistry*, Vol. 2, O. L. Chapman, Ed., Dekker, New York, 1969, p. 63.
59. E. J. Bowen, in *Advances in Photochemistry*, Vol. 1, W. A. Noyes, Jr., G. S. Hammond, and J. N. Pitts, Jr., Eds., Wiley-Interscience, New York, 1963, p. 23.
60. E. J. Bowen and D. W. Tanner, *Trans. Faraday Soc.*, **51**, 475 (1955).
61. E. Paterno and G. Chieffi, *Gazz. Chim. Ital.*, **39**, 341 (1909).
62. G. Buchi, C. G. Inman, and E. S. Lipinsky, *J. Amer. Chem. Soc.*, **76**, 4327 (1954).
63. D. R. Arnold, in *Advances in Photochemistry*, Vol. 6, W. A. Noyes, Jr., G. S. Hammond, and J. N. Pitts, Jr., Eds., Wiley-Interscience, New York, 1968, p. 301.
64. D. R. Arnold, R. L. Hinman, and A. H. Glick, *Tetrahedron Lett.*, **1964**, 1425.
65. N. C. Yang, *Pure Appl. Chem.*, **9**, 591 (1964).
66. D. R. Arnold and A. H. Glick, *Chem. Commun.*, **1966**, 813; H. Gotthardt, R. Steinmetz, and G. S. Hammond, *J. Org. Chem.*, **33**, 2774 (1968).
67. J. Saltiel, R. M. Coates, and W. G. Dauben, *J. Amer. Chem. Soc.*, **88**, 2745 (1966).
68. For leading references, see N. J. Turro and P. A. Wriede, *Ibid.*, **92**, 320 (1970).
69. N. J. Turro, P. Wriede, J. C. Dalton, D. Arnold, and A. Glick, *Ibid.*, **89**, 3950 (1967).
70. N. J. Turro, P. Wriede, and J. C. Dalton, *Ibid.*, **90**, 3274 (1968).
71. N. J. Turro and P. Wriede, *Ibid.*, **92**, 320 (1970).
72. E. H. Gold and D. Ginsburg, *Angew. Chem. Int. Ed. Engl.* **5**, 246 (1966).
73. P. Dowd, A. Gold, and K. Sachdev, *J. Amer. Chem. Soc.* **92**, 5725 (1970).
73a. T. Kubota, K. Shima, S. Toki, and H. Sakurai, *Chem. Commun.*, **1969**, 1462.

74. H. J. F. Angus and D. Bryce-Smith, *Proc. Chem. Soc.*, **1959**, 326; D. Bryce-Smith and A. Gilbert, *J. Chem. Soc.*, **1965**, 918; J. S. Bradshaw, *J. Org. Chem.*, **31**, 3974 (1966).
75. W. M. Hardham and G. S. Hammond, *J. Amer. Chem. Soc.*, **89**, 3200 (1967).
76. D. Bryce-Smith, *Pure Appl. Chem.*, **16**, 47 (1968).
77. R. Srinivasan and K. A. Hill. *J. Amer. Chem. Soc.*, **87**, 4653 (1965).
78. K. E. Wilzbach and L. Kaplan, *Ibid.*, **88**, 2066 (1966).
79. D. Bryce-Smith, A. Gilbert, and B. H. Orger, *Chem. Commun.*, **1966**, 512.
80. N. C. Perrins and J. P. Simmons, *Ibid.*, **1967**, 999.
81. G. Koltzenberg and K. Kraft, *Tetrahedron Lett.*, **1966**, 389.
82. E. Grovenstein, Jr., T. C. Campebll, and T. Shibata, *J. Org. Chem.*, **34**, 2418 (1969).
82a. R. M. Bowman, T. R. Chamberlain, C. W. Huang, and J. J. McCullough, *J. Amer. Chem. Soc.*, **92**, 4106 (1970).
83. L. Kaplan, J. S. Ritscher, and K. E. Wilzbach, *J. Amer. Chem. Soc.*, **88**, 2881 (1966).
84. D. Bryce-Smith and B. Vickery, *Chem. Ind.* (London), **1961**, 429.
85. R. Lapouyade, A. Castellan, and H. Bouas-Laurent, *Tetrahedron Lett.*, **1969**, 3537.
86. R. S. Davidson, *Chem. Commun.*, **1969**, 1450.
87. C. Pac and H. Sakurai, *Tetrahedron Lett.*, **1969**, 3829.
88. G. F. Vesley, Jr., Ph.D. Thesis, California Institute of Technology, 1968; *Diss. Abstr.*, **29**, 1313-B (1968).
89. C. Walling and M. J. Gibian, *J. Amer. Chem. Soc.*, **87**, 3361 (1965).
90. N. C. Yang and R. Dusenbery, *Mol. Photochem.*, **1**, 159 (1969).
91. N. J. Turro, *Molecular Photochemistry*, Benjamin, New York, 1965, p. 137.
92. H.-D. Scharf and F. Korte, *Chem. Ber.*, **97**, 2425 (1964).
93. P. de Mayo, J. B. Stothers, and W. Templeton, *Can. J. Chem.*, **39**, 488 (1961); P. W. Jolly and P. de Mayo, *Ibid.*, **42**, 170 (1964).
94. D. Elad, *Fortschr. Chem. Forsch.*, **7**, 528 (1967); *Ibid.*, in *Organic Photochemistry*, Vol. 2, O. L. Chapman, Ed., Dekker, New York, 1969, p. 168; J. Sperling, *J. Amer. Chem. Soc.*, **91**, 5389 (1969).
95. D. Elad and R. D. Youssefyeh, *Chem. Commun.*, **1965**, 7.
96. D. A. Plank, *Abstracts*, American Chemical Society Meeting, April 1969, Minneapolis, Minn., Paper #145, Organic Division.
97. B. D. Challand, *Can. J. Chem.*, **47**, 687 (1969).
98. W. F. Smith, Jr., *Tetrahedron*, **25**, 2071 (1969).
99. W. F. Smith, Jr., and B. W. Rossiter, *Ibid.*, **25**, 2059 (1969).
100. J. E. Hodgkins and J. A. King, *J. Amer. Chem. Soc.*, **85**, 2679 (1963).
101. R. W. Binkley, *J. Org. Chem.*, **34**, 3218 (1969).
102. K. Sandros, *Acta Chem. Scand.*, **18**, 2355 (1964).
103. K. Sandros and H. L. J. Bäckström, *Acta Chem. Scand.*, **16**, 958 (1962).
104. P. J. Wagner, *Mol. Photochem.*, **1**, 71 (1969).
105. M. Fischer, *Chem. Ber.*, **100**, 3599 (1967).
106. H. Güsten and E. F. Ullman, *Chem. Commun.*, **1970**, 28.
107. A. Beckett and G. Porter, *Trans. Faraday Soc.*, **59**, 2038 (1963).
108. B. M. Monroe and S. A. Weiner, *J. Amer. Chem. Soc.*, **91**, 450 (1969).
109. B. M. Monroe, *Intra-Sci. Chem. Rep.*, **3**, 283 (1969).
110. S. A. Weiner, E. J. Hamilton, Jr., and B. M. Monroe, *J. Amer. Chem. Soc.*, **91**, 6350 (1969).
111. N. C. Yang and S. Murov, *Ibid.*, **88**, 2852 (1966).

112. M. B. Rubin, *Tetrahedron Lett.*, **1969**, 3931.
113. A. Kellmann and J. T. Dubois, *J. Chem. Phys.*, **42**, 2518 (1965).
114. E. Vander Donckt and G. Porter, *Ibid.*, **46**, 1173 (1967).
115. F. Wilkinson and J. T. Dubois, *Ibid.*, **48**, 2651 (1968).
116. G. S. Hammond and P. A. Leermakers, *J. Phys. Chem.*, **66**, 1148 (1962).
117. R. Tanikaga, *Bull. Chem. Soc. Japan*, **41**, 1664, 2151 (1968).
118. B. M. Monroe and C. C. Wamser, *Mol. Photochem.*, **2**, 213 (1970).
119. T. Okada, M. Kawanisi, H. Nozaki, N. Toshima, and H. Hirai, *Tetrahedron Lett.*, **1969**, 927.
120. A. Padwa, W. Bergmark, and D. Pashayan, *J. Amer. Chem. Soc.*, **91**, 2653 (1969).
121. H. C. A. van Beek, P. M. Heertjes, and F. M. Visscher, *J. Soc. Dyers Colour.*, **81**, 400 (1965).
122. W. F. Smith, Jr., and B. W. Rossiter, *J. Amer. Chem. Soc.*, **89**, 717 (1967).
123. J. G. Pacifici and G. Irick, Jr., *Tetrahedron Lett.*, **1969**, 2207.
124. H. Labhart, *Angew. Chem.*, **79**, 826 (1967); *Ibid.*, *Angew. Chem. Int. Ed.*, **6**, 812 (1967).
125. G. R. Seely, *J. Phys. Chem.*, **69**, 2779 (1965).
126. S. Hashimoto, K. Kano, and J. Sunamoto, *Kogyo Kagaku Zasshi*, **71**, 864 (1968).
127. G. R. Seely, *J. Phys. Chem.*, **73**, 117 (1969).
128. T. Matsuura and K. Ogura, *Bull. Chem. Soc. Japan*, **42**, 2970 (1969).
129. H.-D. Becker, *J. Org. Chem.*, **34**, 2472 (1969).
130. N. C. Yang, D. P. C. Tang, D. M. Thap, and J. S. Sallo, *J. Amer. Chem. Soc.*, **88**, 2851 (1966).
131. C. Moussebois and J. Dale, *J. Chem. Soc.*, C, **1966**, 260; J. Dale and C. Moussebois, *Ibid.*, **1966**, 264.
132. E. W. Duck and J. M. Locke, *Chem. Ind.*, **1965**, 507.
133. K. W. Egger and S. W. Benson, *J. Amer. Chem. Soc.*, **87**, 3311, 3314 (1965).
134. D. Elad, in *Organic Photochemistry*, Vol. 2, O. L. Chapman, Ed., Dekker, New York, 1969, p. 181.
135. G. S. Hammond and J. Saltiel, *J. Amer. Chem. Soc.*, **85**, 2516 (1963).
136. G. S. Hammond, J. Saltiel, A. A. Lamola, N. J. Turro, J. S. Bradshaw, D. O. Cowan, R. C. Counsell, V. Vogt, and C. Dalton, *Ibid.*, **86**, 3197 (1964).
137. W. G. Herkstroeter and G. S. Hammond, *Ibid.*, **88**, 4769 (1966).
138. L. M. Coyne, Ph.D. Thesis, California Institute of Technology, 1967; *Diss. Abstr.*, **28**, 515-B (1967).
139. D. E. Wood, Unpublished work, California Institute of Technology, 1966.
140. H. L. Hyndman, B. M. Monroe, and G. S. Hammond, *J. Amer. Chem. Soc.*, **91**, 2852 (1969).
141. J. Saltiel, L. Metts, and M. Wrighton, *Ibid.*, **91**, 5684 (1969).
142. R. S. H. Liu, Ph.D. Thesis, California Institute of Technology, 1965; *Diss. Abstr.*, **26**, 2478 (1965).
143. G. O. Schenck and R. Steinmetz, *Bull. Soc. Chim. Belg.*, **71**, 781 (1962); G. O. Schenck, *Ind. Eng. Chem.*, **55**(6), 40 (1963).
144. N. C. Yang, J. I. Cohen, and A. Shani, *J. Amer. Chem. Soc.*, **90**, 3264 (1968).
145. J. Saltiel, K. R. Neuberger, and M. Wrighton, *Ibid.*, **91**, 3658 (1969).
146. E. K. C. Lee, H. O. Denschlag, and G. A. Haninger, Jr., *J. Chem. Phys.*, **48**, 4547 (1968).
147. R. A. Caldwell and G. W. Sovocool, *J. Amer. Chem. Soc.*, **90**, 7138 (1968).
148. R. A. Caldwell and S. P. James, *Ibid.*, **91**, 5184 (1969); R. A. Caldwell, *Ibid.*, **92**, 1439 (1970).

149. S. M. Japar, M. Pomerantz, and E. W. Abrahamson, *Chem. Phys. Lett.*, **2**, 137 (1968).
150. R. Hurley and A. C. Testa, *J. Amer. Chem. Soc.*, **92**, 211 (1970).
151. G. S. Hammond, N. J. Turro, and R. S. H. Liu, *J. Org. Chem.*, **28**, 3297 (1963).
152. N. J. Turro, *Molecular Photochemistry*, Benjamin, New York, 1965, p. 181.
153. A. Bylina, *Chem. Phys. Lett.*, **1**, 509 (1968).
154a. W. M. Moore, G. S. Hammond, and R. P. Foss, *J. Amer. Chem. Soc.*, **83**, 2789 (1961).
154b. S. Dym and R. M. Hochstrasser, *J. Chem. Phys.*, **51**, 2458 (1969).
155. F. Wilkinson and J. T. Dubois, *Ibid.*, **39**, 377 (1963).
156. A. R. Horrocks and F. Wilkinson, *Proc. Roy. Soc., Ser. A*, **306**, 257 (1968).
157. M. A. El-Sayed, *Accounts Chem. Res.*, **1**, 8 (1968).
158. L. M. Stephenson, D. G. Whitten, G. F. Vesley, and G. S. Hammond, *J. Amer. Chem. Soc.*, **88**, 3665 (1966); L. M. Stephenson, D. G. Whitten, and G. S. Hammond, in *The Chemistry of Ionization and Excitation*, G. R. A. Johnson and G. Scholes, Eds., Taylor and Francis, London, 1967, p. 35; L. M. Stephenson and G. S. Hammond, *Angew. Chem.*, **81**, 279 (1969); *Angew. Chem. Int. Ed.*, **8**, 261 (1969); *Ibid., Pure Appl. Chem.*, **16**, 125 (1968).
159a. J. R. Fox and G. S. Hammond, *J. Amer. Chem. Soc.*, **86**, 4031 (1964).
159b. S. F. Nelsen and P. D. Bartlett, *Ibid.*, **88**, 143 (1966).
160. P. D. Bartlett and P. S. Engel, *Ibid.*, **90**, 2960 (1968).
161. Th. Förster, *Discussions Faraday Soc.*, **27**, 7 (1959).
162. R. G. Bennett and R. E. Kellogg, in *Progress in Reaction Kinetics*, Vol. 4, G. Porter, Ed., Pergamon, Oxford, 1967, p. 221.
163. J. Saltiel and E. D. Megarity, *J. Amer. Chem. Soc.*, **91**, 1265 (1969).
164. For an excellent discussion, see S. P. McGlynn, F. J. Smith, and G. Cilento, *Photochem. Photobiol.*, **3**, 269 (1964).
165. S. P. McGlynn, T. Azumi, and M. Kasha, *J. Chem. Phys.*, **40**, 507 (1964), and references cited therein.
166. G. Fischer, K. A. Muszkat, and E. Fischer, *Israel J. Chem.*, **6**, 965 (1968).
167. W. R. Ware and J. S. Novros, *J. Phys. Chem.*, **70**, 3246 (1966).
168. C. R. Mullin, M. A. Dillon, and M. Burton, *J. Chem. Phys.*, **40**, 3053 (1964).
169. A. R. Horrocks and F. Wilkinson, *Proc. Roy. Soc., Ser. A*, **306**. 257 (1968), and previous papers.
170. P. J. Wagner, *J. Chem. Phys.*, **45**, 2335 (1966).
171. H. Morrison, H. Curtis, and T. McDowell, *J. Amer. Chem. Soc.*, **88**, 5415 (1966).
172. M. A. El-Sayed, *J. Chem. Phys.*, **41**, 2462 (1964).
173. D. O. Cowan and R. L. Drisko, *J. Amer. Chem. Soc.*, **89**, 3068 (1967).
174. A. Lamola and G. S. Hammond, *J. Chem. Phys.*, **43**, 2129 (1965).
175. C. A. Parker and T. A. Joyce, *Trans. Faraday Soc.*, **62**, 2785 (1966).
176. J. E. Baldwin and J. P. Nelsen, *J. Org. Chem.*, **31**, 336 (1966).
177. A. C. Day and T. R. Wright, *Tetrahedron Lett.*, **1969**, 1067.
178. B. S. Solomon, T. F. Thomas, and C. Steel, *J. Amer. Chem. Soc.*, **90**, 2249 (1968).
179. S. L. Murov, R. S. Cole, and G. S. Hammond, *Ibid.*, **90**, 2957 (1968).
180. S. L. Murov and G. S. Hammond, *J. Phys. Chem.*, **72**, 3797 (1968).
181. B. S. Solomon, C. Steel, and A. Weller, *Chem. Commun.*, **1969**, 927.
182. G. S. Hammond, P. Wyatt, C. D. DeBoer, and N. J. Turro, *J. Amer. Chem. Soc.*, **86**, 2532 (1964).
183. G. S. Hammond and R. S. Cole, *Ibid.*, **87**, 3256 (1965).

184. K. Mislow, M. Axelrod, D. R. Rayner, H. Gotthardt, L. M. Coyne, and G. S. Hammond, *Ibid.*, **87**, 4958 (1965).
185. G. S. Hammond, H. Gotthardt, L. M. Coyne, M. Axelrod, D. R. Rayner, and K. Mislow, *Ibid.*, **87**, 4959 (1965).
186. R. S. Cooke and G. S. Hammond, *Ibid.*, **90**, 2958 (1968).
187. H. Leonhardt and A. Weller, in *Luminescence of Organic and Inorganic Materials*, H. Kallmann and G. Spruch, Eds., Wiley, New York, 1962, p. 74.
188. H. Leonhardt and A. Weller, *Ber. Bunsenges. Physik. Chem.*, **67**, 791 (1963).
189. W. R. Ware and H. P. Richter, *J. Chem. Phys.*, **48**, 1595 (1968).
190. N. Mataga, T. Okada, and K. Ezumi, *J. Mol. Phys.*, **10**, 201 (1966).
191. N. Mataga, T. Okada, and N. Yamamoto, *Chem. Phys. Lett.*, **1**, 119 (1967).
192. This phenomenon appears to be quite general since triethylamine quenches acenaphthene singlets in acetonitrile at the rate of $2 \times 10^9 \, M^{-1} \, \text{sec}^{-1}$ as determined by intensity and lifetime measurements. Quenching of triphenylene fluorescence also occurs at this rate in acetonitrile but is two orders of magnitude slower in isooctane. P. S. Engel, Unpublished results.
193. P. S. Engel and P. D. Bartlett, *J. Amer. Chem. Soc.*, **92**, 5883 (1970).
194. I. I. Abram, G. S. Milne, B. S. Solomon, and C. Steel, *Ibid.* **91**, 1220 (1969).
195. P. S. Engel, *Ibid.*, **91**, 6903 (1969).
196. M. Szwarc, Unpublished results.
197. P. D. Bartlett and N. A. Porter, *J. Amer. Chem. Soc.*, **90**, 5317 (1968).
198. N. A. Porter, Ph.D. Thesis, Harvard University, 1969.
199. E. L. Allred and R. L. Smith, *J. Amer. Chem. Soc.*, **89**, 7133 (1967).
200. E. L. Allred and R. L. Smith, *Ibid.*, **91**, 6766 (1969).
201. S. D. Andrews and A. C. Day, *J. Chem. Soc., B*, **1968**, 1271.
202. F. D. Lewis and W. H. Saunders, Jr., *J. Amer. Chem. Soc.*, **90**, 7033 (1968).
203. F. D. Lewis and J. C. Dalton, *Ibid.*, **91**, 5260 (1969).
204. J. S. Swenton, T. J. Ikeler, and B. H. Williams, *Chem. Commun.*, **1969**, 1263.
205. G. Irick, Jr. and J., Pacifici, *Tetrahedron Lett.*, **1969**, 1303.
206. L. B. Jones and G. S. Hammond, *J. Amer. Chem. Soc.*, **87**, 4219 (1965).
207. C. Luner and M. Szwarc, *J. Chem. Phys.*, **23**, 1978 (1955).
208. V. Krongauz and N. Vasil'ev, *Kinetics Catalysis*, **4**, 55 (1963); English translation of *Kinetika i Kataliz*, **4**, 67 (1963).
209. N. Vasil'ev and V. Krongauz, *Ibid.*, **4**, 177 (1963).
210. C. Walling and M. J. Gibian, *J. Amer. Chem. Soc.*, **87**, 3413 (1965).
211. P. S. Engel, Unpublished result.
212. W. M. Hardham and G. S. Hammond, *J. Amer. Chem. Soc.*, **89**, 3200 (1967).
213. H. E. Zimmerman and J. S. Swenton, *Ibid.*, **89**, 906 (1967).
214. J. R. Williams and H. Ziffer, *Tetrahedron*, **24**, 6725 (1968).
215. P. DeMayo, J. P. Pete, and M. Tchir, *Can. J. Chem.*, **46**, 2535 (1968), and earlier papers cited therein.
216. M. A. Golub, *J. Amer. Chem. Soc.*, **91**, 4925 (1969).
217. F. S. Wettack, G. D. Renkes, M. G. Rockley, N. J. Turro, and J. C. Dalton, *Ibid.*, **92**, 1793 (1970).
218. D. I. Schuster and D. J. Patel, *Ibid.*, **90**, 5145 (1968).
219. L. A. Singer, G. A. Davis, and V. P. Muralidharan, *Ibid.*, **91**, 897 (1969).
220. L. A. Singer and G. A. Davis, *Ibid.*, **89**, 158 (1967).
221. L. A. Singer, *Tetrahedron Lett.*, **1969**, 923.
222. S. K. Lower and M. A. El-Sayed, *Chem. Revs.*, **66**, 199 (1966).
223. S. G. Cohen and J. B. Guttenplan, *Tetrahedron Lett.*, **1968**, 5353.

224. T. R. Evans and P. A. Leermakers, *J. Amer. Chem. Soc.*, **90**, 1840 (1968).
225. H. L. J. Bäckström and K. Sandros, *Acta Chem. Scand.*, **14**, 48 (1960).
226. Reference 11, p. 665.
227. E. J. Baum and R. O. C. Norman, *J. Chem. Soc.*, B, **1968**, 227.
228. J. P. Dubois and R. L. Van Hemert, *J. Chem. Phys.*, **40**, 923 (1964), and previous papers.
229. N. J. Turro and R. Engel, *J. Amer. Chem. Soc.*, **91**, 7113 (1969).
230. T. R. Evans and P. A. Leermakers, *Ibid.*, **91**, 5898 (1969).
231. N. C. Yang, R. Loeschen, and D. Mitchell, *Ibid.*, **89**, 5465 (1967).
232. J. Saltiel, H. C. Curtis, L. Metts, J. W. Miley, J. Winterle, and M. Wrighton, *Ibid.* **92**, 410 (1970).
233a. C. A. Parker and T. A. Joyce, *Chem. Commun.*, **1968**, 1421.
233b. P. F. Jones and A. R. Calloway, *J. Amer. Chem. Soc.*, **92**, 4997 (1970).
234. R. S. H. Liu and J. R. Edman, *Ibid.*, **90**, 213 (1968).
235. R. S. H. Liu and J. R. Edman, *Ibid.*, **91**, 1492 (1969).
236. R. S. H. Liu and D. M. Gale, *Ibid.*, **90**, 1897 (1968).
237. R. S. H. Liu, N. J. Turro, and G. S. Hammond, *Ibid.*, **87**, 3406 (1965).
238. R. S. H. Liu, *Ibid.*, **90**, 1899 (1968).
239. R. S. H. Liu and R. E. Kellogg, *Ibid.*, **91**, 250 (1969).
240. Reference 3, p. 43.
240a. P. J. Wagner and D. J. Bucheck, *Ibid.*, **91**, 5090 (1969).
240b. R. L. Cargill, A. C. Miller, D. M. Pond, P. deMayo, M. F. Tchir, K. R. Neuberger, and J. Saltiel, *Mol. Photochem.*, **1**, 301 (1969).
240c. R. O. Loutfry and P. deMayo, *Chem. Commun.*, **1970**, 1040.
241. O. L. Chapman, T. H. Koch, F. Klein, P. J. Nelson, and E. L. Brown, *J. Amer. Chem. Soc.*, **90**, 1657 (1968).
242. N. C. Yang and R. L. Loeschen, *Tetrahedron Lett.*, **1968**, 2571.
243. D. A. Warwick and C. H. J. Wells, *Ibid.*, **1968**, 4401.
244. D. I. Schuster and D. H. Sussman, *Tetrahedron Lett.*, **1970**, 1657.

Photochemical and Spectroscopic Properties of Organic Molecules in Adsorbed or Other Perturbing Polar Environments

COLIN H. NICHOLLS and PETER A. LEERMAKERS, *Hall-Atwater Laboratories, Wesleyan University, Middletown, Connecticut* 06457

CONTENTS

I. Introduction 315
II. Adsorption on Solid Surfaces 316
III. Electronic Absorption Spectra 319
IV. Spectra of Adsorbed Molecules 320
    A. Benzene and Derivatives 320
    B. Polycyclic Hydrocarbons 322
    C. Ketones 322
    D. Charge-Transfer Transitions 323
    E. Carotenoids 324
    F. Luminescence Spectra 325
V. Photochemical Reactions of Adsorbed Molecules 330
    A. *cis-trans* Isomerization 330
    B. Complex and Secondary Photoreactions 332
VI. Summary 334
References 334

I. INTRODUCTION

The classical series of investigations by de Boer and coworkers[1] involving the electronic spectroscopy of adsorbed molecules was the first definitive work concerned with interactions of light with molecular species bound to surfaces. In these early studies of surface field effects, halogens and phenols adsorbed from the gas phase onto sublimed salt films were observed to show often significant perturbations of absorption maxima and molar absorption coefficients.[2] Subsequent investigations have found it convenient to utilize transparent microporous silicate materials with large surface areas (200–600 m²/g) as adsorbents, often immersing them in inert solvents of similar refractive indices in order to increase transparency. Investigations specifically concerned with the spectroscopic behavior in solvent–adsorbant matrices of organic molecules have included those of Robin and Trueblood[3]; their technique of

315

adsorbing aromatic anilines and nitro compounds in silicic acid–cyclohexane slurries produced pronounced shifts in absorption wavelength. Most of the results in this early period have been covered in an excellent review by Terenin.[2]

In recent years an extended series of investigations has been conducted, particularly in these laboratories,[4-9] into the spectra and photochemistry of organic molecules bound to the surfaces of transparent microporous solids such as silica gel. These studies involving a variety of compounds have revealed quite characteristic behaviors that have helped to elucidate the nature of physical adsorption processes, and also something of the configuration of excited electronic states of selected classes of molecules. They have also suggested further utilization of solid surface binding as a means for influencing the course of photochemical processes and probing the reaction mechanisms involved. This report reviews some of the more important findings resulting from these and related investigations.

II. ADSORPTION ON SOLID SURFACES

In order to obtain satisfactory absorption spectra of the substance under investigation it is essential that the beam pass through a large number of adsorbed monomolecular layers. In practice this is most satisfactorily achieved by adsorption on transparent microporous solids with a high surface area to mass ratio (200–600 m^2/g). The solids found most suitable have been silica gel, silicic acid, and microporous glass.

In the granular form these solids strongly scatter the light beam such that very little light is transmitted. The opaque adsorbent can be made relatively transparent by immersion in an inert solvent with a refractive index near that of the solid. There is still some light scattering in this form, but it can be adequately compensated for during spectrophotometric measurements by using an identical slurry in the reference cell.

Only nonpolar solvents, e.g., cyclohexane (most commonly employed with silica gel and silicic acid), methylcyclohexane, methylpentane, and carbon tetrachloride, can be utilized in these slurries since solvents of greater polarity will compete with the intended adsorbate for available binding sites and will result in incomplete sustrate adsorption.

The basic effects on organic molecules of binding to solid surfaces are a function of many possible steric and electronic interactions. However, the ultimate photochemical fate of bound molecules will depend primarily upon their specific physical orientations while adsorbed on the solid surface. Photoreactivity will depend upon which part or parts of the molecules are actually experiencing the adsorptive effects, the inhibitions to rotation and migration, the decreased possibility of making contact with other reagents, and which, if any, remain essentially solvated in the supernatant solution. A consideration of the manner of binding of adsorbate to the surface of the adsorbent, as revealed by infrared spectra, heats of adsorption, adsorption isotherms, and related physical measurements is therefore pertinent to this review. It should be noted at this stage of the review that the authors are not *primarily* concerned with the fundamental physical chemistry of surfaces and adsorbant–adsorbate interactions. Numerous treatments in the areas of catalysis and surface and colloid chemistry are readily available and may be consulted, so such rigorous treatment is omitted in this review. For our purposes the most important types of interactions between surface and substrate are van der Waals or dispersion (induced dipole–induced dipole and dipole–induced dipole), dipole–dipole or static charge–dipole, and hydrogen bonding. For the reader especially interested in surface effects outside of the scope of this report, the paper by Terenin[2] (and references therein) is especially recommended.

Silica gel $(SiO_2)_n$ consists of a three-dimensional network of tetracoordinate oxysilicon tetrahedra, the partially hydrated surface of which generally contains both siloxane (—Si—O—Si—) and silanol (—Si—OH) groups. Some of the silanol groups exist as single (more active) species, while others exist as hydrogen bonded pairs.[10] Interactions of adsorbate with silica gel arise from the nonspecific interaction of the adsorbed molecule with the whole adsorbent and the specific interaction with the active surface hydroxyls. Infrared absorption spectra of silica gel are characterized by the strong —OH stretching band at 3750 cm^{-1}. This absorption band is shifted to lower frequency by the interaction of the silanol–OH groups with adsorbed molecules, the extent of the shift indicating the degree of interaction. At the same time the infrared absorption bands corresponding to the groups of the adsorbate involved in the bonding to the solid surface also show a shift to lower frequency.

Only small shifts (40 cm^{-1}) of the silanol band result from the adsorption of saturated hydrocarbons,[11–14] indicating that only weak dispersion forces are involved. Aromatic hydrocarbons give a significantly larger silanol band shift (125 cm^{-1})[2,14] but this value is small compared with those produced by ethers (400–500 cm^{-1})[14] which are able to form weak hydrogen bonds through the ether oxygen, and by saturated nitrogen compounds (900–1000 cm^{-1})[12,14] which form strong hydrogen bonds. At the same time the —CH vibration bands of the aromatic hydrocarbons are broadened and in the case of benzene shifted 120 cm^{-1} to lower frequency. Ron et al.[15] found that the adsorption of benzene on methylated and sintered porous Vycor gave the same —CH vibration band shift as for adsorption on an untreated surface. Thus it would appear that benzene is bound by dispersion forces to which the π electrons contribute and not, as suggested by Terenin,[2] by hydrogen bonding of the π electrons to the surface silanol groups.

In some cases the introduction of double bonds into a molecule can considerably reduce the interaction at the surface,[14] for example, silanol infrared shift for pyrolidine is 900 cm^{-1} as compared with 325 cm^{-1} for pyrole; tetrahydrofuran causes a shift of 475 cm^{-1}, and furan only 120 cm^{-1}. In each case adsorption is believed to be by means of hydrogen bonding to the hetero-atom. Reduction in the energy of interaction with the unsaturated compounds is presumably due to interaction of the nonbonding electrons with the double bonds, which render the systems aromatic. The shift produced by adsorbed phenol (345 cm^{-1}) probably indicates bonding through the hydroxyl group. This is consistent with the fact that frequencies associated with C—H bonds of the aromatic ring are effected very little.[16]

Heats of adsorption measurements do not lead to very specific interpretation since the isosteric heat of adsorption (ΔH) arises from both nonspecific interactions, which occur in all cases of adsorption, and from specific interactions with the hydroxy groups; nevertheless, valuable conclusions about the binding forces can be deduced. Saturated hydrocarbons, e.g., n-pentane, have a value of $-\Delta H$ of 8.0 kcal/mole, while saturated ethers have values of around 16 kcal/mole.[14] Probably dispersion forces only are involved in the former case and additional specific interaction with the silanol–OH occurs in the second case. On graphite, where there is no specific interaction, the heats of adsorption of hydrocarbons and ethers are very similar.[17] The heat of adsorption of furan (11 kcal/mole) is 5 kcal/mole less than that of tetrahydrofuran; this again indicates the effect that delocalization of electrons by the double bonds has on the binding forces.[14]

The values of $-\Delta H$ for benzene are in the range 10–12 kcal/mole,[15,18–20] being intermediate between values attributed to pure dispersion forces for saturated hydrocarbons and those in which more specific forces are involved. Furthermore, Ron and coworkers calculated the entropies of adsorption for benzene and concluded that the "mobile gas" model of adsorption was applicable, and Whalen[18] found no simple relationship between the hydroxy site content and benzene adsorption. These results confirm the conclusions reached from the infrared data that benzene adsorption is essentially due to dispersion forces which should be greater than with saturated compounds, and that no hydrogen bonding is involved.

Further insight into molecular binding to the solid surface is possible from a determination of the surface area occupied by each molecule in a monolayer. In a detailed study of the adsorption of a range of substances on silica gel (surface area 200 m²/g), Weis and coworkers[9] found that a relatively small number of polynuclear aromatic substances occupied all available binding sites, indicating that the molecules are oriented parallel to the solid surface to maximize interactions. A considerably larger number of molecules of ketones and diketones were accommodated on the surface, implying that in this case the molecule is bound almost solely by the specific forces between the carbonyl and surface hydroxy groups allowing some freedom of motion of the rest of the molecule.

III. ELECTRONIC ABSORPTION SPECTRA

Spectral changes brought about by adsorption onto solid surfaces can be characterized by:

(1) Displacement of the absorption spectra as a whole to higher or lower frequencies, i.e., a spectral shift.
(2) Changes in the extinction coefficients of the absorption bands.
(3) Broadening of the absorption bands.
(4) Appearance of new absorption bands.

Of these four properties, spectral shifts are the most sensitive to environmental changes and also the most readily measured. As a result the majority of investigations into electronic absorption spectral changes resulting from surface adsorption have been confined to measurements of spectral shifts. While the shift of the 0–0 bands is the most meaningful measurement to make, these 0–0 bands are not always discernible, especially when the molecules are adsorbed on polar surfaces, so it has become common practice simply to measure the shift of the absorption maximum. In most cases this measurement would correspond to the shift of the 0–0 band, in others, however, adsorption processes can produce unequal displacement of the ground and excited state potential curves, resulting in a different vibronic band shape.

Electronic absorption spectra are very dependent on the environment of the molecules. In any condensed state the interaction between the molecule and its neighbor affects the energy levels of both the ground and excited states, the greater this interaction the greater the perturbation of the spectra with respect to those obtained in the vapor.

A most comprehensive discussion of the effect of solvent on spectra has been given by Bayliss and McRae.[21] They point out that polarization or dispersion forces are the most general interactions involved in solution and that all solution spectra are subject to a generalized polarization red shift, relative to vapor spectra, due to solvent polarization by the transition dipole. However, these dispersion forces are relatively weak and are easily obscured by the effect of dipole–dipole and dipole–static charge forces *in polar, but not highly polarizable, solvents*. By applying the Franck-Condon principle, they showed that excitation from a solvated ground state, where these stronger forces are operating, gives an excited state with the ground state configuration, but since this is not the equilibrium configuration of the solvated excited state it will have a higher energy.

The direction of a spectral shift, to either higher or lower wavelength, in going from a nonpolar to a polar solvent is dependent on the transition involved; a blue shift almost always occurs with $n \to \pi^*$ transitions and often (with significant exceptions) a red shift with $\pi \to \pi^*$ transitions. Since polar solvents are more strongly hydrogen bonded (or dipole–dipole bonded if H bonding is not possible) to the ground state than to the excited (n,π^*) state (where one less n electron is available), the energy of the ground state is lowered more than that of the excited state. Hence the energy gap between the two states is increased, resulting in a blue spectral shift. In the case of the $\pi \to \pi^*$ transition it has been suggested that the excited (π,π^*) state is more polar and clearly more polarizable than the ground state and is therefore stabilized by interaction with the polar solvent to a greater extent. The latter is clearly not the case if the $\pi \to \pi^*$ transition has significant charge-transfer character of a kind resulting in decreased dipole moment as in the case of several compounds to be mentioned later. Thus, in the usual case of $\pi \to \pi^*$ transitions in hydrocarbons involving little or no change in dipole moment the energy gap between the two states is reduced, hence a red shift in the absorption spectrum.

In most of the cases to be considered in this review the molecules are initially dissolved in a nonpolar solvent (e.g., cyclohexane) in which the original spectrum is recorded, then are adsorbed onto a relatively polar surface. It would, therefore, be anticipated that shifts to lower wavelengths would be observed for the $n \to \pi^*$ transitions, e.g., the low energy transition in carbonyl compounds, while a shift in the reverse direction will occasionally be observed for the $\pi \to \pi^*$ transitions of aromatic *hydrocarbons*.

IV. SPECTRA OF ADSORBED MOLECULES

A. Benzene and Derivatives

Benzene. The ultraviolet absorption spectrum of benzene is characterized by the low intensity L_b band at 256 nm and the more intense L_a band at about 200 nm, both bands being $\pi \to \pi^*$ transitions. Changes in the absorption spectra when benzene is adsorbed on porous silica substrates have been subjected to numerous investigations.[3,7,15,19,20] The spectrum of adsorbed benzene is broader and more diffuse compared to the spectrum of a solution in cyclohexane,[19,20] which in turn is more diffuse than the vapor spectrum.[20] The loss of vibrational structure from the spectrum of adsorbed benzene can be ascribed to the interaction with the surface altering the isolated molecular vibrational motion. The dispersion forces between solvent and solute cause a red shift ($\simeq 200$ cm^{-1}), relative to the vapor spectrum, when benzene is dissolved in cyclohexane,[15,20] while subsequent adsorption from cyclohexane onto silica gel results in a shift of about equal magnitude, but in the reverse direction.[7] The fact that these shifts are of approximately the same magnitude suggests that the same types of forces are involved in each case, hence dispersion forces must be involved in the binding of benzene to the silica gel surface, in accordance with results from infrared and heat of adsorption measurements.

Aniline. The characteristic absorption peaks of aniline resulting from $\pi \to \pi^*$ transitions are found at 288 nm for the ^1L_b band and 234 nm for the ^1L_a band. The nonbonding electrons on the nitrogen occupy a pure p orbital to give maximum overlap with the π electrons of the ring, shifting the $\pi \to \pi^*$ transitions to longer wavelengths than with benzene.[22] Nitrogen substitution by electron donating methyl groups increases the electron density on the nitrogen and the interaction with the π electrons on the ring. The ^1L_a band is shifted from 234 nm for aniline to 243 nm for *N*-methylaniline and 251 nm for *N,N*-dimethylaniline.[22]

Adsorption of aniline onto silicic acid shifts the absorption maxima relative to the cyclohexane solution spectrum about 1800 cm^{-1} to the blue,[3] adsorption of *N,N*-dimethylaniline on silica gel[7] gave a corresponding blue shift of about 3000 cm^{-1}. These quite large *blue* shifts indicate that hydrogen bonding occurs between the silanol sites of the silica surface and the nonbonding electrons of the nitrogen, reducing their interaction with the π electron system of the ring (since these are $\pi \to \pi^*$ transitions, they must be accompanied by decreasing dipole moment, *vide supra*). The blue shift was considerably greater with the *N,N*-dimethylaniline since a greater electron density on the *N* is involved. Confirmation of this hydrogen bonding of aniline to the silanol OH groups has been demonstrated by the red shift (535 cm^{-1}) in the silanol vibrational frequency.[16]

Phenol. In phenol the benzene absorption bands are likewise shifted to the red by the interaction of the nonbonding electrons of oxygen with the π electrons of the aromatic ring. When phenol is adsorbed on slurries of silicic acid,[3] silica gel,[8] and silica–alumina[23] in cyclohexane, relatively small (260–300 cm^{-1}) blue shifts from the spectra in cyclohexane solution are observed. As in the case of aniline the phenol is adsorbed by hydrogen bonding to the silanol group, in this case by the phenolic OH, as indicated by the fact that the vibrational frequency of the silanol OH is shifted while the C—H frequency of the ring remains unchanged.[16] When the phenolic OH group becomes involved in hydrogen bonding, the interaction between the nonbonding electrons and the π electron system of the ring is reduced, hence a blue shift. However, since the shift is small (300 cm^{-1}) compared with those obtained for aniline and derivatives, it is apparent that phenol is not nearly as strongly bound to the silica surface as the amines.

B. Polycyclic Hydrocarbons

Naphthalene, Anthracene. The adsorption of naphthalene or anthracene onto silica gel from cyclohexane causes slight red shifts of around 100–200 cm^{-1}, this effect being much the same as that previously noted for benzene.[7] However, the adsorption of anthracene (and other large polycyclic hydrocarbons) onto silica gel *containing alumina,* in the absence of air, results in color formation.[24] The optical spectra of the polycyclic aromatic hydrocarbon in the adsorbed state and in isooctane solution are nearly identical in the ultraviolet region, but the adsorbed state has strong absorbing bands in the visible region which have been attributed to the presence of a cation radical (confirmed by epr spectra).[25] The one electron transfer responsible for radical formation has been attributed either to undefined oxidation sites on the solid surface[26,27] or to chemisorbed oxygen.[28]

Azulene. The absorption spectrum of azulene, a nonbenzenoid aromatic hydrocarbon with odd-membered rings, can be considered as two distinct spectra, the visible absorption due to the ^1L_b band (0–0 band near 700 nm) and the ultraviolet absorption of the ^1L_a band.[29] This latter band is very similar to the long wavelength bands of benzene and naphthalene (^1L_b) and shows the same 130 cm^{-1} blue shift when adsorbed on silica gel from cyclohexane.[7] As in the case of benzene and naphthalene, this blue shift is due to the fact that the red shift, relative to the vapor spectra, is smaller (305 cm^{-1}) for the adsorbed molecule than in cyclohexane solution (435 cm^{-1}). Thus it would appear that the red shifts of the ^1L_a band are solely due to dispersive forces interacting with the aromatic molecule, in agreement with Weigang's prediction,[29] and dipole–dipole interaction is negligible.

In the case of the visible (^1L_b) band, silica gel adsorption produces a significantly larger blue shift (270 cm^{-1}) relative to cyclohexane,[7] actually a subtractive combination of two blue shifts: relative to the vapor, a 57 cm^{-1} shift when dissolved in cyclohexane, and a 327 cm^{-1} shift when adsorbed on silica gel. These blue shifts are due to dipole–dipole interactions exceeding the dispersion forces, with significantly larger interactions involved between the azulene dipole and silica gel than between azulene and the nonpolar solvent.

C. Ketones

The ultraviolet absorption spectra of ketones are characterized by a low intensity $n \rightarrow \pi^*$ absorption band in the 270–300 nm region, and a high intensity

TABLE I[7]
Spectral Shifts of Adsorbed Ketones

Compound	Transition	Cyclohexane solution	Silica gel–cyclohexane	Spectral chift, cm^{-1}
Acetone	$n \to \pi^*$	2781	2656	1720 (blue)
Benzophenone	$n \to \pi^*$	3365	—	
	$\pi \to \pi^*$	2483	2512	1960 (red)
Benzil	$n \to \pi^*$	3911	3725	1280 (blue)
	$\pi \to \pi^*$	2576	2681	1530 (red)
Tetramethylcyclobutanedione	$n \to \pi^*$	3510	3300	1820 (blue)
	$n \to \pi^*$	3080	2915	1840 (blue)
	$\pi \to \pi^*$	2268	2296	520 (red)

(Wavelength Maximum, Å)

$\pi \to \pi^*$ band at shorter wavelength. The $n \to \pi^*$ transitions are accompanied by a decrease in dipole moment in the excited state, and relative to the *vapor spectra* are displaced to the blue in polar solvents and to the red in nonpolar solvents.[30] The $\pi \to \pi^*$ transitions involve an increase in the dipole moment and are displaced to the red in both polar and nonpolar solvents.

The spectral shifts resulting from the adsorption of various ketones onto silica gel from cyclohexane solution are shown in Table I.[7]

The blue shifts of the $n \to \pi^*$ transitions resulting from adsorption onto silica gel are in the range 1700–2000 cm^{-1} and are much more pronounced than the 400–600 cm^{-1} shifts observed in ethanol.[31]

As well as altering the levels of the singlet states, adsorption onto silica gel can modify the triplet energy levels, in certain cases causing an inversion of the order of energy levels of the lowest triplet states. Acetophenone has a characteristic short-lived n,π^* phosphorescence in a hydrocarbon glass, indicating that the lowest triplet is an n,π^* state. However, in polar solvents and adsorbed onto silic gel, this emission is replaced by a long-lived phosphorescence, typical of a π,π^* state.[32] Inversion of triplet[32a] and singlet states[33] in related systems by silica and other polar perturbing environments has chemical consequences as well (*vide infra*).

D. Charge-Transfer Transitions

Nitrobenzene. The 250 nm primary or ^1L_a band of nitrobenzene has been explained by Nagakura[34] in terms of an "intramolecular charge transfer spectrum." He suggests that interaction between the highest occupied orbital of the donor part of the molecule (the phenyl ring) and lowest unoccupied orbital of the acceptor group (nitro group) forms two new orbitals extending over both donor and acceptor. Excitation occurs by transition between these two orbitals and results in a transfer of charge from donor to acceptor. Accordingly the excited state would be the more polar and have a much greater dipole moment. Bakshiev[35] has calculated that the static dipole of the excited state is 9 D compared to a value of 4 D for the ground state. Dipole–dipole interactions with a polar environment would, therefore, lower the energy of the excited state relative to the ground state, and thus produce a red shift.

Red shifts in nitrobenzene spectra that have been noted (relative to cyclohexane solution) are 1000 cm^{-1} for ethanol,[30] 2200 cm^{-1} for water,[30] 2330 cm^{-1} for silica gel,[7] and 2600 cm^{-1} for silicic acid.[3] Robin and Trueblood[3] proposed that the large red shift observed with adsorption onto silicic acid involved intermolecular hydrogen bonding. However, since neither the NO_2 vibration frequencies[20] nor the silanol–OH frequencies[16] were perturbed by adsorption onto silica gel, yet the C—H vibrational frequencies showed marked shifts,[20] it is apparent that no significant interaction between the silanol and nitro groups occurs. Accordingly, the large red shifts observed for nitrobenzene when adsorbed onto silica gel or silicic acid indicate large dipole–dipole interactions between the adsorbate and solid surface.

Alkylpyridinium Iodides. The strongly solvent-dependent absorption bands of alkylpyridinium iodides have been correlated with a charge-transfer transition. However, in this case the ground state is much more polar than the excited state; according to Kosower,[36] the dipole moment of the ground state is 13.9 D, compared to a value of 8.6 D for the excited state. As a result, dipole–dipole interactions between the ground-state molecule and a polar adsorbent will be greater than the interactions between the excited state and the adsorbent. This has been confirmed by observation of blue shifts of the order of 7000 cm^{-1} (from 4200 to 3200 Å) when 1-methyl, 1-ethyl, or 1-isopropyl-4-carbomethoxypyridinium iodide were adsorbed onto silica gel from chloroform solutions.[7] Kosower[36] has used the transition energy of 1-ethyl-4-carbomethoxypyridinium iodide in various solvents as a measure of the polarity of the solvent (Z value), a value of 3250 Å for the λ_{max} on silica gel gives the adsorbent a Z value of 88, intermediate between that of water (94.6) and methanol (83.6).

E. Carotenoids

In an endeavor to explain the anomolous spectral shifts when certain carotenoids such as astaxanthin and all-*trans* retinal are incorporated into their respective apoproteins, crustacyanin, and opsin, the perturbation of the absorption spectra of the chromophores adsorbed onto silica gel has been examined.[37,38] Buckwald and Jencks[37] found that there was a small red shift

(920 cm^{-1}) with a 20% increase in band width when astaxanthin was adsorbed onto silica gel from cyclohexane. From the size of this shift they were able to conclude that there was little or no difference in polarity between the ground and excited states of astaxanthin.

Irving and Leermakers[38] found that when all-*trans*-retinal was adsorbed onto silica gel from cyclohexane a red shift of 2500 cm^{-1} (40 nm) resulted, a somewhat larger shift than would be expected for a $\pi \to \pi^*$ transition transferring to a polar environment. Considerably larger red shifts (up to 100 nm) were recorded for the retinal Schiff base, possibly due to protonation of the base by the silica gel. However, the long wavelength absorption band of the adsorbed protonated Schiff base was still at a considerably shorter wavelength than the 500 nm absorption band of rhodopsin, suggesting that factors other than those involved in binding to a solid substrate must be involved when retinal Schiff base is bound to its apoprotein.[38a]

F. Luminescence Spectra

Fluorescence and Phosphorescence. Essentially the same effects that influence absorption of radiation are found to perturb the emission processes. Fluorescence spectra are red- or blue-shifted along with the corresponding band of the absorption spectrum.

The results of an extensive study by Eaton[38b] on phosphorescence energies and lifetimes of ketones are given in Tables II–V. For purposes of this review, discussion is limited to this important class of compounds.

The carbonyl compounds in the tables may be classified in one of three groups: (I) those compounds for which the emitting triplet is an n,π^* state in both polar and nonpolar media; (II) those compounds which emit from an n,π^* triplet in low polarity solvent glasses, but which invert to a π,π^* state in high polarity glasses; and (III) carbonyl compounds for which the lowest-lying triplet is π,π^* in all solvents. The classes described represent the permutations of n,π^* and π,π^* triplets possible in solvents of different polarity. Some qualitative observations concerning the relative dispositions of the states can be made in each case. In compounds of Group I the lowest triplet, $^3(n,\pi^*)$, is of considerably lower energy than is T_2, $^3(\pi,\pi^*)$. Thus increasing the polarity of the solvent would not be expected to raise the energy of $^3(n,\pi^*)$ and lower that of $^3(\pi,\pi^*)$ sufficiently to cause an inversion of emitting states. Group II compounds, of which acetophenone is an excellent example, are characterized by a much smaller $T_2 - T_1$ energy difference. Thus it is possible that increased solvent polarity would cause the emitting state to be $^3(\pi,\pi^*)$. No unambiguous statement may be made concerning the energy spacing of the triplets of molecules in Group III. The energy difference may be large or small; however, the

TABLE II
Phosphorescence Maxima and Triplet Energies for Some Ketones in Several Molecular Environments at 77°K

Compound	Solvent[a]	In solvent glass λ_{max}, nm	E_T, kcal	On silica gel λ_{max}, nm	E_T, kcal
Acetone	MCH	444	64.5[b]	440	65.0[b]
1,3-Diphenylacetone	MCH	424	71.5[c]	459	62.3[b]
	EPA	423	70.0[c]	421	67.9[b]
o-Methoxyacetophenone	MCH	423	67.6[b]	440	65.0[b]
	EPA	423	67.6[b]	427	66.8[b]
Acetophenone	MCH	417	73.4[c]	425	70.7[c]
	IP	425	71.5[c]	425	70.7[c]
3,3,5,5-Tetramethylcyclohexanone	MCH	464	66.0[c]	467	64.1[c]
	EPA	459	62.5[c]	430	66.5[c]
Xanthone	MCH	428	74.2[c]	445	68.2[c]
	EPA	404	70.7[b]	403	70.7[b]
Benzophenone	IP	446	68.8[c]	418	72.0[c]
2-Benzoylpyridine	IP	468	65.5[c]	458	62.5[c]
2'-Acetonaphthone	IP	527	56.2[b]	526	57.8[c]
	EPA	516	58.2[c]	515	58.7[c]
	MCH	524	54.6[b]	531	53.9[b]
Di-2-pyridylketone	EPA	465	66.1[c]	460	66.8[b]

[a] MCH = methylcyclohexane. EPA = ether, pentane, ethanol. IP = isopentane.
[b] E_T calculated from phosphorescence maximum.
[c] E_T calculated from 0–0 band of phosphorescence.

emitting state would be expected to be $^3(\pi,\pi^*)$ in solvents of any polarity, with characteristically long lifetimes.

It must be noted that "significant" variations in observed lifetimes, particularly for short-lived states, are presumed to be changes of a factor of 10 or more. The observed phosphorescence lifetime τ_P is governed by the relation $\tau_0 = \tau_P/\phi_P$ (where τ_0 is the true radiative lifetime and ϕ_P is the quantum efficiency of phosphorescence). τ_0 does not vary. However, τ_P and ϕ_P may change significantly, depending on the phase of the medium and implicit variations in the solvent glass. The distribution of the carbonyl compounds among the above three groups is described below.

Group I includes all of the diketones studied and the following monoketones: acetone, 3,3,5,5-tetramethylcyclohexanone, benzophenone, 2-benzoylpyridine, and di-2-pyridyl ketone. This group is characterized by short phosphorescence lifetimes in both the pure solvent glasses and adsorbed in the polar silica gel–solvent matrices, and by an increase in the energy of the emitting

TABLE III
Observed $1/e$ Phosphorescence Lifetimes of the Ketones of Table II

Compound	Solvent	Observed τ_P, sec In solvent glass	On silica gel
Acetone	MCH	2×10^{-4}	8×10^{-4}
1,3-Diphenylacetone	MCH	0.01	0.04
	EPA	0.01	0.04
O-Methoxyacetophenone	MCH	0.01	0.08
	EPA	0.02	0.03
Acetophenone	MCH	1.5×10^{-3}	0.3
	IP	1.7×10^{-3}	0.4
3,3,5,5-Tetramethylcyclohexanone	MCH	6×10^{-4}	2×10^{-4}
	EPA	6×10^{-4}	4×10^{-4}
Xanthone	MCH	0.01	0.04
	EPA	0.05	0.05
Benzophenone	IP	7×10^{-4}	2×10^{-3}
2-Benzoylpyridine	IP	8×10^{-4}	2×10^{-3}
2'-Acetonaphthone	IP	0.05	0.08
	EPA	0.4	0.5
Di-2-pyridylketone	EPA	1.2×10^{-3}	1.8×10^{-3}

TABLE IV
Phosphorescence Maxima and Triplet Energies for Some α-Diketones in Several Molecular Environments at 77°K

Compound	Solvent	In solvent glass λ_{max}, nm	E_T, kcal	On silica gel λ_{max}, nm	E_T, kcal
2,3-Butanedione	MCH	519	55.2[b]	492	58.2[b]
	EPA	500	57.2[b]	495	57.2[b]
2,3-Pentanedione	MCH	506	56.5[b]	462	61.8[a]
	EPA	495	57.7[a]	494	57.9[b]
1-Phenyl-1,2-propanedione	MCH	535	53.5[b]	511	56.0[b]
	EPA	525	54.5[b]	527	56.8[b]
Benzil	IP	526	54.3[b]	501	57.2[b]
4-Methoxybenzil	IP	529	54.1[a]	487	58.7[a]
Anisil	IP	523	54.7[a]	482	59.3[a]
Mesitil	MCH	540	53.0[b]	525	55.4[b]
	EPA	537	53.3[a]	533	54.7[a]
α-Furil	IP	524	54.6[a]	508	56.3[a]
α-Pyridil	IP	491	58.3[a]	481	59.4[a]

[a] E_T calculated from phosphorescence maximum.
[b] E_T calculated from 0–0 band of phosphorescence.

TABLE V
Observed 1/e Phosphorescence Lifetimes of the α-Diketones of Table IV

Compound	Solvent	Observed τ_P, sec In solvent glass	Observed τ_P, sec On silica gel
2,3-Butanedione	MCH	8×10^{-4}	$4. \times 10^{-4}$
	EPA	4×10^{-4}	4×10^{-4}
2,3-Pentanedione	MCH	5×10^{-4}	2×10^{-4}
	EPA	2×10^{-4}	4×10^{-4}
1-Phenyl-1,2-propanedione	MCH	3×10^{-4}	5×10^{-4}
	EPA	7×10^{-4}	6×10^{-4}
Benzil	IP	1.5×10^{-3}	1×10^{-3}
4-Methoxybenzil	IP	1.2×10^{-3}	1.2×10^{-3}
Anisil	IP	1.2×10^{-3}	1.2×10^{-3}
Mesitil	MCH	8×10^{-4}	4×10^{-4}
	EPA	1×10^{-3}	1.2×10^{-3}
α-Furil	IP	6×10^{-4}	1×10^{-3}
α-Pyridil	IP	1.4×10^{-3}	1×10^{-3}

triplet state upon increasing the polarity of the environment. The phosphorescing state is thus $^3(n,\pi^*)$ in all media. Only one compound in the tables is classified in Group II. Acetophenone was observed to phosphoresce from a lower energy triplet when adsorbed on silica gel than it did in the nonpolar glasses. Furthermore, the observed triplet lifetime increased by a factor of 200 upon adsorption, as compared to the radiative lifetime in MCH at 77°K. These observations are consistent with those of Lamola,[32] who concluded that the emitting state in nonpolar solvents was $^3(n,\pi^*)$, but that upon increased solvent polarity the emitting triplet acquired π,π^* characteristics.

The remaining compounds possess π,π^* emitting triplets in all media used (Group III). The members of this class are 1,2-diphenylacetone, o-methoxyacetophenone, xanthone, and 2'-acetonaphthone. This group is characterized by a long-lived phosphorescence in all solvents and by a decrease in the energy of the emitting triplet upon increased solvent polarity.

There is some evidence that xanthone should be classified in Group II. Murmukhametov, Mileshana, and Shigorin[38c] reported that xanthone in n-hexane possesses a short-lived n-π* phosphorescence band, while in alcohol the n,π* and π,π* triplets are inverted, resulting in the presence of a long-lived π-π* band. The mixing of n,π^* and π,π^* triplets in xanthone is explained in terms of the electron donating effects of the bridgehead oxygen causing the π-π* bands to shift to lower energies and the n-π* bands to shift to higher energies. The proximity of the $^3(\pi,\pi^*)$ and $^3(n,\pi^*)$ states will then be affected by changes in the polarity of the medium. Table III indicates that the phosphorescence lifetime of xanthone in MCH is 0.01 sec. It is possible that this lifetime represents a mixed emitting state in MCH, of partial $^3(n,\pi^*)$ character, and predominantly of $^3(\pi,\pi^*)$, since in MCH/silica gel the lifetime increases to 0.04 sec, and the triplet energy decreases from 74.2 kcal mole^{-1} in the pure solvent glass to 68.2 kcal mole^{-1} in MCH/silica gel (Table II).

The emitting triplet of o-methoxyacetophenone is π-π^* in all solvents used. This is in marked contrast to the behavior of acetophenone, which inverts the $^3(\pi,\pi^*)$ and $^3(n,\pi^*)$ states upon increased solvent polarity. The lack of state inversion in o-methoxyacetophenone may be explained in terms of the electronic effect of the methoxy substituent. The electron donating ability of the methoxy group tends to destabilize the n,π^* triplet, in which the excitation is essentially localized on the carbonyl oxygen, and to stabilize the $^3(\pi,\pi^*)$ state through delocalization of the excitation within the aromatic system. Thus it is due to intramolecular perturbations of the electronic states of the parent acetophenone system by the orthomethoxy substituent that the molecule possesses a π,π^* state as its lowest-lying triplet in both polar and nonpolar media, as Pitts[38d] and Yang[36e] have previously reported.

All of the α-diketones and some of the monoketones studied possessed n,π^* emitting triplets (Group I). The triplet energies of the compounds in pure solvent glasses were increased upon adsorption on silica gel. All of the members of Group I, including the monoketones, are characterized by comparable triplet energies, between 53 and 60 kcal mole^{-1} for the diketones, and between 65 and 70 kcal mole^{-1} for the monoketones. The observed phosphorescence lifetimes were short in both solvents of low polarity and in high polarity media. The emission spectra of these compounds were well-defined in nonpolar media, but less sharp in polar solvents, as expected for n-π^* transitions.

To summarize this section, it is apparent that in the media involved some systems are environmentally strongly perturbed, both in terms of triplet energies and measured lifetimes, some are only weakly perturbed. With a few already noted exceptions, it is safe to generalize that triplet lifetimes are slightly longer in the adsorbed state, and that triplet energies of n,π^* states are moderately elevated.

Eximer Fluorescence. Since Forster and Kasper discovered concentration-dependent long-wavelength emission resulting from association of an electronically excited pyrene molecule with another ground state pyrene molecule,[39] the phenomenon of excimer fluorescence has been studied extensively.[40] The mechanism for excimer formation and emission can be represented by

$$A + h\nu_1 \longrightarrow A^*$$
$$A^* + A \longrightarrow (AA)^*$$
$$(AA)^* \longrightarrow (AA) + h\nu_2$$
$$(AA) \longrightarrow A + A$$

Excimer formation has been shown to be a diffusion controlled process[41-43] in which a "sandwich" or "face-on" configuration of the two interacting molecules is required.[44,45] It has been deduced that intermolecular distance in the excimer state is smaller than for the same configuration with both molecules in their electronic ground states.[44,46] Apart from pyrene, excimer-like emission has been observed from a wide range of aromatic compounds including many alkyl derivatives of such hydrocarbons[43,47] and vinyl polymers.[48-50]

When pyrene was adsorbed onto silica gel from cyclohexane solution, the prominent excimer fluorescence band at 4650 Å was no longer observed, but the monomer fluorescence band at 3900 Å remained little changed.[9] This inhibition of the excimer fluorescence indicates that the molecules are so strongly bound to the substrate surface that they become immobilized, thus preventing the conjugate π system overlap required to produce excimer emission.

In contrast to pyrene, the intramolecular excimer fluorescence of 1,3-diphenylpropane was not quenched by adsorption on silica gel.[9] The excimer emission of this molecule is due to conformational changes within the molecule which allow overlap of the conjugated π systems of the two phenyl groups during the lifetime of the excited state. Since adsorption onto silica gel does not quench excimer fluorescence, the forces involved in the intramolecular excimer formation must be sufficiently strong to be little influenced by the dispersion forces responsible for binding to the adsorbent surface. Thus, even this "negative result" illustrates that the adsorption technique distinguishes between the relative free energies of intramolecular excimer formation versus adsorption of the aromatic nuclei by dispersion interactions.

V. PHOTOCHEMICAL REACTIONS OF ADSORBED MOLECULES

A. cis-trans *Isomerization*

Stilbenes. Absorption of light by a compound containing an ethylenic double bond can result in *cis-trans* isomerization. Direct irradiation of either isomer of stilbene results in the formation of a photostationary mixture; the composition of the mixture depending on the conditions. The mechanism of the isomerization process is still a matter of controversy although there is considerable experimental support for the existence of a common triplet metastable intermediate lying just below the individual *cis* and *trans* triplet energy levels.[51-53] On the other hand, Saltiel[54] now believes that there is more compelling evidence for the "singlet mechanism" in which rotation about the central bond occurs in the excited singlet state followed by decay from a common twisted singlet state.

In an attempt to probe further into the complex photochemical system involved, the isomerization of stilbenes adsorbed on silica gel was examined.[9] It was observed that the time required for establishment of the photostationary state was significantly increased (by a factor of 3) and that the composition of the photostationary state changed from 93% cis isomer in cyclohexane solution to 60% cis isomer in the silica gel matrix. Though not definitive, this evidence supports Fischer's "triplet mechanism,"[53] as we have previously reported.[9]

Photosensitized Isomerism. Cis-trans isomerization of olefinic compounds can be achieved by photosensitization, the composition of the photostationary state depending, among other things, on the triplet energy of the sensitizer.[52] In the case of piperylene (1,3-pentadiene), the *trans* form of which has a triplet energy (E_T) of 59 kcal mole^{-1} and the *cis* form a value of 57 kcal mole^{-1}, all sensitizers with E_T values in excess of 65 kcal mole^{-1} give the same photostationary state (58% *trans*) since energy can be transferred with equal facility to either the *cis* or *trans* form. Sensitizers with E_T less than 60 kcal mole^{-1} preferentially transfer energy to the *cis* form and the photostationary state becomes richer in the *trans* isomer. No "normal" sensitization occurs with sensitizers having values of E_T less than about 53 kcal mole^{-1}.

Weis, Bowen, and Leermakers[5] have shown that sensitized isomerism of piperylene in solution can be achieved with the sensitizer adsorbed on silica gel, electronic energy transfer taking place at the solid–liquid interface. High energy sensitizers ($E_T > 65$ kcal mole^{-1}) gave the same photostationary state as that obtained with homogeneous solutions, but for sensitizers having triplet energies in the range 55–59 kcal mole^{-1} the modification of the triplet excitation energy by adsorption onto silica gel was sufficient to alter the composition of the photostationary state to a significant degree. For example, the lowest triplet of biacetyl (E_T 55 kcal mole^{-1}) has an n,π^* configuration, the energy of which is raised by adsorption on silica gel; as a result, the *trans* concentration of the piperylene photostationary state is lowered from 80 to 73%, corresponding to an environmental-induced rise in biacetyl triplet energy of about 3 kcal to 58 kcal mole^{-1}.

Daubendiek et al.[55] have extended this technique to demonstrate electronic energy transfer at the gas–solid interface. They irradiated photosensitizers, in polycrystalline form, in the presence of *cis*-piperylene vapor and found that the extent of isomerization was comparable to values obtained for the same sensitizer in a solution of piperylene. Their results indicated an energy difference of about 1 kcal between triplet states of sensitizers in solution and on the surface of organic crystals.

B. Complex and Secondary Photoreactions

Photochromism of Spiropyrans. The reversible photochromism of spiropyrans has been exhaustively studied by Fischer and his colleagues[56] and shown to be a reversible transformation between the spiropyran form A (colorless) and the merocyanine form B (colored).

In solution, Form A is the more stable and can be converted into the metastable colored form B by irradiation with ultraviolet light. The action of heat or exposure to longer wavelength light reverses the reaction, the rate of which depends on the polarity of the solvent since the more polar form B is relatively stabilized in polar solvents.

Adsorption of the colorless form onto silicic acid[57] or silica gel[6] immediately produces a highly colored matrix which in certain cases, depending on the structure of the pyran, can be reversibly photobleached.[7] The brightly colored matrix indicates that the "open" partly ionic form is the more stable in the highly polar silica gel environment; undoubtedly physical adsorption forces also contribute to the stability of this form.

To explain their results obtained in the study of photochromism of 1,3,3-trimethylindolino-6-nitro-8-methoxybenzospiropyrane adsorbed on silicic acid, Balny et al.[57] have proposed that energy transfer takes place between the singlet states of A (spiropyrane) and B (merocyanine form).

2,4-Cyclohexadienones. The perturbation of the electronic absorption spectra by adsorption on silica gel can have a significant effect on subsequent photochemical reactions. For example, Hart has shown that 2,4-cyclohexadienones photochemically degrade cleanly in nonpolar solvents to a ketene (1) but in highly polar solvents or adsorbed on silica gel bicyclic ketone (2) is the predominant product.[33] The absorption spectra indicate that in nonpolar solvents the lowest singlet state is the n,π^* state, from which the formation of the ketene proceeds. This $n \to \pi^*$ band is obscured by the $\pi \to \pi^*$ band in polar media, inversion of the energy levels of the n,π^*, and the first

π,π^* singlet states occurs and photolysis proceeds from the π,π^* state to give the bicyclic ketone.

Secondary Radical Reactions. Other adsorptive effects including inhibition to rotation and migration and a decreased possibility of making contact with other reagents can also influence the course of secondary photochemical reactions. This has been borne out by the finding that whereas in benzene solution the photolysis of azobisisobutyronitrile produces both dimethyl-*N*-(2-cyano-2-propyl) ketenimine and tetramethylsuccinonitrile,[58] no ketenimine was formed by irradiation in a silica gel–benzene matrix.[7] In this latter case, binding of the cyano groups to the silica surface is believed to prevent migration of the cyanopropyl radicals and only carbon–carbon coupling can occur. It has also been shown[7] that the photodecomposition of tetramethyl-1,3-cyclobutanedione is reduced threefold by adsorption onto silica gel, the radicals formed in the initial step are held in close proximity on the surface so that bond reformation can compete effectively with irreversible photodecarbonylation.

However, attempts by Kiefer and Carlson[59] to prohibit undesired bimolecular reactions by irradiating 2,3,3-trimethyl-1-penten-4-one adsorbed onto silica gel were unsuccessful (due probably to steric inhibition of adsorption); the product composition was the same as that previously obtained in solution. Werbin and Strom[60] attempted to restrain the freedom of movement of the radicals formed from the photolysis of vitamin K_3 (2-methyl-1,4-naphthoquinone) by adsorption onto silica gel, but obtained the same mixture of dimers as that obtained from the irradiation in acetone solution, viz., syn and anticyclobutanes, an oxetane dimer, and a binaphthoquinone dimer. Photolysis of the solid substrate, however, produced only the syn isomer of cyclobutane, in this case no migration of radicals is possible, hence only one product.

VI. SUMMARY

It is apparent that not all molecules are bound to silica gel surfaces strongly enough to prevent migration; reasons for this are manifold and sometimes unclear. However, there exist a large number of organic molecules which can have their photochemical behavior controlled by binding to a solid surface. This technique, by simplifying or selectively channeling product composition, should have wide application with investigations into the primary and secondary processes of photochemical reactions. Clearly the most useful application to date has been spectroscopic rather than photochemical, due to the pronounced spectral shifts in the adsorbed environment. Separation or merging of different electronic transitions facilitate spectroscopic labeling in a variety of molecules of interest, and creation of altogether new bands (as in myrcene[7]) is relevant to geometric structure of the adsorbate. Because many biological pigments are "bound" in one way or another, it is felt that further application of the techniques described in this review may indeed be fruitful in a number of yet unforeseen investigations involving electronic excitation.

Acknowledgment

The authors are greatly indebted to the U.S. Army Research Office (Durham), the Alfred P. Sloan Foundation, the American Chemical Society Petroleum Research Fund, and to the National Institute of Health for generous support. We also wish to thank Drs. H. T. Thomas, T. R. Evans, F. C. James, and A. Weissberger, and A. F. Toth and especially L. D. Weis for advice and technical assistance in support of this manuscript. We are also most grateful to David F. Eaton for permission to cite his unpublished results.

REFERENCES

1. J. H. de Boer, *Z. Phys. Chem.*, **B14**, 163 (1931); **B15**, 281 (1932); **B16**, 397 (1932); **B17**, 161 (1932); **B**18, 49 (1932). J. H. de Boer and C. J. Dippel, *Z. Phys. Chem.*, **B25**, 399, 408 (1943). J. H. de Boer and J. F. H. Custers, *Z. Phys. Chem.*, **B25**, 225 (1934); **B21**, 208, 217 (1933).
2. A. Terenin, *Advan. Catal.*, **15**, 227 (1964).
3. M. Robin and K. N. Trueblood, *J. Amer. Chem. Soc.*, **79**, 5138 (1957).
4. P. A. Leermakers and H. T. Thomas, *Ibid.*, **87**, 1620 (1965).
5. L. D. Weis, B. W. Bowen, and P. A. Leermakers, *Ibid.*, **88**, 3176 (1966).
6. P. A. Leermakers, L. D. Weis, and H. T. Thomas, *Ibid.*, **87**, 4603 (1965).
7. P. A. Leermakers, H. T. Thomas, L. D. Weis, and F. C. James, *Ibid.*, **88**, 5075 (1966).
8. T. R. Evans, A. F. Toth, and P. A. Leermakers, *Ibid.*, **89**, 5060 (1967).
9. L. D. Weis, T. R. Evans, and P. A. Leermakers, *Ibid.*, **90**, 6109 (1968).
10. C. G. Armistead, A. J. Tyler, F. H. Hambleton, S. A. Mitchell, and J. A. Hockey, *J. Phys. Chem.*, **73**, 3947 (1969).
11. L. H. Little, *Infrared Spectra of Adsorbed Species*, Academic Press, London, 1966.

12. A. Davydov, A. V. Kiselev, and B. V. Kuznetson, *Russ. J. Phys. Chem.*, **39**, 1096 (1965).
13. A. V. Kiselev and V. I. Lygin, *Surface Sci.*, **2**, 236 (1964).
14. P. A. Elkington and G. Curthoys, *J. Phys. Chem.*, **72**, 3475 (1968).
15. A. Ron, M. Folman, and O. Schnepp, *J. Chem. Phys.*, **36**, 2449 (1962).
16. V. N. Abramov, A. V. Kiselev, and V. I. Lygin, *Zh. Fiz. Khim.*, **37**, 2783 (1963); cf. *Chem. Abstr.*, **60**, 8786e (1964).
17. A. V. Kiselev, *Discussions Faraday Soc.*, **40**, 205 (1965).
18. J. W. Whalen, *J. Phys. Chem.*, **71**, 1557 (1967).
19. A. N. Pavlova, *C. R. Acad. Sci. USSR*, **49**, 265 (1945); cf. *Chem. Abstr.*, **40**, 5641 (1946).
20. M. Okuda, *Nippon Kagaku Zasshi*, **82**, 1118 (1961); cf. *Chem. Abstr.*, **56**, 2089 (1962).
21. N. B. Bayliss and E. G. McRae, *J. Phys. Chem.*, **58**, 1002 (1954).
22. H. H. Jaffe and M. Orchin, *Theory and Applications of Ultraviolet Spectra*, Wiley, New York, 1968 pp. 407, 410.
23. D. Kobayashi, *Nippon Kagaku Zasshi*, **80**, 1399 (1957).
24. R. N. Roberts, C. Barter, and H. Stone, *J. Phys. Chem.*, **63**, 2077 (1959).
25. W. K. Hall, *J. Catal.*, **1**, 53 (1962).
26. R. L. Hodgson and J. H. Raley, *Ibid.*, **4**, 6 (1965).
27. U. V. Strelko, L. M. Ganyuk, and Z. Z. Vystoskil, *Ukr. Khim. Zh.*, **29**, 363 (1963); cf. *Chem. Abst.*, **59**, 1121 (1963).
28. R. P. Porter and W. K. Hall, *J. Catal.*, **5**, 366 (1966).
29. O. E. Weigang, *J. Chem. Phys.*, **33**, 892 (1960).
30. N. B. Bayliss and E. G. McRae, *J. Phys. Chem.*, **58**, 1006 (1954).
31. E. M. Kosower, *J. Chem. Phys.*, **38**, 2813 (1963).
32. A. A. Lamola, *Ibid.*, **47**, 4810 (1967).
32a. R. D. Rauh and P. A. Leermakers, *J. Amer. Chem. Soc.*, **90**, 2246 (1968).
33. J. Griffith and H. Hart, *Ibid.*, **90**, 5296 (1968).
34. S. Nagakura and J. Tanaka, *J. Chem. Phys.*, **22**, 236 (1954); S. Nagakura, *Ibid.*, **23**, 1441 (1955).
35. N. G. Bakshiev, *Opt. Spectros.*, **13**, 43 (1962).
36. E. M. Kosower, *J. Amer. Chem. Soc.*, **80**, 3253 (1958).
37. M. Buchwald and W. P. Jencks, *Biochemistry*, **7**, 834 (1968).
38. C. S. Irving and P. A. Leermakers, *Photochem. Photobiol.*, **7**, 655 (1968).
38a. C. S. Irving, G. W. Byers, and P. A. Leermakers, *Biochemistry*, **9**, 858 (1970).
38b. D. F. Eaton, Private communication.
38c. R. N. Murmukhametov, L. A. Mileshana, and D. N. Shigorin, *Opt. Spectrosc.*, **22**, 404 (1967).
38d. J. N. Pitts, H. W. Johnson, and T. Kuwana, *J. Phys. Chem.*, **66**, 2456 (1962).
38e. N. C. Yang, D. S. McClure, S. L. Murov, J. J. Houser, and R. Dusenbery, *J. Amer. Chem. Soc.*, **89**, 5466 (1967).
39. T. Forster and K. Kasper, *Z. Phys. Chem.*, **1**, 275 (1954); *Z. Electrochem.*, **59**, 976 (1955).
40. E. L. Wehry and L. B. Rogers, *Fluorescence and Phosphorescence Analysis*, Wiley-Interscience, New York, 1968, pp. 113–118.
41. J. B. Birks, D. T. Dyson, and I. H. Munro, *Proc. Roy. Soc., Ser. A*, **275**, 575 (1963).
42. J. B. Birks, D. T. Dyson, and T. A. King, *Ibid.*, **277**, 270 (1964).
43. J. B. Birks and L. G. Christophorou, *Ibid.*, **277**, 571 (1964).
44. J. Ferguson, *J. Chem. Phys.*, **28**, 765 (1958).

45. E. A. Chandross and J. Ferguson, *Ibid.*, **45**, 3554 (1966).
46. J. Ferguson, *Ibid.*, **44**, 2677 (1966).
47. J. B. Birks, and L. G. Christophorou, *Nature*, **197**, 1064 (1963).
48. S. S. Lehrer and G. D. Fasman, *Biopolymers*, **2**, 199 (1964).
49. J. W. Longworth and F. A. Bovey, *Ibid.*, **4**, 1115 (1966).
50. J. W. Longworth, *Ibid.*, **4**, 1131 (1966).
51. G. S. Hammond and J. Saltiel, *J. Amer. Chem. Soc.*, **84**, 4983 (1962); **85**, 2515 (1963).
52. G. S. Hammond et al., *Ibid.*, **86**, 3197 (1964).
53. D. Gegiou, K. A. Muszkat, and E. Fischer, *Ibid.*, **90**, 12, 3907 (1968).
54. J. Saltiel, *Ibid.*, **91**, 1265 (1969).
55. R. L. Daubendiek, H. Magid, and F. R. McMillan, *Chem. Commun.*, **1968**, 218.
56. R. Heiligman-Rim, Y. Hirshberg, and E. Fischer, *J. Phys. Chem.*, **66**, 2470 (1962), and preceding papers.
57. C. Balny and P. Douzou, *C. R. Acad. Sci. Paris*, **264**, 477 (1967); C. Balny, D. Djaparidze, and P. Douzou, *Ibid.*, **265**, 1148 (1967).
58. P. Smith and A. M. Rosenberg, *J. Amer. Chem. Soc.*, **81**, 2037 (1959).
59. E. F. Kiefer and D. A. Carlson, *Tetrahedron*, **1967**, 1617.
60. H. Werbin and E. J. Strom, *J. Amer. Chem. Soc.*, **90**, 7296 (1968).

Author Index

Numbers in parentheses are reference numbers and show that an author's work is referred to although his name is not mentioned in the text. Numbers in *italics* indicate the pages on which the full references appear.

A

Abrahamson, E. W., 10(23), 20, 21(23), 25(23), 34, 54, *71*, 271, *311*
Abram, I. I., 282(194), *312*
Abramov, V. N., 318(16), 321(16), 324(16), *335*
Adalberon, M., 241, *241*
Adams, W. P., 144(95), *159*
Adamson, A. W., 228(5), *243*
Aditya, S., 60, 61, *75*
Adler, S. E., 70(VI), *70*
Adman, E., 230(20), *243*
Airey, J. R., 7(17), 15(17), 16(17), 35(17), 45(17), *71*
Aladekomo, J. B., 170(19), 184(39), 185(39), 186(39), 202(39), *223*
Alder, K., 79(22), *105*
Alexander, N., 305(r), *306*
Algar, B. E., 190(54,56), *224*, 298(t), 299(s, t), *300*
Allred, E. L., 283(199,200), *312*
Almgren, M., 103(105), *108*, 305(w), *306*
Amata, C. D., 297(h), 298(h), 299(h), *300*
Anderson, J. C., 109, 116(40), 122(40), 123(40), 126, *155, 157*
Anderson, L. C., 29(99,100), *73*
André, J., 84(60), *107*, 252(41), *308*
Andreev, E. A., 39, *39*
Andrews, S. D., 285(201), *312*
Angus, H. J. F., 257(74), *309*
Applequist, D. E., 207(97), *225*
Appleyard, J. H., 207(92), *225*
Applied Optics, Supp. 2., 14(34), 15(34), *72*
Armistead, C. G., 317(10), *334*
Armstrong, A. T., 196(68), 197, 198(68), *224*, 228, *243*
Arnold, D., 256(69), *308*
Arnold, D. R., 256, 256(64,66), 302, *308*

Arnold, S. J., 35(116), *74*
Ashmore, P. G., 10(20), 40, *71, 74*
Assay, J., 31(108), *73*
Atherton, N. M., 119(48), *157*
Augenstein, L., 216(133), *226*
Axelrod, M., 279(184,185), 280(185), *312*
Ayad, K. N., 113(26), *156*
Azumi, T., 196(68), 197, 197(67), 198(68), 219(67), *224*, 275(165), 299(mm), 305(V), *301, 306, 311*

B

Bach, F. L., 113(21), *156*
Bäckström, H. L. J., 80(30), 82(49), *106*, 193(62), *224*, 262(103), 291(225), *309, 313*
Bader, D. W., 18(45), 19, 51, 52(45), 68(45), *72*
Badger, B., 164(8), 221, *222*
Baker Co., (J. T.), 228(8), 234(8), *243*
Bakshiev, N. G., 324, *335*
Baldwin, B. A., 182(36), *223*
Baldwin, J. E., 113(23,24), 155(24), *156*, 278, *311*
Balny, C., 332, *336*
Balzani, V., 233, 242, *243, 244*
Ban, M. I., 184(11), 185(11), 186(11), 204(83), 218(83), *222, 224*
Banthorpe, D. V., 113(27), *157*
Barclay, J. C., 113(21), *156*
Barltrop, J. A., 139(86), *159*
Barnard, M., 88(69), *107*
Barnes, R. L., 185(41), 186(41), 207(41), *223*
Barter, C., 322(24), *335*
Bartlett, P. D., 274(159,160), 281, 281(193), 283, 283(159b,160,193), 285(193), 286(160), 296(160), *311, 312*

Barton, D. H. R., 125(61), *157, 158*
Basila, M. R., 53(142), *74*
Basinski, A., 125(58), *157*
Bass, A. M., 8(29), 10(29), 26(29), 32(29), 34(29), 35(29), 39(29), 46(29), 47(29), 48(29), *71*
Bates, D. R., 5(10), *71*
Bates, J. R., 29(99,100), *73*
Baum, E. J., 82(55), 83(55), *106,* 125(58), 139(86), *157, 159,* 249(25), 251(25), 252(25,44), 291(227), *307, 308, 313*
Bäuminger, B., 78, 97(11), *105*
Bayliss, N. B., 319, 323(30), 324(30), *335*
Bayliss, N. S., 25(73), *73*
Beard, C., 113(26), *156*
Becker, H.-D., 268(129), *310*
Becker, R. S., 228, 229, 233, *243*
Beckett, A., 86(64), *107,* 250(32), 263, 264(107), 265(107), *307, 309*
Bednar, T. W., 164(3), 177, 204(3), *222*
Beens, H., 174, 177(28), 200(28), *223*
Beer, M., 170(13), *222*
Beinoravichute, 241, *244*
Bell, E. R., 114(30), 121(30), 126, 130(30), 131, 133(30), 137(30), *157*
Bell, J., 170(21), *223*
Bell, J. A., 234(30), *244*
Bellůs, D., 109(2), 110(4), 116(4,41), 117(41), 118(41), 125(41,59), 126(41,59), 127(41), 128(41), 129(59), 130(59,71), 133(79,83), 136(71), 138(71,83), 139(71), 141(11,41), 142(11), 145(11), 148(107), 155(4), *155-159*
Belorit, A., 79(28), *106*
Beltram-Lopez, V., 17(41), *72*
Ben-Bassat, J. M., 90(73), *107*
Bennett, C., 235(33), *244*
Bennett, R. G., 204(85), *224,* 275(162), 297(uu), 298(uu), *301, 311*
Benrath, A., 78, 85(3), *105*
Benson, S. W., 58(146), *75,* 268(133), *310*
Bentrude, W. G., 80(44), 82(44), 83(56), 84(44), 86(44), *106*
Berg, H., 90(72), *107*
Bergmark, W., 267, *310*
Berlman, I. B., 181(33), 222(143), *223, 226,* 297(d), 298(d), 299(d), 304(c), 305(c), *300, 305*
Berson, J. A., 128(67), *158*

Bertele, E., 152(20), 153(20), *156*
Binkley, R. W., 259(101), *309*
Bird, R. B., 43(129), 44(129), 46(129), *74*
Birks, J. B., 163, 164(10), 167, 168(10), 169(1,37,42,43), 170(16,19), 178, 179, 184(37,39,40), 185(39,41,42,43), 186(16,37,39,40,41,42,43), 193, 194(37), 197(71), 202(16,39,40,42,43), 204(40,43), 207, 207(92), 218(120, 121,122), *167, 222-225,* 255, 255(56, 57), 297(ee), 298(ee), 299(ee), 304(o), 305(o), *300, 306, 308,* 330(41,42,43, 47), *335, 336*
Bishop, R., 104(114), *108*
Blackwell, J. E., 102(103), *108*
Blanksma, J. J., 113(26), *156*
Blethen, M. L., 70(XIV), *70*
Bloomfield, J. J., 93(78), *107*
Boggus, J. D., 200(76), *224*
Boos, H., 152(20), 153(20), *156,* 176(27), *223*
Booth, M. H., 18, *72*
Borkman, R. F., 304(k), *306*
Bouas-Laurent, H., 207(98), *225,* 257(85), *309*
Bouchy, M., 84(60), *107,* 252(41), *308*
Bovey, F. A., 330(49), *336*
Bowen, B. W., 316(5), 331, *334*
Bowen, E. J., 78, 82(6), *105,* 207(90), 208(90), *225,* 255, 255(60), 297(e,cc), 298(e,cc), 299(e,cc), *300, 308*
Bowers, K. D., 17(41), *72*
Bowers, P., 297(p), 298(p), 299(p), 305(u), *300, 306*
Bowers, P. G., 299(w), *300*
Bowman, R. M., 257(82a), *309*
Boykin, D. W., Jr., 113(21), *156*
Bozak, R. E., 115(38), *157,* 232(24), 239(45,46), 240(46), *243, 244*
Bradshaw, J. S., 80(37), *106,* 125(60), 126, 134(60), 135(60), *157,* 257(74), 269(136), 270(136), 273(136), *309, 310*
Braga, C. L., 170(16), 186(16), 202(16), *223*
Brand, J. C. D., 229, 235(19), 242(19), *243*
Brauman, J., 235(35), *244*
Braun, C. L., 116(42), *157*
Braun, W., 8(29), 10(29), 26(29), 32(29), 34(29), 35(29), 39(29), 46(29), 47(29), 48(29), *71*

AUTHOR INDEX

Briegleb, G., 200(76), *224*
Briggs, A. G., 22, 25(59), 53, *72*
Brinen, J. S., 297(f), 298(f), 299(f), *300*
Britton, D., 19(54), 19(57), 26(85), 53(54), 53(57), *72*
Broadbent, A. D., 249(24), 270(24), 302(m), *303, 307*
Broadbent, T. W., 26(77), *73*
Brocklehurst, B., 16(38), *72,* 164(8), 221, 221(130), *222, 225*
Brodowski, W., 250(34), *307*
Brown, E. L., 294(241), *313*
Brown, R. F. C., 95(84), *107*
Brown, W., 26(83), 27, *73*
Brown, W. G., 26(91), *73*
Bruce, J. M., 133(78), *158*
Brudka, M., 125(57), *157*
Bryce-Smith, D., 94(79), *107,* 257(74,76,79,84), *309*
Bucheck, D. J., 294(240a), *313*
Buchi, G., 256(62), *308*
Buchwald, M., 324, *335*
Bunbury, D. L., 85(63), 87(63), *107*
Bunker, D. L., 53(139), *74*
Burkoth, T. L., 104(112), *108*
Burley, D. R., 249(24), 270(24), 302(m), *303, 307*
Burton, C. S., 84(59), *107,* 252(42), *308*
Burton, M., 58(147), *75,* 276(168), 297(h), 298(h,jj), 299(h), *300, 301, 311*
Busch, G. E., 69(IV,V), 70(VII, XIII), *70*
Buschhoff, M., 113(23), *156*
Byers, G. W., 325(38a), *335*
Bykhovskii, V. K., 41(126), *74*
Bylina, A., 273, *311*

C

Cadman, P., 7(15), 15(15,36), 16, 16(15), 30(15,36), 34(15,36), 35(15,36), 37(15,36), 44(15,36), 49(15), *71, 72*
Cadogan, J. I. G., 113(28), *157*
Cais, M., 228(9), 229, 231, 232(9), 234(9), *243*
Calas, C. A., 207, *225*
Caldwell, R. A., 271, *310*
Callear, A. B., 23, 24(65,66), 26, 38(66), 40, 40(123), 57, 58(65,66), 61, *24, 72-74*
Calloway, A. R., 292, *313*
Calvert, J., 302(1), 303

Calvert, J. G., 23(62), 30(62), 58(62), 67(62), *72,* 80(42), *106,* 114(32a), 114(32b), 115(32a), 119(32b), 135(32c), 139(32d), 143(32d), 147(32d), *157,* 248(11), 249(17), 291(226), *307, 313*
Campbell, T. C., 257(82), *309*
Cantrell, T. S., 148(108), *159*
Carabetta, R. A., 19(55), 53(55), *72*
Carassiti, V., 233, *243*
Cargill, R. L., 294(240b), *313*
Carlson, D. A., 333, *336*
Caronna, G., 78(12), 87(12), *105*
Carrington, A., 17, *72*
Carrington, T., 53(139), *74*
Carter, J. G., 297(c), *300*
Case, W. A., 252(45), 302(e), 303(e), *303, 308*
Cashion, J. K., 7(16), 16, 35, 35(40), 36(40), *71, 72*
Castellan, A., 257(85), *309*
Castro, G., 203(82), *224*
Cava, M. P., 155(117), *159*
Chakrabarti, S. K., 203(82), *224*
Challand, B. D., 259, *309*
Chamberlain, T. R., 257(82a), *309*
Chambers, R. W., 305(q), *306*
Chandra, A. K., 196(73), 198(73), 199, 200(73), 201, *224*
Chandross, E. A., 164(7), 195, 195(65), 198, 208(65), 219, *222, 224, 225,* 330(45), *336*
Chapman, O. L., 109(2), 144(95,96), *155, 159,* 246(3,8), 247(3), 253, 254, 255(48,58), 259(94), 268(134), 294, 294(240), 295(3,94), *307-310, 313*
Charney, E., 103(106), *108*
Chattaway, F. D., 113(26), *156*
Chaudet, J. H., 133(79), *158*
Chen, R. F., 305(r), *306*
Cherkasov, A. S., 207(93), 208(93), *225,* 299(kk), *301*
Chieffi, G., 256(61), *308*
Chien, J. C. W., 53(141), *74*
Chow, Y. L., 125(61), *157, 158*
Christie, M. L., 53, 53(139), 63(151), *74, 75*
Christophorou, L. G., 164(10), 167, 168(10), *167, 222,* 297(c), *300,* 330(43,47), *335, 336*
Chuang, T. T., 85(63), 87(63), *107*

Ciamician, G., 78, 85(2), *105*
Cilento, G., 275(164), *311*
Claesson, S., 30, *73,* 173(24), 175(24), 176(24), 187(24), 209(24), 210(24), 211(24), 220(24), *223*
Clar, E., 297(tt), 298(aa, tt), 299(tt), *300, 301*
Clark, T. C., 51(135), *74*
Cleveland, P. G., 144(96), *159*
Clyne, M. A. A., 18(44,46,48,49), 19, 25, 26(49,93), 29(49,93), 35(117), 51, 51(46,48), 52, 52(93), 65, *72–74*
Coates, R. M., 256(67), *308*
Coche, A., 193(63), *224*
Cohen, J. I., 271(144), *310*
Cohen, S. G., 290, *312*
Cohen, W. D., 78, 82(4), 85, *105*
Cole, R. S., 278(179), 279(179,183), *311*
Compton, L. E., 30, *30*
Condon, E. U., 4(7), 6, *71*
Conte, J. C., 169(37), 184(37), 186(37), 193(37), 194(37), *223*
Cook, C. F., 228, 231, 235, *243*
Cooke, R. S., 280(186), *312*
Cookson, R. C., 150(112), *159*
Cooper, R., 222(141), *226*
Coppinger, G. M., 114(30), 121(30), 126, 130(30), 131, 133(30,78), 137(30), *157, 158*
Cordes, H., 26(80,81,82), *73*
Corliss, C. H., 9(154), *75*
Cornelius, J. F., 69(IV), *70*
Cotton, F. A., 227(1), *243*
Coulson, C. A., 207(96), *225*
Coulson, J., 113(26), *157*
Counsell, R. C., 80(37), *106,* 269(136), 270(136), 273(136), *310*
Cowan, D. O., 80(37), *106,* 269(136), 270(136), 273(136), 277, *310, 311*
Coyle, J. D., 139(86), *159*
Coyne, L. M., 80(38), *106,* 270(138), 279(184,185), 280(185), *310, 312*
Cox, A., 125(61), *157, 158*
Cox, M., 193(59), *224*
Coxon, J. A., 18(44,46,49), 19(49), 25, 26(49,93), 29(49,93), 51, 51(46), 52, 52(93), *72, 73*
Cuingnet, E., 241, *241*
Cundall, R., *301*(qq)
Cundall, R. B., 103(107), *108,* 255(50), *308*

Cunningham, P. T., 190(57), *224*
Curthoys, G., 317(14), 318(14), *335*
Curtis, H., 277(171), *311*
Curtis, H. C., 292(232), 294(232), *313*
Curtiss, C. F., 43(129), 44(129), 46(129), *74*
Custers, J. F. H., 315(1), *334*
Cutts, E., 133(78), *158*
Cvetanovic, R. J., 50(133), *74*
Czekalla, J., 200(76), *224*

D

Daehne, W. v., 113(21), *156*
Dainton, F. S., 10(20), 40, *71, 74*
Dale, J., 268(131), *310*
D'Allessio, J. T., 298(jj), *301*
Dalton, C., 80(37), *106,* 269(136), 270(136), 273(136), *310*
Dalton J. C., 246(3), 247(3), 256(69,70), 286(203), 289(217), 294(240), 295(3), *307, 308, 312, 313*
Dannenberg, J. J., 231, 234(21), *243*
Darnall, K. R., 80(44), 82(44), 83(56), 84(44), 86(44), *106*
Dauben, W. G., 246(6), 256(67), *307, 308*
Daubendiek, R. L., 331, *336*
Davidson, N., 19(57), 53(57,139), *72, 74*
Davidson, R. S., 257(86), *309*
Davis, A., 118(46), 125(46), *157*
Davis, D. D., 8(29), 10(29), 26(29), 32(29), 34(29), 35(29), 39(29), 46(29), 47(29), 48(29), *71*
Davis, D. R., 31(108), *73*
Davis, G. A., 289(219,220), 290, 290(219), *312*
Davydov, A., 317(12), *335*
Day, A. C., 278(177), 285(201), *311, 312*
DeBoer, C., 255(51), *308*
DeBoer, C. D., 279(182), *311*
de Boer, J. H., 315, *334*
de Groot, Ae., 94(81), *107*
De Mayo, P., 82(54), 87(66), 99(54), *106, 107,* 150(110), *159,* 258(93), 288, 294, 294(240b), *309, 312, 313*
Deglise, X., 84(60), *107,* 252(41), *308*
Deleo, E., 78, 87(12,13), *105*
Denschlag, H. O., 272(146), *310*
Derible P., 149(18), 151(18), *156*
Derwent, R. G., 35, *35*

AUTHOR INDEX

Dickens, P. G., 49(130), *74*
Dickinson, T., 170(18), 171(18), 207(94), 208(94), 213, 213(94,111), 217(111), *223, 225*
Diesen, R. W., 54(144), 70(VI), *70, 75*
Dilling, W. L., 104(115), 246(4), *108, 307*
Dillon, M. A., 276(168), *311*
Dippel, C. J., 315(1), *334*
Dirania, M. K. M., 113(21), *156*
Dirscherl, W., 78, 97(9,10), 98(10), *105*
Djaparidze, D., 332(57), *336*
Döller, E., 169(47), 185(45), 186(45), 202(45), 204(45), *223*
Domrachev, G. A., 242(57), *244*
Donovan, R. J., 8(27,29,30,75,76), 10(27,29), 11, 12(27), 16(37), 21(27), 24(67), 25(75,76), 26(26,90), 27, 29(90), 31(37,76), 32(29), 34, 34(27,29,30, 76,114,115), 35(29,76), 36(75), 37(27,114), 38(31,75,76,114,115), 39(29), 42(27), 43(27,37,67), 44(27,37,67,76), 46, 46(27,29,31,37, 67,75,76,115), 47(27,29,36,67,75), 48, 48(29), 49(37,76,131), 53, 53(76), 57, 57(31), 59, 59(131), 61(31), 63(31), 64(31,75,90), *48, 71-74*
Dougherty, T. J., 250(28), *307*
Douzou, P., 332(57), *336*
Dowd, P., 257, *308*
Drisko, R. L., 277, *311*
Dubois, J. P., 291, *313*
Dubois, J. T., 79(26), *106*, 167, 170(14), 193, 193(59), *167, 223, 224*, 265(113), 266(115), 273 (155), 277(115), 291(113,155), 304(m), 305(j), *306 310, 311*
Duck, E. W., 268(132), *310*
Dunitz, J. D., 152(20), 153(20), *156*
Durie, R. A., 67(152), *75*
Ďurišinová, Ľ., 133(83), 138(83), *158*
Dürr, H. G., 115(36), *157*
Dusenberry, R., 101(99), *108*
Dusenbery, R., 329 (38e), *335*
Dusenbery, R. L., 253(49), 254(49), 258(49,90), *308, 309*
Dusenbury, R., 302(a), 303(a), *303*
Dym, S., 273(154), 291(154), *311*
Dyson, D. J., 255(57), *308*, 163(1), 169(1,42), 178, 179, 185(42), 186(42), 202(42), *222, 223*, 255(56), *308*

Dyson, D. T., 330(41,42), *335*

E

Eastman, R. H., 249(20), *307*
Eaton, D. F., 325, *335*
Eaton, P. E., 227(3), *243*
Edman, J. R., 292(234,235), 299(nn), 305(d), *301, 305, 313*
Edwards, A. G., 150(112), *159*
Egger, K. W., 268(133), *310*
Eicken, R. R., 152(116), 153, 154(116), 155(116), *159*
Eisinger, J., 215(131), 216(132,134), 217(136), *226*
El-Sayed, M. A., 273(157), 277(172), 290, *311, 312*
Elad, D., 141(88,89), *159*, 259, 259(95), 268(134), *309, 310*
Elder, E., 200(76), *224*, 228, *243*
Eliel, E. L., 147(104), *159*
Elkington, P. A., 317(14), 318(14), *335*
Elliott, S. P., 249(22), 250(29), *307*
Elsinger, F., 152(20), 153(20), *156*
Emmons, S. A., 207(95), 213(95), *225*
Engel, P. S., 249(19), 274(160), 281, 281(192,193), 282, 283(160,193), 285(193), 286(160), 287(211), 289(211), 296(160), *307, 311, 312*
Engel, R., 82(50), 85(50), 103(108,109), *106, 108*, 291(229), 305(t), *306, 313*
Engleman, J., 53(139), *74*
Ermolaev, V. L., 297(o), 298(o), 302(d), 303(d), 304(g), 305(g), *300, 303, 305*
Ernest, I., 124(56), *157*
Eschenmoser, A., 152(20), 153, 153(20), *156*
Esser, H., 79(22), *105*
Evans, D. F., 298(z), 299(z), 304(f), 305(f), *300, 305*
Evans, G. B., 103(107), *108*
Evans, T. R., 79(34,40), 80(34), 81(40), 98(34), *106*, 291(224,230), *313* 316(8,9), 318(9), 321(8), 330(9), 331(9), *334*
Eyring, H., 4, *71*
Ezumi, K., 173(25), 176(27), 280(190), *223 312*

F

Falconer, W. E., 25(71), 47(71), *73*
Fasman, G. D., 170(17), 184(17), *223, 330(48), 336*
Feit, E. D., 249(16), *307*
Feldkimel–Gorodetsky, M., 147(102), *159*
Felmlee, W. J., 54(144), *75*
Felner, I., 152(20), 153(20), *156*
Ferguson, J., 195, 198, 211(105), *224, 225,* 330(44,45,46), *335, 336*
Fettis, G. C., 2, *70*
Filseth, S. V., 23, 26, *72*
Fink, P., 33(110), *73*
Finlayson, N., 35(116), *74*
Finnegan, R. A., 115(37), 116(37), 123, 125(62), 126(37,64), 129(70), 134, 134(37,62,64,81), 135(64), 139(64), 147(16), *156–158, 236, 244*
Fischer, E., 275(166), *311,* 330(53), 331(53), 332, *336*
Fischer, E. O., 240(49), 241, *244*
Fischer, G., 275, *311*
Fischer, M., 141(13,14), 145(13,14,101), *156, 159,* 262, 267(105), *309*
Fish, R. W., 235(33), *244*
Flowers, M. C., 48(146), *75*
Folman, M., 317(15), 318(15), 320(15), 321(15), *335*
Fonken, G. J., 246(8), *307*
Foote, R. S., 115(7), 141(7), 142(7), *156*
Forster, L. S., 79(24), 80(24), *105,* 249(15), *307*
Förster, T., 225, *308,* 329, *335*
Förster, Th., 164, *222,* 173(22), 185(44,45), 186(45), 187(44), 192(58), 193 (58), 202(45), 204(45), 218(58), *223, 224,* 275, *311*
Foss, R. P., 273(154), 291(154), *311*
Foster, W. R., 113(28), *157*
Fox, J. R., 274(159), 282, 283 (159a), *311*
Fraenkel, G. K., 221(129), *225*
Frankevich, Ye L., 56(145), 57(145), 65(145), *75*
Freeman, C. G., 18, *72*
Friedrich, E. C., 207(97), *225*
Fry, A. J., 115(35), *157,* 234, *243*
Fujisawa, T., 97(90), 98(90), *107*
Furey, R. L., 141(12), 143(12), *156*

G

Gale, D. M., 292(236), *313*
Ganyuk, L. M., 322(27), *335*
Garstang, R. H., 4, 5, 5(10), 6, 31, 45, *71*
Garwood, R. F., 113(26), *156*
Gegiou, D., 330(53), 331(53), *336*
Gehman, W., 19(57), 53(57), *72*
Gerasimenko, A. V., 238(41), *244*
Geresc, A., 110(3), 124(3), *155*
Giacalone, A., 78, 87(13), *105*
Gibbs, D. E., 18(43), 25(43), 52, *72*
Gibian, M. J., 86(65), *107,* 258(89), 287(210), *309, 312*
Gibson, G. E., 26(91), *73*
Gilbert, A., 94(79), *107,* 257(74,79), *309*
Ginsburg, D., 257(72), *308*
Giuliano, C. R., 15, *15*
Givens, R. S., 115(36), 128(68), *157, 158*
Givens, W. G., 53(142), *74*
Glass, L., 200(75), 203(75), *224*
Glick, A., 256(69), *308*
Glick, A. H., 256(64,66), *308*
Go, R. M., 209(100), *225*
Godfrey, T. S., 139(85a), *159*
Godtfredsen, W. O., 113(21), *156*
Goeth, H., 133(78), *158*
Gold, A., 257(73), *308*
Gold, E. H., 257(72), *308*
Golden, J. H., 118(46), 125(46), *157*
Goldschmidt, C. R., 222(143), *226*
Golub, M. A., 288(216), 291, *312*
Goodeve, C. F., 29(97,100), 33, *73, 74*
Gorodetsky, M., 147(102), 148(17,102, 105), 149(105), *156, 159*
Gotthardt, H., 256(66), 279(185,185), 280(185), *308, 312*
Gould, E. S., 239, *244*
Grabowski, Z. R., 125(57), *157*
Graf, P. E., 53(141), *74*
Gream, G. E., 92(75), *107*
Greenberg, S. A., 79(24), 80(24), *105,* 249(15), *307*
Grellmann, K. H., 145(99), *159*
Gribi, H. P., 152(20), 153(20), *156*
Griffith, J., 323(33), 332(33), *335*
Griffiths, J. G. A., 78, *105*
Grossweiner, L. I., 250(31), *307*
Grovenstein, E., Jr., 257(82), *309*

AUTHOR INDEX

Grubert, H., 240(49), *244*
Gschwend, H., 152(20), 153(20), *156*
Guéron, M., 215(131), 216(132), *226*
Guillory, J. P., 228, 231, 235, *243*
Gurvich, L. V., 56(145), 57(145), 58(145), 65(145), *75*
Guseva, L. N., 133(78), *158*
Güsten, H., 262(106), *309*
Gutsche, C. D., 123(54), 124(54), *157*
Guttenplan, J. B., 290, *312*

H

Haddon, W. F., 96(86), *107*
Haebig, J., 215(115), *225*
Hageman, H. J., 113(21), *156*
Hagen, A. W., 147(16), *156,* 236, *244*
Halford, J. O., 29(99,100), *73*
Hall, D. W., 232(23), *243*
Hall, W. K., 322(25,28), *335*
Haller, W. S., 148(108), *159*
Hambleton, F. H., 317(10), *334*
Hamer, N. K., 104(114), *108*
Hamilton, E. J., Jr., 82(52), 86(52), 87(52), 92(52,76), 93(52,76), *106, 107,* 263(110), *309*
Hammond, G. S., 58(149), *75,* 78(19), 79(25,27,35), 80(32, 33, 35,36,37), 81(35,45), 82(32,53), 84(59), 87(53), 89(70), 90(71), 92(76), 93(76), 97(89,90), 98(70,90), 99(33), *105–107,* 114(31), 115(35), 142(92), *157, 159,* 190(50,51), 206, *224,* 234, 234(29), *243, 244,* 246(1), 247(1), 249(14), 250(27,37), 251, 252(42), 252(59), 256(63,66), 257(75), 258(1), 266(116), 269(135,136,137), 270(136,140), 272,272(151), 273(1,135,136,137,154), 274(158,159), 277, 278, 279(158,179), 180, 182, 183, 184, 185), 280(185, 186), 282, 283(159a), 286(206), 287, 288(158,174), 291, 291(154), 292(237), 293(237), 296(158), 297(a,e,g,l), 298(a,e,g,j,1), 299(a,e, 1), 301, 302(63,b,c), 303(b,c,f), 304(a,e,i), 305(a,e,i), *300, 303, 305– 313,* 330(51,52), 331(52), *336*
Haninger, G. A., Jr., 272(146), *310*

Haranath, P. V. B., 26(88,89), *73*
Hardham, W. M., 257(75), 287, *309, 312*
Härdtl, K. H., 189(48), *223*
Hardwick, R., 54(143), *75*
Harrison, A. J., 53(140), *74*
Hart, H., 323(33), 332(33), *335*
Hartley, D. B., 58(146), *75*
Harvey, J. S. M., 17(41), *72*
Hashimoto, S., 267(126), *310*
Hatchard, C. G., 164(6), 218(6), *222,* 218(118), *225,* 298(v), *300*
Hathorn, F. G. M., 10(28), 11, 34(114,115), 37(114), 38(31,114,115), 46(28,31, 115), 50(28), 57(31), 59, 61(31), 63(31), 64(31), 65(28), 66(28), 67(28), 68(28), *71,74*
Havinga, E., 121(50), 122, 140(9), *156, 157*
Havlicek, S. C., 155(117), *159*
Hazeldine, R. N., 67(153), *75*
Hedaya, E., 115(34), 120(34), 121(34), 122(34), 126(34), 127, *157*
Heertjes, P. M., 267(121), *310*
Heiligman–Rim, R., 332(56), *336*
Heinzelmann, W., 297(u), 298(u), 299(u), *300*
Helman, W. P., 297(h), 298(h), 299(h), *300*
Henderson, W. A., Jr., 250(35), *307*
Herberhold, M., 241(55), *244*
Herkstroeter, W. G., 79(35), 80(34), 81(35), *106,* 234(29), *244,*249(14), 251, 269(137), 273(137), 297(a,g,l), 298(a,g,l), 299(a,l), 302(c), 303(c,f), 304(a,e), 305(a,e), *300, 303, 305, 307, 308, 310*
Herr, K. C., 10(24), *71*
Herre, W., 200(76), *224*
Herzberg, G., 5(9), 10, 33, 41(125), *71, 74*
Herzfeld, K. F., 40(122), *74*
Hess, L. D., 15, *15,*139(86), *159*
Hickinbottom, W. J., 113(26), *156*
Hill, J., 113(21), *156*
Hill, K. A., 257(77), *309*
Hillier, I. H., 198(74), 200(75), 203(75), *224*
Hinman, R. L., 256(64), *308*
Hirai, H., 267(119), *310*
Hirayama, F., 169(140), 184(140), 186, 201, 213, 213(113), 214, 215(114), *225, 226*
Hiroaka, H,. 54(143), *75*

Hirota, Y., 105, *108*
Hirschfelder, J. O., 43(129), 44(129), 46(129), *74*
Hirshberg, Y., 332 (56), *336*
Hirt, R. C., 241, *244*
Hochstrasser, R. M., 78(16), *105,* 117(44), *157,* 203(81,82), 213, *224, 225,* 273(154), 291(154), *311*
Hockey, J. A., 317(10), *334*
Hodges, F. W., 113(26), *156*
Hodgkins, J. E., 259(100), *309*
Hodgson, R. L., 322(26), *335*
Hoffmann, R., 127(65), *158*
Hoffmann, R. W., 152(116), 153, 154(116), 155(116), *159*
Hoffsommer, R. D., 133(78), *158*
Hoigné, J., 116(41), 117(41), 118(41), 125(41), 126(41), 127(41), 128(41), 141(41), *157*
Holdy, K. E., 70(XV), *70*
Hori, Y., 151(113), 152(113), *159*
Horii, Z., 151(113), 152(113), *159*
Hornig, D. F., 19, 53(53), *72*
Horrocks, A. R., 205(89), *225,* 273(156), 276(169), 277(156), 279(156), 297(i), 298(i), 299(i,y), *300, 311*
Horspool, W. M., 105(119), *108,* 126(5,63), *155, 158*
Horton, A. T., 78, 82(6), *105*
Hostettler, H. U., 150(112), *159*
Houser, J. J., 101(99), *108,* 253(49), 254(49) 258(49), 302(a), 303(a), *303, 308,* 329(38e), *335*
Howarth, O. W., 221(129), *225*
Howells, W. G., 229(17), 232(17), *243*
Hoytink, G. J., 170, 195(66), 203(20), *223, 224*
Hrdlovic, P., 109(2), 125(59), 126(59), 129(59), 130(59,71), 133(79,83), 136(71), 138(71,83), 139(71), *155, 157, 158*
Huang, C. W., 257(82a), *309*
Hudec, J., 150(112), *159*
Hudlický, M., 124(56), *157*
Huffman, K. R., 250(33), *307*
Hughes, V. W., 17(41), *72*
Hulthen, E., 26(92), *73*
Humphrey, J. S., 121(50), 132(6), 138(6), *156, 157*
Hurley, R., 272(150), *311*

Husain, D., 4(11), 5, 7(11), 8(27, 29, 30, 75, 76), 10(23, 27,28,29), 11, 12, 12 (27), 15, 16(37), 20, 21(23,27), 24(67), 25(23,75,76), 26(29,90), 27, 29(90), 31, 31(37,76), 32(29), 34, 34(27,29, 30,76,114,115), 35(29,76), 36(75), 37(27,114,157), 38(31,75,76,114, 115), 39(29), 42(27), 43(11,27,37,67), 44, 44(27,37,67,76), 45(11), 46, 46(11,27,28,29,31,37,67,75,76,115), 47(11,27,29,37,67,75), 48, 48(11,29), 49(11,37,67,131), 50(28), 53, 53(76), 54, 57 57(31), 59, 59(131), 61(31), 63(31), 64(31,75,90), 65(28), 66(28), 67(28), 68(28), *48, 71–75*
Hutton, E., 18(47), 19, 25(47), 51, 52(47), *72,* 164(2,5), 218(116), *222, 225*
Huyser, E. S., 99(96), 101(96), *108*
Hyndman, H. L., 270(140), *310*

I

Ikeler, T. J., 286(204), *312*
Imoto, E., 133(79), *158*
Inman, C. G., 256(62), *308*
Ipaktschi, J., 95(83), *107*
Irick, G., Jr., 267(123), 286(205), *310, 312*
Irving, C. S., 324, 325, 324(38a), *335*
Ivanova, T. V., 170(15), *223,* 297(dd,ff), 298(dd), *300*
Iwata, S., 200(77), 205(77), *224*
Izawa, Y., 249(21), *307*
Izzo, P. T., 152(114), *159*

J

Jackson, J. M., 40(121), *74*
Jaff, H. H., 117(45), 129(69), *157, 158,* 321, (22), *335*
James, F. C., 316(7), 320(7), 321(7), 322(7), 323(7), 324(7), 332(7), 333(7), 334(7), *334*
James S. P., 271(148), *310*
Japar, S. M., 271, *311*
Jarlsater, N., 26(92), *73*
Jencks, W. P., 324, *335*
Jennings, K. R., 16(38), *72*
Johnson, C. D., 19(56), 53(56), *72*
Johnson, G. R. A., 274(158), 277(158), 279(158), 288(158), 296(158), *311*

AUTHOR INDEX

Johnson, H. W., 139(85a), *159*, 329(38d), *335*
Johnston, K. M., 113(26), *157*
Jolly, J. E., 78(14), 80(14), 81(14), *105*, 248(10), *307*
Jolly, P. W., 82(54), 99(54), *106*, 258(93), *309*
Jones, J., 204(84), *224*
Jones, L. B., 251, 286(206), *308, 312*
Jones, P. F., 213(107), *225*, 292, *313*
Jortner, F., *158*(75)
Jortner, J., 198(74), *224*
Joschek, H. I., 113(22), 125(22), 139(87), *156, 159*
Joyce, T., 219, 219(124), *225*
Joyce, T. A., 277(175), 292(233a), 297(m), 298(m), 299(m,pp), *300, 301, 311, 313*

K

Kagen, J., 248(12), *307*
Kallmann, H., 280(187), 281(187), *312*
Kallmann, H. P., 79(26), *106*, 193(59), *224*
Kamper, R. A., 17(41), *72*
Kan, R. O., 141(12), 143(12), *156*
Kano, K., 267(126), *310*
Kaplan, L., 257(78,83), *309*
Karl, G., 16(39), *72*
Kasha, M., 170(13), *222*, 275(165), 297(bb), 298(bb), 299(bb,mm), 302(k), 303(k), 304(s), 304(s,v), *300, 301, 303, 306, 311*
Kasper, J. V. V., 3(3), 7(3), 12(3), 14(3,32), 21(3), 33(3,32), 36(3), 37(3), 38(3,32), 42(3), *71*
Kasper, K., 164, *222*, 329, *335*
Kato, S., 116(42), *157*
Kaufman, F., 16(38), *72*
Kaufman, V., 9(156), *75*
Kawanisi, M., 146(15), 149(109), 150(109), 151(109), *156, 159*, 267(119), *310*
Kealy, T. J., 227(1), *243*
Kearns, D. R., 35, *35*, 148(106,107), *159*, 252(45,47), 297(ss), 298(ss), 302(e), 303(e,j), 304(k,n), 305(n,q), *301, 303, 306, 308*
Kearvell, A., 205(89), *225*, 299(y), *300*
Kellmann, A., 265(113), 291(113), *310*

Kellogg, R. E., 275(162), 293(239), 297(uu), 298(uu), *301, 311, 313*
Kelly, D. P., 113(21), *156*
Kelso, P. A., 249(23), *307*
Kemball, G. E., 4(4), *71*
Kemble, E. C., 4(6), *71*
Kendall, D. S., 96(88), 97(88), *107*
Kende, A. S., 152(114), *159*
Khandelwal, G. D., 105(119), *108*
Kharasch, M. S., 113(21), *156*
Kharasch, N., 141(90), *159*, 252(46), *308*
Kiefer, E. F., 333, *336*
Kiess, C. C., 9(154), *75*
Kikuchi, N., 113(21), *156*
Kim, B., 249(22), *307*
King, J. A., 259(100), *309*
King, T. A. 169(42), 184(40), 185(42), 186(40,42), 202(40,42), *223*, 255(56), *308*, 330(42), *335*
Kingsland, M., 150(112), *159*
Kirby, G. W., 125(61), *157, 158*
Kirkbridge, F. W., 78, *105*
Kirmse, W., 113(23), *156*
Kiselev, A. V., 317(12,13), 318(16,17), 321(16), 324(16), *335*
Kitaura, Y., 133(78), *158*
Klappmeier, F. H., 80(31), 82(31), 90(31), *106*
Klein, F., 294(241), *313*
Klemperer, W., 20(58), *72*
Klenert, M., 37(119), *74*
Klevens, H. B., 297(rr), 298(rr), 299(rr), *301*
Kling, O., 137(82), *158*
Klingele, H. O., 90(74), *107*
Klinger, H., 78, 85(1), *105*
Klopffer, W., 215, *226*
Klotz, L. C., 70(XV), *70*
Knibbe, H., 174, 175(26), 176, 177(28), 187, 188(30), 200(26,28), 203(31), 209, 210(31,101), *223, 225*
Knox, J. H., 2, *70*
Knutson, D., 125(62), 126(64), 129(70), 134, 134(64,81), 135(64), 139(64), *158*
Kobayashi, D., 321(23), *335*
Koblitz, W., 207(91), *225*
Kobsa, H., 114(29), 120, 124, 126(29), 129, 130(29), *157*, 236, *244*
Koch, T. H., 294(241), *313*
Kochetkova, N. S., 228(10), 231(10), *243*

Koda, T., 213(108), *225*
Koelsch, C. F., 93(77), *107*
Koffman, L., 26(92), *73*
Koga, G., 113(21), *156*
Koga, N., 113(21), *156*
Koltzenberg, G., 257(81), *309*
Kondelka, J., 216(133), *226*
Kondratiev, V., 53(137), *74*
Kondratiev, V. N., 56(145), 57(145), 58(145), 65(145), *75*
Kondratjew, E., 25(70), 47(70), *73*
Konijnenberg, E., 197, 198(70), *224*
Korte, F., 258(92), *309*
Kosower, E. M., 323(31), 324, *335*
Koutecky, J., 198(72), *224*
Kovář, J., 124(56), *157*
Kraft, K., 257(81), *309*
Kresge, A. J., 79(20), *105*
Krieger, H., 79(22), *105*
Kroening R. D., 104(115), *108*
Krongauz, V., 287(208,209), *312*
Krotoszynski, B. K., 235(36), *244*
Kruss, P., 16(39), *72*
Kubota, T., 257(73a), *308*
Kudryashov, P. I., 297(dd), 298(dd), *300*
Kuo, C. H., 133(78), *158*
Kupperman, A., 30, 31(108), *73*
Kuwana, T., 139(85a), *159*, 302(i), 303(i), *303*, 329(38d), *335*
Kuzmin, M. G., 133(78), *158*
Kuznetson, B. V., 317(12), *335*
Kwei, K. P. S., 241, *244*

L

La Barge, R. G., 79(23), 90(23,73), *105, 107*
Labhart, H., 267(124), 297(u), 298(u), 299(u,x), *300, 310*
Lalande, R., 207, *225*
Lamberti, V., 132(73), 141(73), 143(73), *158*
Lamboy, A. M. F., 170(20), 203 (20), *223*
Lamola, A. A., 79(35), 80(32,35,37), 81(35), 82(32), *106*, 114(31), 133(74), 139(74), *157, 158*, 216(135), 217, 217(135,136), *226*, 247(9), 249(14), 250(36), 255, 269(136), 270(136), 273(9,136), 277, 288(174), 291, 297(a,b,g,j), 298(a,b,g,j), 299(a,b), 301, 302(b,c), 303(b,c,f), 304(a,b,e,i),

Lamola, A. A. *(continued)*
305(a,b,e,i), *300, 303, 305-308, 310, 311*, 323(32), 328, *335*
Land, E. J., 103(107), *108*, 119(48), *157*
Landau, L., 4, 27(94), 41(94), 71, *73*
Langelaar, J., 170(20), 203(20), *223*
Laposa, J. D., 204(86), *225*
Lapouyade, R., 257(85), *309*
Larsh, A. E., 53(139), *74*
Latif, N., 98(92), *107*
Latowska, E., 125(57,58), *157*
Latowski, T., 125(57,58), *157*
Laustriat, G., 193(63), *224*
Lawrence, G. M., 9, 33(18), 39, *71*
Le Claire, C. D., 93(77), *107*
Lea, K. R., 17(41), *72*
Lee, E. K. C., 272(146), *310*
Lee, H. K., 26(77), *73*
Lee, T.-J., 103(111), *108*
Leermakers, P. A., 79(34,40), 80(33,34,36), 81(40), 95(85), 96(87,88), 97(87,88, 91), 98(34,87), 99(33,93,94), 101(94), *106, 107,* 139(86), 142(92), *159.* 246(2), 247(9), 250(37), 255(9), 262(2), 266(116), 273(9), 291(224, 230), *307, 308, 310, 313,* 316(4,5,6, 7,8,9), 318(9), 320(7), 321(7,8), 322(7), 323(7,32a), 324, 324(7), 325, 325(38a), 330(9), 331, 331(9), 332(6,7), 333(7), 334(7), *334, 335*
Legay, F., 67(152), *75*
Lehrer, S. S., 170(17), 184(17), *223,* 330(48), *336*
Leighton, P. A., 23(63), 58(63), *72*
Leipunsky, A., 53(137), *74*
Lemaire, J., 84(58,59,60), 95(58), *107,* 252(40,41,42), *308*
Lenz, G. R., 133(76), 152(115), 155(76), *158, 159*
Leonard, N. J., 79(20), *105*
Leonhardt, H., 182, 210(35), *223,* 280(187, 188), 281(187), *312*
Letsinger, R. L., 82(47), 90(47), *106*
Levy, D. H., 17(42), *72*
Lewis, F. D., 249(26), *307,* 285(202), 286(203), 298(l l), 299(l l), 303(g), *301, 303, 312*
Lewis, G. N., 297(bb), 298(bb), 299(bb), *300,* 302(k), 303(k), 304(s), 305(s), *303, 306*

Lewis, H. B., 141(90), *159*
Lewis, I. C., 221(129), *225*
Lewis, R. G., 115(36), *157*
Lieben, F., 78, 97(11), *105*
Lieber, C. O., 173(22), *223*
Lifshitz, E. M., 4(8), *71*
Lim, E. C., 164(4), 169(140), 184(4,140), 186(140), 190(52), 192(52), 196(73), 198(73), 199, 200(73), 201, 201(140), 203(82), 204(86), 206(52), *222, 224–226*
Linnett, J. W., 18, 49, *72, 74*
Linschitz, H., 234(30), *244*
Linshitz, H., 170(21), *223*
Lipinsky, E. S., 256(62), *308*
Lippert, E., 176(27), *223*
Lipsky, S., 116(42), *157*, 169(140), 184(140), 186, 201, *226*
Litovitz, T. A., 40(122), *74*
Little, J. C., 104(115), *108*
Little, L. H., 317(11)
Liu, R. S. H., 115(35), *157*, 234, *243*, 270(142), 272(151), 292, 292(235, 236,237), 293(237,238,239), 299(nn), 305(d), *301, 305, 310, 311, 313*
Livingstone, R., 82(51), 83(51), *106*, 209, *225*
Locke, J. M., 268(132), *310*
Loeschen, R., 291(231), 295(242), *313*
Longuet-Higgins, H. C., 170(13), *222*
Longworth, J. W., 164(7), 219, *222, 225*, 330(49,50), *336*
Loutfry, R. O., 294, *313*
Loveridge, E. L., 125(60), 134(60), 135(60), *157*
Löwe, L., 78, 97(11), *105*
Lower, S. K., 290(222), *312*
Loyola International Conference on Molecular Luminescence, 164(4), 184(4), *222*
Luder, W., 176(27), *223*
Ludwig, P., 298(jj), *301*
Ludwig, P. K., 297(h), 298(h), 299(h), *300*
Lumb, M. D., 169(43), 170(16), 185(43), 186(16,43), 202(16,43), 204(43), *223*
Lumry, R., 164(3), 177, 204(3), *222*
Lundquist, R. T., 228(9), 229, 231, 232(9), 234(9), *243*
Luner, C., 287(207), *312*
Lustig, C. D., 17(41), *72*

Lutz, R. E., 113(21), *156*
Lygin, V. I., 317(13), 318(16), 321(16), 324(16), *335*

M

Maerov, S. B., 133(79), *158*
Mager, K. J., 200(76), *224*
Magid, H., 331(55), *336*
Mahoney, R. T., 69(IV,V), 70(VII, VIII, X, XI, XII, XIII, XIV), *70*
Majer, J. R., 58(149), *75*
Malliaris, A., 203(81), *224*
Maňásek, Z., 133(79,83), 138(83), *158*
Mani, J. C., 249(24), 270(24), 302(n), *303, 307*
Marchand, A. P., 93(78), *107*
Marchetti, A. P., 252(47), 297(ss), 298(ss), 303(j), 304(n), 305(n), *301, 303, 306, 308*
Margulis, T. N., 230(20), *243*
Marsh, G., 148(106), *159*
Marshall, R., 53(139), *74*
Martin, R. B., 82(47), 90(47), *106*
Martin, R. M., 29, 30, *30, 73*
Mataga, N., 173, 176, 176(27), 209(102), 210(102), *223, 225*, 280(190), 281(191), 298(gg), *301, 312*
Matheson, M. S., 80(43), 81(43), *106*, 249(13), *307*
Mathews, J. H., 113(26), *156*
Mathieson, L., 25, 40(68), *73*
Matsuura, T., 133(78), *158*, 268(128), *310*
Mattheus, A., 141(14), 145(14), *156*
Mattice, J. J., 115(37), 116(37), 123, 126(37), 134(37), 157, 236, *244*
Mauret, P., 207, *225*
Mazur, Y., 147(102), 148(17,105), 149(19, 105), 150(19), *156, 159*
McCartin, P. J., 204(85), *224*
McClure, D. S., 101(99), *108*, 190(53), *224*, 253(49), 254(49), 258(49), 302(a,h), 303(a,h), *303, 308*, 329(38e), *335*
McCullough, J. J., 257(82a), *309*
McDaniel, R. S., 113(21), *156*
McDowell, T., 277(171), *311*
McGlynn, S. P., 196(68), 197, 197(67,68), 198(68), 200(76,78), 219(67), *224*, 228, *243*, 275(164,165), 299(mm), 305(v), *301, 306, 311*

McGregor, D. E., 113(21), *156*
McIntosh, C. L., 147, 148(103), 150(110), *159*
McIntosh, J. S. E., 79(29), *106*
McLafferty, F. W., 96(86), *107*
McMillan, F. R., 331(55), *336*
McRae, E. G., 319, 323(30), 324(30), *335*
McRae, J. A., 120(49), 121(49), *157*
Medinger, T., 205(88,89), 206, *225*
Medvedev, V. A., 56(145), 57(145), 58(145), 65(145), *75*
Megarity, E. D., 275, *311*
Meinwald, J., 90(74), *107*
Metts, L., 270(141), 292(232), 294(232), *310, 313*
Meyer, E. F., 152(20), 153(20), *156*
Meyer, R. T., 22, 22(60) 37(60), 43(60), 59, 61, *72*
Michejev, J. A., 133(78), *158*
Michio, H., 113(25), *156*
Migdalf, B. H., 252(39), *308*
Mileshana, L. A., 328, *335*
Miley, J. W., 292(232), 294(232), *313*
Miller, A. C., 294(24,b), *313*
Miller, A. L., 117(45), *157*
Miller, S. I., 113(22), 125(22), 139(87), *156, 159*
Miller, T. A., 17(42), *72*
Milne, G. S., 282(194), *312*
Minomura, S., 213(108), *225*
Mislow, K., 279, 279(185), 280(185), *312*
Mitchell, A. C. G., 42(128), *74*
Mitchell, D., 291(231), *313*
Mitchell, S. A., 317(10), *334*
Mokeeva, G. A., 170(15), *223*, 297(ff), *300*
Molchanov, V. A., 299(kk), *301*
Monot, M. R., 113(21), *156*
Monroe, B. M., 79(39), 80(39), 82(52), 86(52), 87(52), 90(39,71), 92(39,52), 93(52), 97(90), 98(90), *106, 107,* 252(43), 263, 264, 265(108), 266(118), 270(140), *308–310*
Moore, C. E., 4(19), 9, 56(19), 57(19), 58(19), 65(19), 68(19), *71*
Moore, G. F., 218(119,121), *225*
Moore, W. M., 273(154), 291(154), *311*
Morduchowitz, A., 99(97), *108*
Morrison, H., 277(171), *311*
Morrison, H. A., 252(39), *308*
Morse, R. I., 69(IV), 70(VII, XIII), *70*

Mott, N. F., 40(121), *74*
Moubasher, R., 98(92), *107*
Moussebois, C., 268(131), *310*
Mulac, W. A., 80(43), 81(43), *106,* 249(13), *307*
Mulliken, R. S., 25(74), 26(84), 29, 33, 48(104), *73,* 190(55), 200(80), *224*
Mullin, C. R., 276(168), *311*
Munro, I. H., 163(1), 169(11,43), 178, 179, 185(43), 186(43), 202(43), 204(43), 218(119,121), *222, 223, 225,* 330(41), *335*
Munro, I. M., 255(57), 297(ee), 298(ee), 299(ee), 304(o), 305(o), *300, 306, 308*
Muralidharan, V. P., 289(219), 290(219), *312*
Murmukhametov, R. N., 328, *335*
Murov, S., 253(49), 254(49), 258(49), 265, *308, 309*
Murov, S. L., 101(99), *108,* 253(49), 254(49), 258(49), 278, 279(179,180), 297(a,n), 298(a,n), 299(a,n), 302(a), 303(a,f), 304(a), 305(a), *300, 303, 305, 311,* 329(38e), *335*
Murrell, J. N., 197, 198(69), 200(79), *224*
Mustafa, A., 144(98), *159*
Muszkat, K. A., 275(166), *311,* 330(53), 331(53), *336*

N

Nagakura, S., 200(77), 205(77), *224,* 323, *335*
Nakadaira, Y., 105, *108*
Nakahara, A., 228(13,14), *243*
Nakamura, H., 233(25), *243*
Nakanishi, K., 105, *108*
Neckers, D. C., 99(96), 101(96,98), *108,* 132(72), *158*
Neidel, F., 250(34), *307*
Nelsen, J. P., 278(176), *311*
Nelsen, S. F., 274(159), 283(159b), *311*
Nelson, P. J., 294(241), *313*
Nesmeyanov, A. N., 228(10), 229, 231(10), 323(10), 238(41,42,43,44), 240(43), *243, 244*
Neuberger, K. R., 271(145), 294(240b), *310, 313*
Newland, G. C., 133(79), *158*
Niclause, M., 84(60), *107,* 252(41), *308*

Nicol, M., 213(107), *225*
Niemann, E. G., 37(119), *74*
Niemeyer, D., 143(93), *159*
Nikitin, E. E., 39, 41, 41(126), *39, 74*
Nikolaiski, E., 137(82), *158*
Nishikawa, H., 233(26), *243*
Nishmura, H., 298(gg), *301*
Norman, H. W., 302(i), 303(i), *303*
Norman, R. O. C., 82(55), 83(55), *106*, 291(227), *313*
Norrish, R. G. W., 10(20,21,22), 22, 25(59), 53, 53(139,140), *71, 72, 74, 78, 105*
Novros, J. S., 276(167), *311*
Noyes, W. A., 23(63), 58(63, 149), *72, 75*
Noyes, W. A., Jr., 78(14,17), 79(25,27), 80(14,43), 81(14,43,45), 84(59), *105–107*, 246(1), 247(1), 248(10), 249(13), 252(42), 255(59), 256(63), 258(1), 273(1), 297(e), 298(e), 299(e), 302(63), *300, 307, 308*
Noyori, R., 146(15), 149(109,111), 150(109), 151(109), *156, 159*
Nozaki, K., 146(15), 149(109,111), 150(109), 151(109), *156, 159*, 267(119), *310*
Nudenberg, W., 113(21), *156*
Nurmukhametov, R., 297(r), 298(r), 299(r), *300*

O

Obara, H., 144(97), *159*
O'Connell, E. J., 139(85c), *159*
Odaira, Y., 99(95), 102(100,101,102), *108*, 113(25), *156*
Offen, H. W., 213(109), *225*
Ogata, Y., 249(21), *307*
Ogryzlo, E. A., 18, 18(43,45), 19, 19(51), 25(43), 35, 51, 52, 52(45), 68(45), *72, 74*
Ogura, K., 268(128), *310*
Okada, T., 146(15), 149(109), 150(109), 151(109), 156, *159*, 173(25), 176, 209(102), 210(102), *223, 225*, 267(119), 280(190), 281(191), *310, 312*
Okawara, M., 133(79), *158*
Oki, M., 79(20), *105*
Okuda, M., 318(20), 320(20), 321(20), 324(20), *335*

Olah, G. A., 110(3), 124(3), *155*
Oldman, R. J., 40(123), 70(VIII,X,XI), *70, 74*
Oliveri-Mandala, E., 78, 87(12,13), *105*
Onsager, L., 176(27), *223*
Opitz, K., 249(18), *307*
Orchin, M., 321(22), *335*
Orgel, L. E., 207(96), *225*
Orger, B. H., 257(79), *309*
Orloff, M. K., 297(f), 298(f), 299(f), *300*
Orton, K. J., 113(26), *156*
Osborn, C. L., 115(7), 141(7), 142(7), *156*
Osborne, A. D., 86(64), *107*
O'Sullivan, M., 255(52), 282(52), 289(52), 304(l), *306, 308*
Oude Alink, B. A. M., 123(53,54), 124(54), *157*
Oudman, D., 94(81), *107*
Ourisson, G., 79(21), *105*

P

Pac, C., 133(77), *158*, 257(87), *309*
Pacey, P. D., 7(17) 15(17), 16(17), 35(17), 45(17), *71*
Pacifici, J., 286(205), *312*
Pacifici, J. G., 267(123), *310*
Padwa, A., 267, *310*
Paice, J. C., 92(75), *107*
Paillous, N., 94(82), *107*
Paldus, J., 198(72), *224*
Palmer, H. B., 3(2), 19, 19(55), 53(2,53,55), *70, 72*
Pande, C. D., 133(78), *158*
Pappas, B. C., 102(103), *108*
Pappas, S. P., 102(103), 103(104), *108*
Parker, C. A., 164(6), 218(6,117,118), 219, 219(124), 220(127), *222, 225*, 277(175), 292(233a), *313*, 297(m), 298(m,v), 299(m,pp), *300, 301, 311*
Parker, J. H., 14(32, 33(32), 38(32), *71*
Parmenter, C. S., 80(41), *106*
Pashayan, D., 267, *310*
Patel, D. J., 289(218), *312*
Paterno, E., 256(61), *308*
Patterson, J. M., 82(47), 90(47), *106*
Patton, H. W., 133(79), *158*
Pauson, P. L., 126(5,63), *155, 158*, 227(1), *243*
Pavlova, A. N., 318(19), 320(19), *335*

Perrins, N. C., 257(80), *309*
Persson, G., 84(60), *107,* 252(41), *308*
Pesaro, M., 152(20), 153(20), *156*
Pete, J. P., 288(215), 294(215), *312*
Phillips, D., 84(59), *107,* 252(42), *308*
Phillips, L. F., 18, *72*
Pimentel, G. C., 3(3), 7(3), 10(24), 12(3), 14, 21(3), 33(3,32), 36, 36(118), 37(3), 38(3,32), 42(3), *71, 74*
Pinhey, J. T., 113(21), 125(57), 149(111), *156, 157, 159*
Pisker-Trifunac, N., 240, *244*
Pitts, J., Jr., 302(l), *303*
Pitts, J. N., 23(62), 30(62), 58(62), 67(62), 70(IX), *70, 72,* 302(i,m), 303(i), *303, 329, 335*
Pitts, J. N., Jr., 58(149), *75,* 78(18,19), 79(25,27), 80(18,42), 81(45), 82(47), 84(59), 90(47), *105–107,* 114(32a), 114(32b), 115(32a), 119(32b), 125(58), 135(32c), 139(32d,85a,86), 143(32d), 147(32d), *157–159,* 246(1), 247(1), 248(11), 249(17,24,25), 251(25), 252(25,42,44), 255(59), 256(63), 258(1), 270(24), 273(1), 291(226), 297(e), 298(e), 299(e), 302(63), *300, 307, 308, 313*
Plank, D. A., 116(39), 118, 123, 124, 126(39), *157, 259, 309*
Platt, J. R., 297(rr), 298(rr), 299(rr), *301*
Plumley, H. J., 29(102), *73*
Polak, L., 25(70), 47(70), *73*
Polanyi, J. C., 7(15,16,17), 15, 15(15,17, 36), 16, 16(15), 30,34,34(35), 35, 35(17,40), 36(40), 37(15,36), 44(15, 36), 45, 49(15), *71, 72*
Pollack, M. A., 14, 15, 33(33), 37(33), *72*
Pomerantz, M., 271, *311*
Pond, D. M., 294(240b), *313*
Pope, R., 207(92), *225*
Porret, D., 33, *74*
Porter, C. W., 78, 86(5), *105,* 113(26), *156*
Porter, G., 10(20,21,22) 53(139,140), *71, 74,* 86(64), *107,* 119(48), 139, 139(85b), *157, 159,* 170(21), 193(60), 222(142), *223, 224, 226,* 250(32), 263, 264(107), 265(107, 114), 275(162), 297(p,ee), 298(p,ee), 299(p,w,ee), 304(o), 305(o,u), *300, 306, 307, 309–311*

Porter, G. B., 78(14), 79(29), 80(14), 81(14), *105, 106,* 117(44), *157,* 248(10), *307*
Porter, G. W., 78(16), *105*
Porter, N. A., 283, 283(198), *312*
Porter, R. P., 322(28), *335*
Powell, G. L., 297(q), 299(q), *300*
Price, W. C., 32(109), *73*
Przybytek, J. T., 248(12), *307*

Q

Quinkert, G., 124(55), *157,* 249(18), *307*

R

Radford, H. E., 17(41), *72*
Radziemski, L. J., 9(156), *75*
Raley, J. H., 29(101), *73,* 322(26), *335*
Ramsay, D. A., 67(152), *75*
Ramsey, C. C. R., 92(75), *107*
Ramsperger, H. C., 78, 86(5), *105*
Rao, D. V., 132(73), 141(73,89), 143(73), *158, 159*
Rao, P. J., 26(88,89), *73*
Rauh, R. D., 323(32a), *335*
Rausch, M., 115(38), *157*
Rausch, M. D., 227(2), 233(2), 239, 240(49), *243, 244*
Rayner, D. R., 279(184,185), 280(185), *312*
Razuvaev, G. A., 242, *244*
Recktenwald, G., 82(47), 90(47), *106*
Rees, A. L. G., 25, 25(72,73), 40(68), *73*
Reese, C. B., 109, 116(40), 122(40), 123(40), 126, *155, 157*
Rehm, D., 175(26), 176, 187, 188(30), 200(26), 210(101), *223, 225*
Reisch, J., 143(93), *159*
Renkes, G. D., 255(53), 289(53,217), *308, 312*
Rettschnick, R. P. H., 170(20), 203(20), *223*
Reubke, R., 79(22), *105*
Rice, S. A., 198(74), 200(75), 203, 215, *224, 225*
Richards, J. H., 231, 232(231), 234(21, 32), 240, *243, 244*
Richter, H. P., 176, 181, 182(34), 187, 189, 203(34), *223,* 280(189), 281(189), *312*

Richtol, H. H., 79(28), 80(31), 82(31), 90(31), *106*
Rigaudy, J., 149(18), 151(18), *156*
Rigby, R. D. G., 113(21), 125(57), *156, 157*
Riguady, J., 94(82), *107*
Rinehart, K. L., 232(24), *243*
Ritchie, C. D., 129(69), *158*
Ritscher, J. S., 257(83), *309*
Rivas, C., 250(30), 251(30), *307*
Robbins, W., 249(20), *307*
Roberts, R. N., 322(24), *335*
Robertson, J. M., 211(104), 212(106), *225*
Robin, M., 315, 320(3), 321(3), 324, *334*
Robinson, H. G., 17(41), *72*
Rockley, M. G., 289(217), *312*
Rodemeyer, S. A., 297(h), 298(h), 299(h), *300*
Rodionova, N. A., 238(42), *244*
Rogers, L. B., 329(40), *335*
Rogers, M. T., 207(97), *225*
Rollefson, G. K., 58(147), *75*
Röllig, K., 176, 203(31), 209, 210(31), *223*
Romand, J., 29(98), 31, *73*
Romanenko, V. I., 238(42,43,44), 240(43), *244*
Ron, A., 317, 318(15), 320(15), 321(15), *335*
Rosenberg, A. M., 333(58), *336*
Rosenberg, H., 115(38), *157*, 239, *244*
Rosenblum, M., 228(7), 231(7), *243*, 229(17), 230(20), 232(17), 234(31), 235, 235(33), *235, 243, 244*
Ross, M. E., 99(93), *107*
Rossiter, B. W., 259(99), 267(122), *309, 310*
Roy, R. S., 63(151), *75*
Rubin, M. B., 79(23), 90(23,73), 105, *105, 107, 108*, 265, *310*
Russell, K. E., 53(139), *74*
Russell, R. D., 164(8), 221(8,130), *222, 225*
Rust, F. F., 29(101), *73*
Ryang, H.-S., 104(116), *108*

S

Sachdev, K., 257(73), *308*
Sager, W. F., 129(69), *158*
Sakurai, H., 104(116), *108*, 257(73a,87), *308, 309*
Sallo, J. S., 268(130), *310*
Saltiel. J., 80(37), *106*, 256(67), 269(135, 136), 270(136,141), 271, 272, 273(135,136), 275, 292, 294(232, 240b), *308, 310, 311, 313*, 330, 330(51), *336*
Sander, R. K., 70(VIII,X,XI), *70*
Sandner, M. R., 115(33,34), 120(34), 121, 122(33,34), 126, 126(34), 127, *157*
Sandris, G., 79(21), *105*
Sandros, K., 80(30), 82(49), *106*, 193(62), *224*, 262, 291(102,225), 297(vv), 298(vv), *301, 309, 313*
Santer, J. O., 229(17), 232(17), *243*
Sato, S., 50(133), *74*
Saunders, W. H., Jr., 285(202), 298(ll), 299(ll), 303(g), *301, 303, 312*
Sazonova, V. A., 238(41,42,43,44), 240(43), *244*
Scandola, F., 233, *243*
Schafer, F. P., 176, 203(31), 209, 210(31), *223*
Schäfer, H. K., 79(22), *105*
Schaffner, K., 113(21), 116(41), 117(41), 118(41), 125(41), 126(41), 127(41), 128(41), 141(ll,41), 142(11), 145(11), 148(106,107), 149(111), *156, 157, 159*
Scharf, H.-D., 258(92), *309*
Scharmann, A., 189(48), *223*
Scheffold, R., 152(20), 153(20), *156*
Schenck, G. O., 270(143), *310*
Schiff, L. I., 4(5), *71*
Schläfer, H. L., 137(82), *158*
Schlosser, D. W., 69(IV), *70*
Schmid, H., 113(21), *156*
Schmid, K., 113(21), *156*
Schmitt, R. C., 241, *244*
Schnepp, O., 317(15), 318(15), 320(15), 321(15), *335*
Scholes, G., 274(158), 277(158), 279(158), 288(158), 296(158), *311*
Schönberg, A., 87, 98(92), *107*, 227(4), *243*
Schott, G., 19(57), 53(57), *72*
Schuck, A. E., 139(86), *159*
Schumacher, H. J., 207(91), *225*
Schuster, D. I., 289(218), 295, *312, 313*
Schutte, L., 121(50), 122, *157*
Scott, A. R., 103(110), *108*
Scott, D. R., 228, 229, 231, 233, 235, *243*

Scribe, P., 113(21), *156*
Seebach, D., 127(66), 128(66), *158*
Seely, G. R., 267(125,127), *310*
Seery, D. J., 19(54), 26(85), 53(54), *72, 73*
Seidel, H. P., 172, 173, *223*
Selin, L. E., 26(86,87), *73*
Selinger, B. K., 169(38,46), 172, 173, 184(38), 185(38,46), 186(38,46), 187, 201(38), 202(38,46), 204(38, 46), 207(46), 208, *223*
Setlow, R. B., 217(137), *226*
Shani, A., 133(76), 155(76), *158*, 271(144), *310*
Sharma, R. K., 252(46), *308*
Sharp, L. J., 133(74), 139(74), *158*, 250(36), *308*
Sharpe, R. R., 207(94, 95), 208(94), 213(94,95), *225*
Shibata, T., 257(82), *309*
Shigorin, D. N., 328, *335*
Shima, K., 104(116), *108*, 257(73a), *308*
Shionaya, S., 213(108), *225*
Shizuka, H., 142(91), *159*
Shoosmith, J., 10, *71*
Short, G. D., 219, 220(127), *225*
Shortley, G. H., 4(7), 5, 6, *71*
Shulman, R. G., 215(131), 216(132), *226*
Sidman, J. W., 78(15), *105*
Siedel, H. P., 173(22), *223*
Siga, K., 29(102), *73*
Silber, P., 78, 85(2), *105*
Simmons, J. P., 257(80), *309*
Simons, J., 53(139), *74*
Simons, J. P., 58(149), *75*, 209(99), *225*
Sina, A., 98(92), *107*
Singer, L. A., 289(219,220), 290, 290(219), 291, *312*
Singer, L. S., 221(129, *225*
Singh, A., 103(110), *108*
Singh, S. P., 248(12), *307*
Sláma, P., 125(59), 126(59), 129(59), 130(59,71), 133(79,83), 136(71), 138(71,83), 139(71), *157, 158*
Slates, H. L., 133(78), *158*
Smiley Irelan, J. R., 93(78), *107*
Smith, B. H., 145(100), *159*
Smith, F., 288, *243*
Smith, F. J., 275(164), *311*
Smith, I. W. M., 7(15), 15(15), 16(15,39), 30(15), 34(15), 35(15), 37(15), 44(15), 49(15), *71, 72*

Smith, J. A., 53(139), *74*
Smith, P., 333(58), *336*
Smith, R. L., 283(199,200), *312*
Smith, W. F., Jr., 259(98,99), 267(122), *309, 310*
Sneddon, W., 229, 235(19), 242(19), *243*
Snell, B. K., 141(10), *156*
Söderborg, B., 26(87), *73*
Solly, R. K., 95(84), *107*
Solomon, B. S., 278, 278(178), 282(194), *311, 312*
Solomon, J., 69, *70*
Sopchyshyn, F., 103(110), *108*
Sovers, O., 49(130), *74*
Sovocool, G. W., 271(147), *310*
Sperling, J., 259(94), *309*
Sponer, H., 26(80,82), *73*
Spruch, G., 280(187), 281(187), *312*
Spruch, G. M., 193(59), *224*
Spruck, G. M., 79(26), *106*
Srinivasan, R., 81(45), *106*, 257(77), *309*
Staab, H. A., 95(83), *107*
Stampa, G., 113(21), *156*
Steacie, E. W. R., 58(148), *75*
Stedman, D. H., 18(48), 19(48), 35(117), 51, 51(48), 52, 65, *72, 74*
Steel, C., 78, 86(5), *105*, 170(21), *223*, 278, 278(178), 282, *311, 312*
Steinfeld, J. I., 20(58), *72*
Steinmetz, R., 256(66), 270(143), *308, 310*
Stenberg, V. I., 109(2), 141(89), *155, 159*
Stephenson, L. M., 97(89), *107*, 190(50, 51), 206(51), *224*, 274(158), 277(158), 279(158), 288, 296(158), *311*
Stermitz, F. R., 155(118), *159*
Stevens, B., 79(26), *106*, 164(2,5), 167, 170(14,18), 171(18), 184(11), 185(11), 186(11), 190(52,54,56), 192, 192(52), 193(59,61), 204(83,84, 87), 206(52), 207, 208(94), 211(103), 212(103), 213(94,95,103,111), 217(111), 218(83,116,123), *167, 222–225*, 298(t), 299(s,t), *300*
Stevenson, C. D., 48, *48*
Stobbe, H., 144(94), *159*
Stoessl, A., 87(66), *107*
Stone, H., 322(24), *335*
Stone, T. J., 119(47), *157*
Stoner, M. R., 133(78), *158*
Stothers, J. B., 258(93), *309*

AUTHOR INDEX

Stratenus, J. L., 114(8), 140(8,9), *156*
Strating, J., 94(80), *107*
Strelko, U. V., 322(27), *335*
Strom, E. J., 333, *336*
Strong, R. L., 53, 53(142), *74*
Sugden, T. M., 10(20), 40, *71, 74*
Sugihara, T., 102(100), *108*
Sullivan, S., 230(20), *243*
Sunamoto, J., 267(126), *310*
Suppan, P., 139, 139(85b), *159*
Suprunchuk, T., 241, *244*
Sussman, D. H., 295, *313*
Sveshnikov, B., 297(dd,ff), 298(dd), *300*
Sveshnikov, B. Y., 170(15), *223*
Swenton, J. S., 246(7), 286, 287, 288(213), *307, 312*
Symons, M. C. R., 120(49), 121(49), *157*
Szabo, L. G., 53(139), *74*
Szwarc, M., 85(62), *107,* 283(196), 287(207), *312*

T

Takagi, K., 249(21), *307*
Takahashi, H., 144(97), *159*
Takeshi, F., 113(25), *156*
Tamblyn, J. W., 133(79), *158*
Tanaka, C., 164(5), 218(116), *222, 225*
Tanaka, I., 142(91), *159*
Tanaka, J., 164(5), 197, 198(69), 200(77), 205(77), 213(108,110), 218(116), *222, 224, 225,* 323(34), *335*
Tang, D. P. C., 268(130), *310*
Tani, S., 133(79), *158*
Tanikaga, R., 266(117), *310*
Tanner, D. W., 207(90), 208(90), *225,* 255(60), *308*
Tarr, A. M., 228, 232(15), 240, 241, *243*
Taub, D., 133(78), *158*
Tauer, E., 145(99), *159*
Taylor, A. W. C., 29(97,100), *73*
Taylor, R. P., 82(47), 90(47), *106*
Taylor, W., 207(96), *225*
Tchir, M., 288(215), 294(215), *312*
Tchir, M. F., 294(240b), *313*
Teale, F. W. J., 297(k), 298(k), 299(k), 305(h), *300, 306*
Tech, J. L., 9(155), *75*
Templeton, W., 258(93), *309*
Teranishi, H., 58(146), *75*
Terenin, A., 315(2), 316, 317, 318, *334*

Testa, A. C., 255(52), 272(150), 282(52), 289(52), 304(l), *306, 308, 311*
Thap, D. M., 268(130), *310*
Thoai, Nguyen, 113(21), *156*
Thomas, H. T., 316(4,6,7), 320(7), 321(7), 322(7), 323(7), 324(7), 332(6,7), 333(7), 334(7), *334*
Thomas, J. A., 113(27), *157*
Thomas, J. K., 222(141), *226*
Thomas, T. F., 278(178), *311*
Thomaz, M. F., 190(52), 192, 204(84,87), 206(52), *224, 225*
Thrush, B. A., 10, 10(20,21,22), 16(38), 35, 63, *35, 71, 72, 75*
Thyagarajan, B. S., 141(90), *159*
Tickle, K., 205(89), *225,* 299(y), *300*
Tiffany, W. B., 26, *73*
Toki, S., 257(73a), *308*
Tominaga, T., 102(100, 101, 102), *108*
Tominga, T., 99(95), *108*
Tomkiewicz, Y., 222(143), *226*
Tomura, N., 298(gg), *301*
Topp, M. R., 222(142), *226*
Torihashi, Y., 176(27), *223*
Toshima, N., 267(119), *310*
Toth, A. F., 316(8), 32(8), *334*
Townsend, M. G., 53(139), *74*
Trecker, D. J., 82(46), 87(46,68), 90(68), *106, 107,* 115(7,33,34), 120(34), 121, 122(33,34), 126, 126(34), 127, 141(7), 142(7), *156, 157,* 255, *308*
Tripathi, B. N., 133(78), *158*
Trueblood, K. N., 315, 320(3), 321(3), 324, *334*
Truscott, T. G., 209(100), *225*
Tsai, L., 103(106), *108*
Tsubomura, H., 190(55), *224*
Tsuchida, R., 228(13,14), 233(25,26), *243*
Tsutsumi, S., 99(95), 102(100,101,102), *108,* 113(25), 133(77), *156, 158*
Tuck, A. F., 70(XIV), *70*
Turro, N. J., 80(33,37), 82(48,50), 85(50), 96(86), 99(33), 103(108,109,111), *106-108,* 117(43a), 118(43a), 132(43b), 142(92), *157, 159,*227(4), *243,* 246(2,3,5), 247(3), 249(26), 250(37), 256, 256(68), 257(71), 258(91), 262(2), 269(136), 271, 272(5,71,151), 273(136, 152), 279(182), 289, 291, 292(237), 293(237), 294(240), 295(3), 305(t), *306-313*

AUTHOR INDEX

Tyler, A. J., 317(10), *334*

U

Uchida, Y., 53(138), *74*
Udding. A. C., 94(80), *107*
Ullman, E. F., 104(112), *108,* 250(33,35), 262(106), *307, 309*
Unger, I., 78(17), 79(27), *105, 106*
Urry, W. H., 82(46), 87(46,68), 90(68), *106, 107*

V

Vahlensieck, H. J., 200(76), *224*
Vala, M. T., 198(74), 215(115), *224, 225*
Valentine, D., Jr., 228(6), 234, *243*
van Beek, H. C.A., 267(121), *310*
VanderDonckt, E., 265(114), *310*
Van Duuren, B. L., 299(hh), *301*
Vangedal, S., 113(21), *156*
Van Hemert, R. L., 193(59), *224,* 291(228), *313*
Van Thiele, M., 19(54), 53(54), *72*
Van Vleck, J. H., 25(69), 26, 47(69), 52(69), *73*
Vasil'ev, N., 287, *312*
Vaughan, W. E., 29(101), *73*
Vedeneyev, V. I., 56(145), 57(145), 58(145), 65(145), *75*
Vember, T. A., 207(93), 208(93), *225*
Vember, T. M., 299(kk), *301*
Venkataramani, B., 133(78), *158*
Vesley, G., 139(86), *159*
Vesley, G. F., 95(85), 96(87), 97(87,91), 98(87), 99(93,94), 101(94), *107,* 190(51), 206(51), *224,* 274(158), 277(158), 279(158), 288(158), 296(158), *311*
Vesley, G. F., Jr., 257(88), *309*
Vickery, B., 257(84), *309*
Vinje, M. G., 113(21), *156*
Visco, R. E., 164(7), 219, *222, 225*
Visscher, F. M., 267(121), *310*
Viswanath, G., 170(13), *222*
Vogel, M., 115(38), *157,* 239, *244*
Vogt, V., 80(37), *106,* 269(136), 270(136), 273(136), *310*
Voldaiking, K. G., 299(kk), *301*
Voltz, R., 193(63), *224*
von Gustorf, E. A. K., 235(34), *244*

Vorek, G. G., 305(r), *306*
Voss, A. J. R., 255(50), *308*
Vystoskil, Z. Z., 322(27), *335*

W

Wagenaar, A., 94(80), *107*
Wagner, O. E., 297(c), *300*
Wagner, P. J., 78(19), 89(70), 98(70), 104(113), *105, 107, 108,* 246(1), 247(1), 249(23), 250(27), 258(1), 262, 273(1), 277(170), 294, 297(oo), *301, 307, 309, 311, 313*
Wahr, J. C., 70(VI), *70*
Walker, L. E., 113(23,24), 155(24), *156*
Walker, M. S., 164(3), 177, 193(61), 204(3), *222, 224*
Walling, C., 86(65), *107,* 258(89), 287(210), *309, 312*
Walsh, A. D., 33, *74*
Walter, J., 4(4), *71*
Wampfler, G., 253, 254, 255(48), *308*
Wamser, C. C., 266(118), *310*
Wan, J. K. S., 78(18), 80(18), *105,* 139(86), *159,* 249(25), 251(25), 252(25), *307*
Wang, C. T., 85(63), 87(63), *107*
Ware, W. R., 176, 181, 182(34), 187, 189, 190(57), 203(34), *223, 224,* 276(167), 280(189), 281, 299(ii), *301, 311, 312*
Ware, W. R., 186(36), *223*
Warren, P. C., 99(93,94), 101(94), *107*
Warwick, D. A., 295, *313*
Wasserman, E., 25(71), 47, *73*
Waters, W. A., 119(47), *157*
Weber, G., 297(k), 298(k), 299(k), 305(h), *300, 306*
Wehry, E. L., 329(40), *335*
Weigang, O. E., 322(29), *335*
Weiner, S. A., 79(39), 80(39), 82(52,53), 86(52), 87(52,53), 90(39,71), 92(39, 52,76), 93(52,76), *106, 107,* 263, 264, 265(108), *309*
Weinlich, J., 249(18), *307*
Weinmayr, V., 232(22), *243*
Weinreb, A., 181(33), *223*
Weir, D. S., 84(57), *106*
Weis, L. D., 316(5,6,7,9), 318, 320(7), 321(7), 322(7), 323(7), 324(7), 330(9), 331, 331(9), 332(6,7), 333(7), 334(7), *334*
Weisbuch, F., 113(21), *156*

AUTHOR INDEX

Weiss, D. S., 96(86), *107*, 246(3), 247(3), 294(240), 295(3), *307, 313*
Weiss, J., 207(96), *225*
Weissberger, A., 246(2), 247(9), 255(9), 262(2), 273(9), *307*
Weliky, N., 239, *244*
Weller, A., 85(61), *107*, 173, 173(22), *223*, 174, 175, 175(26), 176, 177(28), 182, 187, 188(30), 190(49), 191(49), 200(26,28), 203(31), 209, 209(24), 210, 211, 220, *223, 225*, 278, 280, 280(187), 281(187), *311, 312*
Weller, A., 182, 210, *223*
Wells, C. H. J., 295, *313*
Wendler, N. L., 133(78), *158*
Werbin, H., 333, *336*
Wettack, F. S., 255(53), 289, 289(53), *308, 312*
Whalen, J. W., 318, *335*
White, J. G., 212(106), *225*
White, J. M., 31(108), *73*
White, L., 125(60), 126(60), 134(60), 135(60), *157*
Whittemore, I. M., 85(62), *107*
Whitten, D. G., 190(50,51), 206(51), *224*, 274(158), 277(158), 279(158), 288(158), 296(158), *311*
Wiemann, J., 113(21), *156*
Wiersdorff, W., 249(18), *307*
Wiesenfeld, J. R., 4(11), 5, 7(11), 10(23), 12, 15, 20, 21(23), 25(23), 31, 34, 37(157), 43(11), 44, 45(11), 46(11), 47(11), 48(11), 49(11), 54, *71, 75*
Wigner, E., 49, *74*
Wilairat, P., 169(46), 185(46), 186(46), 187(46), 202(46), 204(46), 207(46), 208(46), *223*
Wilbur, P., 113(26), *156*
Wiles, D. M., 228, 232(15), 240, 241, *243, 244*
Wilkinson, F., 79(25), *106*, 193(60), 205, 205(88), 206, *224, 225*, 266(115), 273(155,156), 276(169), 277, 277(115), 279(156), 291(155), 297(i), 298(i), 299(i,y), 304(m), 305(j), *300, 306, 310, 311*
Wilkinson, G., 227(1), 229(18), *243*
Willard, J. E., 23, 26, 29, 53(141, 142), 60, 61, *72-75*

Williams, A. H., 297(cc), 298(cc), 299(cc), *300*
Williams, B. H., 286(204), *312*
Williams, G. H., 113(26), *157*
Williams, J. C., 148(108), *159*
Williams, J. R., 288(214), *312*
Williamson, R. V., 113(26), *156*
Wilson, J. H., 23, 24, 24(65,66), 26, 38(66), 57, 58(65,66), 61, *24, 72*
Wilson, K. R., 69, 70(I,II,VII,VIII,IX,X,XI, XII, XIII,XIV,XV), *70*
Wilzbach, K. E., 257(78,83), *309*
Windsor, M. W., 170(21), *223*
Winey, D. A., 87(68), 90(68), *107*
Winterle, J., 292(232), 294(232), *313*
Wipke, W. T., 246(6), *307*
Wolf, W., 141(90), *159*
Wood, D. E., 270(139), *310*
Woodward, R. B., 127(65), *158*
Wriede, P. A., 256(68,69,70), 257(71), 271, 272(71), *308*
Wright, M., 18(47), 19, 25(47), 51, 52(47), *72*
Wright, T. R., 278(177), *311*
Wrighton, M., 270(141), 271(145), 292(232), 294(232), *310, 313*
Wyatt, P., 279(182), *311*
Wynberg, H., 94(81), *107*

Y

Ya, J. M. H., 204(86), *225*
Yager, W. A., 25(71), 47(71), *73*
Yamada, S., 228(13,14), 233, *243*
Yamaguti, Z., 149(109,111), 150(109), 151(109), *159*
Yamamoto, N., 176, 209(102), 210(102), *223, 225*, 281(191), *312*
Yamane, T., 215(131), 217, *226*
Yamazaki, H., 233(26), *243*
Yang, N. C., 88(69), 99(97), 101(99), *107, 108*, 133(76), 152(115), 155(76), *158, 159*, 249(16,22), 250(29,30,31), 251(30), 253(49), 254(49), 256(65), 258(49,90), 265, 268(130), 271, 291, 295, 302(a), 303(a), *303, 307-310, 313,* 329, *335*
Yavorskii, B. M., 228(10), 229(10), 231(10), 232(10), *243*
Yip, R. W., 150(110), *159*

Yogev, A., 148(105), 149(19,105), 150(19), *156, 159*
Youssefyeh, R. D., 259(95), *309*
Ywata, C., 151(113), 152(113), *159*

Z

Zachariasse, K., 220, *225*
Zahlan, A. B., 200(77), 205(77), 216(132), *224, 226*
Zander, M., 298(aa), *300*
Zaslavskaya, G. B., 228(10), 229(10), 231(10), 232(10), *243*
Zeldes, H., 82(51), 83(51), *106*
Zemansky, M. W., 42(128), *74*
Zener, C., 27(95,96), 40, 41(95,96), *73, 74*
Ziffer, H., 288(214), *312*
Zimmerman, H. E., 115(36), 287, 288(213), *157, 312*
Zolnikova, G. P., 238(42,44), *244*
Zwanenburg, B., 94(80), *107*
Zwicker, E. F., 250(31), *307*

Subject Index

A

Acetanilide, photorearrangement, 142
2'-acetonaphthone, phosphorescence liftime and spectra on solid surfaces, 326, 327
Acetone, adsorption spectra on solid surfaces, 323
  phosphorescence lifetime and spectra on solid surfaces, 326, 327
Acetophenone, phosphorescence lifetime and spectra on solid surfaces, 326–329
3-Acetoxycyclohexenone, photorearrangement, 148
Acridine, 265
Adsorption interactions, 317
Adsorption on solid surfaces, 316–319
Adsorption spectra of adsorbed molecules, acetone, 323
  alkylpyridinium iodides, 324
  aniline, 321
  anthracene, 322
  azulene, 322
  benzene, 320
  benzil, 323
  benzophenone, 323
  carotenoids, 324
  N,N-dimethylaniline, 321
  N-methylaniline, 321
  naphthalene, 322
  nitrobenzene, 323
  phenol, 321
  tetramethylcyclobutanedione, 323
Alkylpyridinium iodides, adsorption spectra on solid surfaces, 32
  change-transfer transactions, 324
  dipole moment, 324
Aniline, adsorption spectra on solid surfaces, 321

Anisil, phosphorescence lifetime and spectra on solid surfaces, 327–328
  triplet energy, 79
Anthracene, adsorption spectra on solid surfaces, 322
2-Azidobiphenyl, 286
Azobisisobutyronitrile, radical reactions of, 333
Azo-1-cyanocyclohexane, 282
Azo-2-methyl-2-propane, 281–282
Azoxybenzene, 266
Azulene, adsorption spectra on solid surfaces, 322
  self-quenching, 170

B

Benzahilide, photorearrangement, 141
Benzene, adsorption spectra on solid surfaces, 320
Benzil, adsorption spectra on solid surfaces, 323
  fluorescence yield, 79
  hydrogen abstraction reaction, 86
  intersystem crossing yield, 80
  phosphorescence lifetime and spectra on solid surfaces, 327–328
  phosphorescence yield, 80
  photochemical reactions, 85–87
  triplet energy, 79
3,4-Benzocumarin, photorearrangement, 123
Benzocyclobutene-1,2-dione, 95
Benzophenone, adsorption spectra on solid surfaces, 323
  phosphorescence lifetime and spectra on solid surfaces, 326–327
Benzoylferrocene, 228, 232, 239–240
2-Benzoylpyridine, phosphorescence lifetime and spectra on solid surfaces, 326–327

SUBJECT INDEX

o-Benzylbenzophenone, 251
Benzyl-o-benzyloxyphenylglyoxylate, 102
Biacetyl, energy transfer rate constants,
    table, 261
  fluorescence yield, 79
  hydrogen abstraction reaction, 80–82, 84
  intersystem crossing yield, 80
  isomerization, 84
  phosphorescence, 90
  photochemical reactions, 80–85, 252
  quantum yield, 80, 84
  triplet energy, 79
bis-Benzenechromium(I), 228, 233, 242
p-Bromobutyrophenone, 252
2,3-Butanedione, phosphorescence lifetime
    and spectra on solid surfaces, 327–328
γ-Butyrolactone, photochemical
    reaction, 259

C

Camphorquinone, hydrogen abstraction
    reaction, 90
  photochemical reactions, 90–91,
    104–105, 263
  triplet energy, 79
Carotenoids, adsorption spectra on solid
    surfaces, 324
Change-resonance excimer states, 196–197
Change-transfer transactions,
    alkylpyridinium iodides, 324
  nitrobenzene, 323
Chemical sensitized photoreduction
    reactions, 262–268
Chemiluminescence, 220
Cis,cis-cyclotetradecane-1,8-diene, 268
Cis-trans isomerization by radical
    reactions, 268–270
Complexes, ferrocene, 229–231, 235
Criteria of photoassociation, 170, 208, 222
1,2-Cyclodecanedione, 87–88
2,3-Cyclohexadienones, photochromism
    of, 332
Cyclopentene, 258
Cymantrene, 233, 235, 241

D

5,6-Decanedione, 87–89
Decarboxylation by photorearrangement,
    134–136

Delayed fluorescence, 218–219
2,3-Diazabicyclo[2.2.1]-2-heptene, 282
1,2-Dicyanoethylene, 256
Diethyl oxalate, hydrogen abstraction
    reactions, 102
  photochemical reactions, 102
Diffusion coefficients for I($5^2 P_{1/2}$) in noble
    gases, table, 44
3,4-Dihydrocumarin, photorearrange-
    ment, 123
Dimer cation neutralization, 220–221
Dimerization of isoprene, table, 251
N,N-Dimethylaniline, adsorption spectra on
    solid surfaces, 321
3,3-Dimethyl-1,2-indandione, 93
2,7-Dimethyl-4,5-octanedione, 87
1,3-Diphenylacetone, phosphorescence
    lifetime and spectra on solid
    surfaces, 326–328
1,2-Diphenylcyclopropane, 279
2,2-Diphenyl-1,3-indanedione,
    photorearrangement, 151
3,3-Diphenyl-1,2-indanedione, 94
1,3-Diphenylpropane, excimer fluorescence
    on solid surfaces, 330
Dipole-dipole interactions, 317, 319,
    322, 324
Dipole moments, alkylpyridinium
    iodides, 324
  nitrobenzene, 323
Di-2-pyridylketone, phosphorescence life-
    time and spectra on solid
    surfaces, 326–327
Dispersion forces, 317, 321–322
3,4-Di-t-butylcyclobutanedione, 94
4,4-Dimethyl-2-cyclohexenone, 253
DNA luminescnce, 215–216

E

Electrochemiluminescence, 219–221
Electronic absorption spectral shifts on
    absorbed surfaces, 319–320
Electronically excited halogen atoms,
    atomic transitions, 6–8
  collisional quenching, 37–38, 45–50
  detection methods, atomic emission in
    flow systems, 15–17
  electron spin resonance, 17
  molecular halogen emission in flow
    discharge system, 17–19

Electrically excited halogen
atoms *(continued)*
  molecular halogen emission in shock-
    heated gases, 19–20
  photochemical studies, 23
  stimulated emission from excited
    halogen atoms, 14–15
  time-resolved adsorption
    spectroscopy, 10–12
  time-resolved atomic emission, 12–13
  time-resolved mass spectrometric
    studies, 22
  time-resolved molecular emission, 20–22
Einstein A factors, 4
mean radiative lifetimes, 4, 44–45
population inversion with lasers, 36–39
reactions, hydrogen abstraction, 57–58
  iodine atom abstractions, 58–61
  with halogens, 61–65
  with polyatomic molecules, 65–68
recombination, in flash photolysis
    experiments, 53–55
  in flow discharges, 51–52
  in shock tubes, 52–53
spin orbit coupling, 4
spin orbit relaxation, 39–50
spontaneous emission, 5
Encounter efficiencies for
    photoassociation, 169, 181
Energy transfer, exciplex intermediate
    in, 192
  in photoassociating solvents, 193–194
Energy transfer from sensitizer singlet
    states, 273–292
Energy transfer from sensitizer upper triplet
    states, 292–296
Enthalpies of photoassociation, 184–188
Entropies of photoassociation, 184–188
Equilibrium constants for photoassociation,
    176–177, 184–186
Ethylphenylcarbonate, photorearrange-
    ment, 133
Ethylphenylglyoxylate, hydrogen abstrac-
    tion reaction, 100–101
  Norrish Type II elimination reactions, 101
  phosphorescence yield, 80
  photochemical reactions, 99–100
  triplet energy, 79
Ethylpyruvate, hydrogen abstraction
    reaction, 100–101
  photochemical reactions, 99–100

Excimer, binding energies, 184–185,
    194–200
  configuration, 194–195, 201
  definition of, 164
  destabilization energy of 'ground'
    state, 184–186
  dissociation frequency factors of, 185
  relaxation scheme for, 163
  triplet state, 171, 200, 203, 205
Excimer fluorescence, compounds
    exhibiting, 167
  decay characteristics, 178–182
  delayed, 208
  in crystalline state, 207, 211–213
  in polyphenyl alkanes, 213–215
  lifetimes, 201, 202
  of pure liquids, 171
  origin of, 166, 203
  polarization of, 203
  pressure dependence of, 172, 213
  quantum yields of, 166, 168, 201–202
Excimer fluorescence of adsorbed molecules,
    1,3,-diphenylpropane, 330
  pyrene, 329–330
Exciplex, as charge transfer state, 189
  binding energies, 188, 200
  definition of, 164, 177
  destabilization energy of 'ground'
    state, 188
  dipole moment of, 174–177
  relaxation scheme for, 163
Exciplex fluorescence, and solvent
    polarity, 176, 209
  decay characteristics of, 181
  in chemiluminescence, 220
  in DNA, 216
Excited states of α-dicarbonyl compounds,
    78–80
Extinction coefficients for hydrogen halides,
    table, 31

F

Ferricenium ion, 229–230, 235, 240
Ferrocene, as a quencher, 234
  as a sensitizer, 234
  derivatives as protective UV
    absorbers, 241
  electronic absorption spectrum of,
    228–229, 232
  electronic absorption spectrum of
    derivatives, 232

Ferrocene *(continued)*
  photolability, 235
  various complexes, 229–231
Fluorescence yield, of benzil, 79
  of biacetyl, 79
Formation of excited halogen atoms by atomic metathetical reactions, 34–36
Fries reaction, photochemical, 236
α-Furil, phosphorescence lifetime and spectra on solid surfaces, 327–328

G

Glyoxal, 80

H

Halogen reactions, with Br ($4^2 P_{1/2}$) and Cl ($3^2 P_{1/2}$), 64–65
  with I ($5^2 P_{1/2}$) and I ($5^2 P_{3/2}$), 61–64
Heats of adsorption, 318
2,3-Heptanedione, intersystem crossing yield, 80
  phosphorescence, 90
  quantum yield, 104
Hexafluorobiacetyl, 85
3,4-Hexanedione, 87, 88
Hot hydrogen and deuterium atom formation, from HI and DI photolysis, 30
Hydrogen atom abstraction reactions, of benzil, 86
  of biacetyl, 80–82, 84
  of γ-butyrolactone, 259
  of camphorquinone, 90
  of cyclopentene, 258
  of ethylphenylglyoxylate, 100–101
  of ethylpyruvate, 100–101
  of pyruvic acid, 96
  with I ($5^2 P_{1/2}$), 57–58
Hydrogen bonding, 317–318, 321, 324
2-Hydroxybenzophenone, 251
o-Hydroxybutyrophenone, 251

I

Intermolecular reactions, 78
Intersystem crossing yields, of benzil, 80
  of biacetyl, 80
  of 2,3-heptanedione, 80

Intersystem crossing yields *(continued)*
  of 1-naphthylacetate, 144
  of 2,3-octanedione, 80
  of 2,3-pentanedione, 80
Intramolecular reactions, 78, 80
Iodine atom abstraction reactions, with I ($5^2 P_{1/2}$), 58–61
Isomerization, of biacetyl, 84
  of piperylene, 331
  of stilbenes, 330–331
Isoprene, dimerization of, table, 251

K

Ketenimine, 282
α-Ketodecanoic acid, Norrish Type II elimination reaction, 98
  phosphorescence, 80
  photochemical reactions, 98
  triplet energy, 79
α-Ketopentanoic acid, Norrish Type II elimination reaction, 98
  photochemical reactions, 98

L

Luminescence in metallocene systems, 228

M

M-C bands, 229, 231, 234
Mesitil, phosphorescence lifetime and spectra on solid surfaces, 327–328
o-Methoxyacetophenone, phosphorescence lifetime and spectra on solid surfaces, 326–329
4-Methoxybenzil, phosphorescence lifetime and spectra on solid surfaces, 327–328
4-Methoxyphenylacetate, photorearrangement, 122
N-Methylaniline, adsorption spectra on solid surfaces, 321
Methyl-o-benzyloxyphenylglyoxylate, 102
5-Methyl-2,3-hexanedione, quantum yield, 104
2-Methyl-1,4-naphthoquinone, radical reactions of, 333
3-Methyl-2-pentene, photochemical reaction, 271
  quantum yields, table, 271

SUBJECT INDEX

4-Methylphenylacetate, photorearrangement, 115–116, 120–121
  quantum yields, table, 121
4-Methylphenylbenzoate, photorearrangement, 115, 120
  quantum yields, 138
4-Methylphenyl-N-methylcarbamate, photorearrangement, 115

N

Naphtalene, adsorption spectra on solid surfaces, 322
1-Naphthylacetate, intersystem crossing yield, 14
  photorearrangement, 114
  triplet energy, 114
Nickelocene, 233
Nitrobenzene, adsorption spectra on solid surfaces, 323
  change-transfer transitions, 323
  dipole moment, 323
Norrish Type II elimination reactions, of ethylphenylglyoxylate, 101
  of α-ketodecanoic acid, 98
  of α-ketopentanoic acid, 98

O

2,3-Octanedione, intersystem crossing yield, 80
  phosphorescence, 90
4,5-Octanedione, 87–89
Oxetane formation reaction, 256–257

P

2,3-Pentanedione, intersystem crossing yield, 80
  phosphorescence, 90
  phosphorescence lifetime and spectra on solid surfaces, 327–328
  photochemical reactions, 87
  quantum yield, 104
  triplet energy, 79
Phenol, adsorption spectra on solid surfaces, 321
Phenylbenzoate, photorearrangement, 115, 118
  UV and γ-irradiation, table, 116

Phenylglyoxal, triplet energy, 79
Phenylglyoxylic acid, 97–98
1-Phenyl-1,2-propanedione,
  phosphorescence lifetime and spectra on solid surfaces, 327–328
  triplet energy, 79
Phenyl-p-toluenesulfonate, photorearrangement, 140
Phenylsalicylate, photorearrangement, 132
Phosphorescence, of biacetyl, 84, 90
  of ethyl phenylglyoxylate, 80
  of 2,3-heptanedione, 90
  of α-ketodecanoic acid, 80
  of 2,3-octanedione, 90
  of 2,3-pentanedione, 90
  of pyruvic acid, 80
Phosphorescence lifetime and spectra on solid surfaces, 2'-acetonaphthone, 326–328
  acetone, 326–327
  anisil, 327–328
  benzil, 327–328
  benzophenone, 326–327
  2-benzoylpyridine, 326–327
  2,3-butanedione, 327–328
  1,3-diphenylacetone, 326–327
  di-2-pyridylketone, 326–327
  α-furil, 327-328
  mesitil, 327–328
  o-methoxyacetophenone, 326–329
  4-methoxybenzil, 327–328
  2,3-pentanedione, 327–328
  1-phenyl-1,2-propanedione, 327
  α-pyridil, 327
  3,3,5,5-tetramethylcyclohexanone, 326
  xanthone, 326, 328
Phosphorescence yield, of benzil, 80, 84
Photochemical addition to aromatic hydrocarbons, 257–258
Photochemical reactions of adsorbed molecules, 330–333
Photochemistry of substituted metallocenes, 236–241
Photochromism, of 2,4-cyclohexadienones, 332
  of spiropyrans, 332
Photodissociation, of homonuclear halides, 25–26
  of hydrogen halides, 29–32
  of interhalogen molecules, 26–29

SUBJECT INDEX

Photodissociation *(continued)*
  of organic halides, 32–34
Photodissociation in ferrocenyl esters, 236–237
Photoenolization reactions, 250–252
Photorearrangement, of N–aryl amides of carboxylic acid, 141–145
  of N–aryl amides of sulfonic acid, 146
  of ary esters of carboxylic acid, 114–140
  of aryl esters of sulfonic acid, 140–141
  of aryl lactams, 145
  of enamides of carboxylic acids, 152–155
  of enol esters of carboxylic acid, 146–149
  of enol lactones of carboxylic acid, 149–152
Piperylene, isomerizations of, 331
Pivalil, triplet energy, 79
Polyatomic molecular reactions, with I ($5^2P_{1/2}$), 65–68
N-1-Propenyl-N-1-propylbenzamide, photorearrangement, 152
[4.4.2] Propella-3,8-diene-11,12-dione, 93, 94
Properties of aromatic carbonyl sensitizers, table, 302–303
Properties of aromatic hydrocarbon sensitizers, table, 297–301
Properties of miscellaneous sensitizers, table, 304–306
Pyrene, eximer fluorescence on solid surfaces, 329–330
α-Pyridil, phosphorescence lifetime and spectra on solid surfaces, 327–328
Pyruvic acid, hydrogen abstraction reaction, 98
  phosphorescence yield, 80
  photochemical reactions, 95–97

Q

Quadricyclene, 278
Quantum yields, of biacetyl, 80–84
  of 2,3-heptanedione, 104
  of 5-methyl-2,3-hexanedione, 104
  of 4-methylphenylacetate, 121
  of 4-methylphenylbenzoate, 138
  of 2,3-pentanedione, 104
Quantum yield calculations for photorearrangement, 136–139

Quenching of the singlet state of aromatic hydrocarbons, 277–288
Quenching of the singlet state of carbonyl compounds, 288–292

R

Radical reactions, of azobisisobutyronitride, 333
  of 2-methyl-1,4-naphthoquinone, 333
  of tetramethylsuccinonitrile, 333
  of 2,3,3-trimethyl-1-penten-4-one, 333
Reversible energy transfer between sensitizer and acceptor, 259–262

S

Salicylanilide, photorearrangement, 132
Schenck mechanism for photosensitized olefin reactions, 270–272
Sensitizer self-quenching reactions, 253–255
Singlet molecular oxygen reactions with iodine atoms, 35
Spriopyranes, photochromism of, 332
Stilbenes, isomerization of, 330–331
Styrylferrocene, 240–241

T

Tetramethylcyclobutanedione, adsorption spectra on solid surfaces, 323
3,3,5,5-Tetramethylcyclohexanone, phosphorescence lifetime and spectra on solid surfaces, 326–327
1,1,4,4-Tetramethyl-2,3-dioxotetralin, 92
Tetramethylsuccinonitrile, radical reactions of, 333
Thermodynamic constants for some excited halogen atom reactions, table, 56
2,3,3-Trimethyl-l-penten-4-one, radical reactions of, 333
Triplet energies, table, 79
Triplet energy, of 1-naphthylacetate, 114
Triplet-triplet annihilation, and eximer formation, 218–219
  in chemiluminescence, 220–221
  mixed, 219, 221

U

Unimolecular photosensitized reactions, 248–252
UV and γ-irradiation, of phenylbenzoate, 116

V

Volume change in photoassociation, 173

X

Xanthane, phosphorescence lifetime and spectra on solid surfaces, 326–329

Cumulative Index, Volumes 1–8

	VOL.	PAGE
Addition of Atoms to Olefins, in Gas Phase (Cvetanovic)	1	115
Alkanes and Alkyl Radicals, Unimolecular Decomposition and Isotope Effects of (Rabinovitch and Setser)	3	1
Aromatic Hydrocarbon Solutions, Photochemistry of (Bower)	1	23
Carbonyl Compounds, The Photocycloaddition of, to Unsaturated Systems: The Syntheses of Oxetanes (Arnold)	6	301
Cobalt (III) and Chromium (III) Complexes, the Photochemistry of, in Solution (Valentine, Jr.)	6	123
Cyclic Ketones, Photochemistry of (Srinivasan)	1	83
a-Dicarbonyl Compounds, The Photochemistry of (Monroe)	8	
Electronic Energy Transfer between Organic Molecules in Solution (Wilkinson)	3	241
Electronically Excited Halogen Atoms (Husain and Donovan)	8	
Electron Transfer Luminescence in Solution (Zweig)	6	425
Free Radical and Molecule Reactions in Gas Phase, Problems of Structure and Reactivity (Benson)	2	1
Gas Phase, Addition of Atoms of Olefins in (Cvetanovic)	1	115
Gas Phase Reactions Involving Hydroxyl and Oxygen Atoms, Mechanisms and Rate Constants of (Avramenko and Kolesnika)	2	25
Gas Phase Reactions, Photochemical, in Hydrogen-Oxygen System (Volman)	1	43
Halogenated Compounds, Photochemical Processes in (Majer and Simons)	2	137
Hydrogen-Oxygen System, Photochemical Gas Phase Reactions in (Volman)	1	43
Hydroxyl and Oxygen Atoms, Mechanisms and Rate Constants of Elementary Gas Phase Reactions Involving (Avramenko and Kolesnikova)	2	25
Hypophalites, Developments in Photochemistry of (Akhtar)	2	263
Ionic States, in Solid Saturated Hydrocarbons, Chemistry of (Kevan and Libby)	2	183
Isotopic Effects, in Mercury Photosensitization (Gunning and Strausz)	1	209
Mechanism of Energy Transfer, in Mercury Photosensitization (Gunning and Strausz)	1	209
Mechanistic Organic Photochemistry, A New Approach to (Zimmerman)	1	183
Mercury Photosensitization, Isotopic Effects and the Mechanism of Energy Transfer in (Gunning and Strausz)	1	209
Metallocenes, Photochemistry in the (Bozak)	8	
Methylene, Preparation, Properties, and Reactivities of (DeMore and Benson)	2	219
Nitric Oxide, Role in Photochemistry (Heicklen and Cohen)	5	157
Nucleic Acid Derivatives, Advances in the Photochemistry of (Burr)	6	193
Organic Molecules in Adsorbed or Other Perturbing Polar Environments, Photochemical and Spectroscopic Properties of (Nicholls and Leermakers)	8	

CUMULATIVE INDEX

Organic Molecules, Photochemical Rearrangements of (Chapman)	1	323
Organic Molecules in their Triplet States, Properties and Reactions of (Wagner and Hammond)	5	21
Organic Nitrites, Developments in Photochemistry of (Akhtar)	2	263
Perhalocarbons, Gas Phase Oxidation of (Heicklen)	7	57
Phosphorescence and Delayed Fluorescence from Solutions (Parker)	2	305
Photoassociation in Aromatic Systems (Stevens)	8	
Photochemical Mechanisms, Highly Complex (Johnston and Cramarossa) .	4	1
Photochemical Oxidation of Alkehydes by Molecular Oxygen, Kinetics and Mechanism of (Niclause, Lemaire, and Letort)	4	25
Photochemical Reactivity, Reflections on (Hammond)	7	373
Photochemical Rearrangements of Conjugated Cyclic Ketones; The Present State of Investigations (Schaffner)	4	81
Photochemical Transformations of Polyenic Compounds (Mousseron) . . .	4	195
Photochemistry of Conjugated Dienes and Trienes (Srinivasan)	4	113
Photochemistry, Vocabulary of (Pitts, Wilkinson, Hammond)	1	1
Photochromism (Dessauer and Paris)	1	275
Photo-Fries Rearrangement and Related Photochemical (l,j)−Shifts of (j=3,5,7) of Carbonyl and Sulfonyl Groups (Bellus, Geigy and Basel) . . .	8	
Photoionization and Photodissociation of Aromatic Molecules, by Ultraviolet Radiation (Terenin and Vilessov)	2	385
Photolysis of the Diazirines (Frey)	4	225
Photooxidation Reactions, Gaseous (Hoare and Pearson)	3	83
Photooxygenation Reactions, Type II, in Solution (Gollnick)	6	1
Photosensitized Reactions, Complications in (Engel and Monroe)	8	
Radiationless Transitions, Isomerization as a Route for (Phillips, Lemaire, Burton, and Noyes, Jr.)	5	329
Radiationless Transitions in Photochemistry (Jortner and Rice)	7	149
Singlet Molecular Oxygen (Wayne)	7	311
Singlet and Triplet States: Benzene and Simple Aromatic Compounds (Noyes and Unger)	4	49
Solid Saturated Hydrocarbons, Chemistry of Ionic States in (Kevan and Libby)	2	183
Spin Conservation (Matsen and Klein)	7	1
Sulfur Atoms, Reactions of (Gunning and Strausz)	4	143
Sulfur and Nitrogen Heteroatomic Organic Compounds, Photochemical Reactions of (Mustafa)	2	63
Triatomic Free Radicals, Spectra and Structures of (Herzberg)	5	1
Ultraviolet Photochemistry, Vacuum (McNesby and Okabe)	3	157
Ultraviolet Radiation, Photoionization and Photodissociation of Aromatic Molecules by (Terenin and Vilessov)	2	385